The Digital Health in the Pandemic Era

The Digital Health in the Pandemic Era

Editor

Daniele Giansanti

MDPI • Basel • Beijing • Wuhan • Barcelona • Belgrade • Manchester • Tokyo • Cluj • Tianjin

Editor
Daniele Giansanti
Centro Nazionale per le
Tecnologie Innovative in
Sanità Pubblica
ISS
Rome
Italy

Editorial Office
MDPI
St. Alban-Anlage 66
4052 Basel, Switzerland

This is a reprint of articles from the Special Issue published online in the open access journal *Life* (ISSN 2075-1729) (available at: www.mdpi.com/journal/life/special_issues/DigitalHealth_Pademic).

For citation purposes, cite each article independently as indicated on the article page online and as indicated below:

LastName, A.A.; LastName, B.B.; LastName, C.C. Article Title. *Journal Name* **Year**, *Volume Number*, Page Range.

ISBN 978-3-0365-7733-3 (Hbk)
ISBN 978-3-0365-7732-6 (PDF)

Contents

About the Editor

Daniele Giansanti

Dr. Giansanti received an MD in Electronic Engineering at Sapienza University, Rome, in 1991; a PHD in Telecommunications and Microelectronics Engineering at Tor Vergata University, Rome, in 1997; and an Academic Specialization in Cognitive Psychology and Neural Networks at Sapienza University, Rome, in 1997. His Academic Specialization was in Medical Physics at Sapienza University, Rome, in 2005. Dr. Giansanti was in charge of the Design of VLSI Asics for DSP in the Civil Field (1991–1997) during his MD and PHD, and he served as a CAE-CAD-CAM system manager and Design Engineer in the project of electronic systems (Boards and VLSI) for the Warfare at Elettronica spa (1992–2000), one of the leaders in the military field. More importantly, he also conducts varied research at ISS (the Italian NIH) (2000–today) in the following fields:

1) Biomedical engineering and medical physics with the development of wearable and portable devices (three national patents).

2) Telemedicine and e-Health: technology assessment and the integration of new systems in the field of telerehabilitation, domiciliary monitoring, digital pathology, and digital radiology.

3) Mhealth: recent interest in the integration of smartphones and tablet technology in health care with particular interest in the opportunities and the relevant problems of risks, abuse, and regulation.

4) Acceptance of and consensus in the use of robots for assistance and rehabilitation.

5) Challenges and acceptance of the use of Artificial Intelligence in Digital Radiology and Digital Pathology.

6) Cybersecurity in the health domain.

Dr. Giansanti is a Professor at Sapienza and Catholic University in Rome and a tutor of theses. He is a Board Editor and reviewer for several journals. He has 152 publications indexed on Scopus and more than 200 contributions, such as monographies, chapters, and congress papers.

Preface to "The Digital Health in the Pandemic Era"

Digital health, virtual assistance, and telemedicine are terms often used interchangeably to refer to remote medical assistance, monitoring, and care. Several studies and insights have developed these issues, analyzing the advantages and disadvantages and successes and failures and offering reflections on the implications and issues surrounding these technologies in the health domain. The results of these investigations are affecting the redesign of hospital and outpatient management based on digital innovation using eHealth and mHealth. Digital health encompasses a broad spectrum of technologies, including wearable personal devices and internal devices, as well as various types of sensors and innovative solutions. Digital health can help identify risks and correct assistance in the diagnosis, treatment, and monitoring of health conditions, offering new potential both to the population and insiders in the health domain. During the COVID-19 pandemic, this approach made it possible to offer assistance and continue care at home, protecting patients, preserving health workers, limiting the spread of the virus, and reducing the need for hospitalization. For example, the opportunity to make digital measurements of oxygen saturation at home has been used to make fundamental decisions regarding the health of patients, such as the choice between hospitalization and respiratory support. It has also become possible to monitor frail patients from home (e.g., with diabetes or cardiovascular or oncological problems), improving the continuity of care and reducing the pressure on hospitals. Digital Health (DH) also contributed to the fight against the pandemic in various new ways, such as in the management of digital contact tracing and vaccination processes through smart technology. This reprint, which deals with the development of DH during the COVID-19 pandemic, contains contributions from various experts in different fields.

Emerging topics in the Special Issue were:

·Digital contact tracing (DCT) and its impact on the spread of the pandemic in different populations, together with the factors that influenced its use.

·The use of DH in life sciences, including anatomy, bioinformatics, cell biology, neuroscience, physiology, and population biology, among others.

·Artificial intelligence (AI) applications in DH, including COVID-19-specific diagnostics, physical activity monitoring, obesity diagnosis, the detection of abnormalities in chest X-rays, and mental health monitoring.

·Large-scale population surveys analyzing the impact of biomedical parameters, health determinants, and digital literacy on the population.

·Remote healthcare interventions and their impact on self-care in chronic patients.

·Chatbots in the health domain and their increasing use during the pandemic.

My sincere thanks to Shane Zheng, who provided exceptional support in every phase of the creation of this collection.

Daniele Giansanti
Editor

life

Editorial

The Digital Health: From the Experience of the COVID-19 Pandemic Onwards

Daniele Giansanti

Centro Tisp ISS, 00161 Rome, Italy; gianslele@gmail.com or daniele.giansanti@iss.it

Digital health has a long history of development and is particularly resonant in the last two years, due to the pandemic [1,2]. A recent study [2] revised the definitions associated with digital health. The authors undertook a quantitative analysis and term mapping of the published definitions of digital health. They analyzed 95 unique definitions of digital health, from both scholarly and general sources. The findings showed that digital health, as has been used in the literature, is more concerned about the provision of healthcare rather than the use of technology. The wellbeing of people, both at population and individual levels, have been emphasized more than the care of patients suffering from diseases. Furthermore, the use of data and information for the care of patients was highlighted. A dominant concept in digital health appeared to be mobile health (mHealth), which is related to other concepts, such as telehealth, eHealth, and artificial intelligence in healthcare.

Even the World Health Organization (WHO) entered into this discussion [3]. The WHO is harnessing the power of digital technologies and health innovation to accelerate the global attainment of health and well-being. WHO has three key objectives [4] to promote the adoption and scale-up of digital health and innovation:

- Translating the latest data, research, and evidence into action: this means promoting standards for interoperability and data sharing, and supporting the implementation of digital solutions that contribute to informed decision making;
- Enhancing knowledge through scientific communities of practice;
- Systematically assessing and linking the needs of the country with the supply of innovations.

Citation: Giansanti, D. The Digital Health: From the Experience of the COVID-19 Pandemic Onwards. *Life* **2022**, *12*, 78. https://doi.org/10.3390/life12010078

Received: 28 December 2021
Accepted: 4 January 2022
Published: 6 January 2022

National and supranational entities are also clearly addressing this issue. For example, the Food and Drug Administration (FDA) [5] in the U.S.A., states that digital health includes categories, such as mHealth, health information technology (IT), wearable devices, telehealth and telemedicine, and personalized medicine. Furthermore, the FDA recognizes that (a) from mobile medical apps and software that support the clinical decisions that doctors make every day to artificial intelligence and machine learning, digital technology has been driving a revolution in health care. (b) Digital health tools have the vast potential to improve our ability to accurately diagnose and treat disease and to enhance the delivery of healthcare for the individual. (c) Digital health technologies use computing platforms, connectivity, software, and sensors for healthcare and related uses; these technologies span a wide range of uses, from applications in general wellness to applications as a medical device.

The potential benefits have been identified:

- Reduce inefficiencies,
- Improve access,
- Reduce costs,
- Increase quality, and
- Make medicine more personalized for patients.

The following FDA action topics were also highlighted in the context of digital health, to provide clarity using practical approaches that balance the benefits and risks:

- Software as a Medical Device (SaMD);
- Artificial Intelligence and Machine Learning in Software as a Medical Device;
- Cybersecurity;
- Device Software Functions, including Mobile Medical Applications;
- Health IT;
- Medical Device Data Systems;
- Medical Device Interoperability;
- Telemedicine;
- Wireless Medical Devices.

As a supranational entity, the European (Eu) Commission has also addressed the issue [6,7]. It stated that digital health is based on tools and services that use information and communication technologies (ICTs) to improve prevention, diagnosis, treatment, monitoring, and management of health-related issues, and to monitor and manage lifestyle habits that impact health. Digital health and care is innovative and can improve access to care and the quality of that care, as well as increase the overall efficiency of the health sector. It identified the following Eu pillars:

- Pillar 1: Secure data access and sharing;
- Pillar 2: Connecting and sharing health data for research, faster diagnosis, and improved health;
- Pillar 3: Strengthening citizen empowerment and individual care through digital services.

WHO's objectives in launching the integration of digital health, the FDA actions topics, and the EU pillars are shareable and are all elements put to the test in the last two years, due to the COVD-19 pandemic.

From a quick overview on Pubmed using the search key "digital health" [8], we observe that, at the date of writing this editorial (28 December 2022), there are 41,165 studies that have addressed this issue.

We also observe that, in the last two years, marked by the pandemic, we have seen the publication of 18446 works, equal to 44.32% of the total.

Several studies and insights published during 2020–2021 have developed these issues, analyzing the advantages and disadvantages, successes and failures, and offering reflections on the implications and issues of these technologies in the health domain. The results of these investigations will affect the redesign of hospital and outpatient management, based on digital innovation using eHealth and mHealth. It has been highlighted by the WHO, FDA, and EU [4–7], that digital health encompasses a broad spectrum of technologies, including wearable personal devices and internal devices, as well as various types of sensors and innovative solutions. Digital health can help in the diagnosis, treatment, and monitoring of health conditions, offering new potential to both the population and the insiders of the health domain. During the pandemic, this approach made it possible to offer assistance and continue care at home, protecting patients, preserving health workers, limiting the spread of the virus, and reducing the need for hospitalization [9]. For example, the opportunity to make digital measurements of oxygen saturation [10] at home has been used to make fundamental decisions for the health of patients, such as the choice between hospitalization and respiratory support. It has also become possible to monitor frail patients from home (e.g., with diabetes or cardiovascular or oncological problems) [11], thus improving the continuity of care and reducing the pressure on the hospitals. Digital health also continues to contribute to the fight against the pandemic in various new ways, such as the management of digital contact tracing [12] and vaccination processes through smart technology [13]. Limitations were also clearly shown, which mainly concerned the following points:

- The digital divide;
- Organizational aspects (both with regard to administrative and technological aspects).

The digital divide has two important components. The first component is represented by the difficulty in accessing to the infrastructures; to date, this also remains a problem in the richest and most technologically advanced countries in the world [14]. The second component is represented by the literacy [15]. These two components of the digital divide were particularly visible during the COVID-19 pandemic [16–21]. They can depend on cultural, ethnic, social, national, and political factors [19,20]; furthermore, they can exacerbate the disparities [21].

In regards to the organizational aspects, it can be highlighted that they are broad-spectrum. Undoubtedly, developing countries start from a base of technological need that must exist before they can fully apply digital health in the health domain. However, even the most developed countries, such as Italy, found themselves facing difficulties, especially at an early stage, which, according to some authors [22], were caused by the following factors:

- The scattered distribution and heterogeneity of available tools;
- The lack of integration with the electronic health record of the national health system;
- The poor interconnection between telemedicine services operating at different levels;
- The lack of a real multidisciplinary approach to the patient's management;
- The heavy privacy regulations and lack of clear guidelines, together with the lack of reimbursement.

From the Editorial, it emerges that the pandemic was an important driver for digital health. This concerned: (a) the improvement of medical and technological knowledge; (b) the stimulus for the implementation of solutions in the health domain; (c) the stimulus for the resolution of long-standing problems related to management and organizational aspects (e.g., reimbursement, and introduction into the treatment process); And (d) the proposition of new solutions (e.g., contact tracing). The analysis of how well it worked and how much it needs to be improved, is important to both improve the offer from the point of view of technology and quality, and to focus the interest of the scholars at 360°. All that is important to both continue the battle against COVID-19 and to prepare new effective and functioning stable health models for the post-pandemic future. For this purpose, the Special Issue, "The Digital Health in the Pandemic Era" (https://www.mdpi.com/journal/life/special_issues/DigitalHealth_Pademic (accessed on 28 December 2021)) [23], has been prepared with the aim of creating a meeting of experiences of the experts of the health domain.

Funding: This research received no external funding.

Institutional Review Board Statement: Not applicable.

Informed Consent Statement: Not applicable.

Conflicts of Interest: The authors declare no conflict of interest.

References

1. Available online: https://medium.com/that-medic-network/a-brief-history-of-digital-health-b238f1f5883c (accessed on 28 December 2021).
2. Fatehi, F.; Samadbeik, M.; Kazemi, A. What is Digital Health? Review of Definitions. *Stud. Health Technol. Inform.* **2020**, *275*, 67–71. [CrossRef] [PubMed]
3. Available online: https://www.who.int/health-topics/digital-health#tab=tab_1 (accessed on 28 December 2021).
4. Available online: https://www.who.int/health-topics/digital-health#tab=tab_2 (accessed on 28 December 2021).
5. Available online: https://www.fda.gov/medical-devices/digital-health-center-excellence/what-digital-health (accessed on 28 December 2021).
6. Available online: https://ec.europa.eu/health/ehealth/home_en (accessed on 28 December 2021).
7. Available online: https://ec.europa.eu/health/sites/default/files/ehealth/docs/2018_ehealth_infographic_en.pdf (accessed on 28 December 2021).
8. Available online: https://pubmed.ncbi.nlm.nih.gov/?term=digital+health&sort=date&size=200 (accessed on 28 December 2021).
9. Giansanti, D. The Role of the mHealth in the Fight against the Covid-19: Successes and Failures. *Healthcare* **2021**, *9*, 58. [CrossRef] [PubMed]

10. Wurzer, D.; Spielhagen, P.; Siegmann, A.; Gercekcioglu, A.; Gorgass, J.; Henze, S.; Kolar, Y.; Koneberg, F.; Kukkonen, S.; McGowan, H.; et al. Remote monitoring of COVID-19 positive high-risk patients in domestic isolation: A feasibility study. *PLoS ONE* **2021**, *16*, e0257095. [CrossRef]
11. Brahmbhatt, D.H.; Ross, H.J.; Moayedi, Y. Digital Technology Application for Improved Responses to Health Care Challenges: Lessons Learned From COVID-19. *Can. J. Cardiol.* **2021**. [CrossRef] [PubMed]
12. Kolasa, K.; Mazzi, F.; Leszczuk-Czubkowska, E.; Zrubka, Z.; Péntek, M. State of the Art in Adoption of Contact Tracing Apps and Recommendations Regarding Privacy Protection and Public Health: Systematic Review. *JMIR mHealth uHealth* **2021**, *9*, e23250. [CrossRef] [PubMed]
13. Simeoni, R.; Maccioni, G.; Giansanti, D. The Vaccination Process against the COVID-19: Opportunities, Problems and *mHealth* Support. *Healthcare* **2021**, *9*, 1165. [CrossRef]
14. Van Deursen, A.J.; van Dijk, J.A. The first-level digital divide shifts from inequalities in physical access to inequalities in ma-terial access. *New Media Soc.* **2019**, *21*, 354–375. [CrossRef] [PubMed]
15. Neter, E.; Brainin, E.; Baron-Epel, O. Group differences in health literacy are ameliorated in ehealth literacy. *Health Psychol. Behav. Med.* **2021**, *9*, 480–497. [CrossRef]
16. Van Deursen, A.J.; Helsper, E.J. Collateral benefits of Internet use: Explaining the diverse outcomes of engaging with the In-ternet. *New Media Soc.* **2018**, *20*, 2333–2351. [CrossRef]
17. Gabbiadini, A.; Baldissarri, C.; Durante, F.; Valtorta, R.R.; De Rosa, M.; Gallucci, M. Together Apart: The Mitigating Role of Digital Communication Technologies on Negative Affect During the COVID-19 Outbreak in Italy. *Front. Psychol.* **2020**, *11*, 554678. [CrossRef]
18. Shah, S.G.S.; Nogueras, D.; Van Woerden, H.C.; Kiparoglou, V. The COVID-19 Pandemic—A pandemic of lockdown loneliness and the role of digital technology: A viewpoint (Preprint). *J. Med. Internet Res.* **2020**, *22*, e22287. [CrossRef]
19. Lai, J.; Widmar, N.O. Revisiting the Digital Divide in the COVID-19 Era. *Appl. Econ. Perspect. Policy* **2021**, *43*, 458–464. [CrossRef]
20. Shek, D.T.L. COVID-19 and Quality of Life: Twelve Reflections. *Appl. Res. Qual. Life* **2021**, *16*, 1–11. [CrossRef]
21. Eberly, L.A.; Kallan, M.J.; Julien, H.M.; Haynes, N.; Khatana, S.A.M.; Nathan, A.S.; Snider, C.; Chokshi, N.P.; Eneanya, N.D.; Takvorian, S.U.; et al. Patient Characteristics Associated with Telemedicine Access for Primary and Specialty Ambulatory Care During the COVID-19 Pandemic. *JAMA Netw. Open* **2020**, *3*, e2031640. [CrossRef] [PubMed]
22. Omboni, S. Telemedicine During the COVID-19 in Italy: A Missed Opportunity? *Telemed. e-Health* **2020**, *26*, 973–975. [CrossRef] [PubMed]
23. Available online: https://www.mdpi.com/journal/life/special_issues/DigitalHealth_Pademic (accessed on 28 December 2021).

life

MDPI

Opinion

The Chatbots Are Invading Us: A Map Point on the Evolution, Applications, Opportunities, and Emerging Problems in the *Health Domain*

Daniele Giansanti

Centre Tisp, Istituto Superiore di Sanità, 00161 Roma, Italy; daniele.giansanti@iss.it; Tel.: +39-06-49902701

Abstract: The inclusion of chatbots is potentially disruptive in society, introducing opportunities, but also important implications that need to be addressed on different domains. The aim of this *study* is to examine chatbots in-depth, by mapping out their technological evolution, current usage, and potential applications, opportunities, and emerging problems within the *health domain*. The study examined three *points of view*. The *first point of view* traces the technological evolution of chatbots. The *second point of view* reports the fields of application of the chatbots, giving space to the expectations of use and the expected benefits from a cross-domain point of view, also affecting the *health domain*. The *third and main point of view* is that of the analysis of the state of use of chatbots in the health domain based on the scientific literature represented by systematic reviews. The overview identified the topics of greatest interest with the opportunities. The analysis revealed the need for initiatives that simultaneously evaluate multiple domains all together in a synergistic way. Concerted efforts to achieve this are recommended. It is also believed to monitor both the process of osmosis between other sectors and the *health domain*, as well as the chatbots that can create psychological and behavioural problems with an impact on the *health domain*.

Keywords: chatbot; health; health domain; artificial intelligence

Citation: Giansanti, D. The Chatbots Are Invading Us: A Map Point on the Evolution, Applications, Opportunities, and Emerging Problems in the Health Domain. *Life* **2023**, *13*, 1130. https://doi.org/10.3390/life13051130

Academic Editor: Tao Huang

Received: 29 March 2023
Revised: 26 April 2023
Accepted: 27 April 2023
Published: 5 May 2023

1. Introduction

The world has recently witnessed the diffusion of the technological phenomenon of chatbots [1–4]. This phenomenon is simultaneously attracting and worrying public opinion, scholars and stakeholders. The attraction is due to the rapid diffusion, the easy accessibility, and the opportunities that chatbots, increasingly integrated with artificial intelligence, seems to offer. However, it is worrying that such rapid diffusion and easy accessibility have not been adequately accompanied by robust reflections on the impact they have on many domains of public life, from the social to the ethical and regulatory.

A chatbot can be defined as:

1. *A computer program designed to have a conversation with a human being, especially over the internet* (https://dictionary.cambridge.org/dictionary/english/chatbot [1]);
2. *A computer program in the form of a virtual e-mail correspondent that can reply to messages from computer users* (https://www.dictionary.com/browse/chatbot [2]; https://www.collinsdictionary.com/dictionary/english/chatbot [3]);
3. A bot that is designed to converse with human beings. Bot: computer program or character (as in a game) designed to mimic the actions of a person (https://www.merriam-webster.com/dictionary/chatbot [4]).

The introduction of artificial intelligence (AI) has inspired further stimulating scientific debates regarding its large-scale application, including in the world of healthcare [5]. An example of this is represented by the chatGPT tool [6], which has recently become widespread, rapidly attracting scientific attention to its potential and implications in its applications in social life [7] and in the health domain [8]. On the other hand, facing the challenges of integrating technologies with AI is an open and current challenge with opportunities, challenges, and bottlenecks to overcome, affecting several domains [9,10], and also affecting learning processes and ethics [11].

The inclusion of chatbots is potentially disruptive in society, introducing opportunities but also important implications that need to be addressed on different domains.

The aim of this analysis is to examine chatbots in depth by mapping out their technological evolution, current usage, and potential applications within the health domain.

The sub-aims are:

- To determine the historical development of chatbot technology and its evolution over time;
- To analyze the current usage patterns of chatbots within different fields;
- To identify the key features, applications and opportunities of chatbots that are specific to the *health domain;*
- To identify the potential problems and bottlenecks in the *health domain.*

2. Methods

The study was arranged into three points of view.

The *first point of view* traces the technological evolution of chatbots starting from the first pioneering experiences.

The *second point of view* reports the fields of application of the chatbots both from the point of view of categorization and of the sector of use, also giving space to the expectations of use and the expected benefits from a cross-domain point of view, also impacting the *health domain.*

The *third and main point* of view is that of the analysis of the state of use of chatbots in the health domain based on the scientific literature represented by systematic reviews. We decided to analyze the systematic reviews, the highest level of evidence in healthcare, because they are capable of providing a comprehensive evaluation of a particular topic by identifying and analyzing all the available primary research studies. This summary of evidence can help detect the principal patterns of interest and highlight areas where additional research is needed or where current research is insufficient to support clinical decisions, in this case, in the clinical domain.

The overview related to the first and second point of views was based on targeted searches on Google and Google Scholar.

The overview related to the third point of view followed a targeted search on PubMed by means of a properly settled composite key.

The overview, as a whole followed the ANDJ checklist, a standardized checklist for the structure of a narrative review.

This overview was carefully crafted with a consideration of five parameters (N1–N5) that have been evaluated on a scale ranging from one (minimum) to five (maximum). The parameters are as follows:

N1: Clarity of introduction and rationale for the review.

N2: Appropriateness of review design.

N3: Clear description of methods.

N4: Clear presentation of results.

N5: Justification of conclusions based on results.

N6: Full disclosure of potential conflicts of interest by authors.

These parameters have been thoughtfully selected to ensure the comprehensiveness and quality of this overview. All selected elements must receive a score of at least three on all parameters in order to be included.

3. Results

3.1. An Overview of the Evolution of Technology

The origin of chatbots [12] can probably be attributed to Alan Turing's 1950s vision of intelligent machines. Artificial intelligence, the basis of chatbots, has therefore, developed these tools. We can summarize this evolution in brief [12]. The first chatbot named *ELIZA*, created in 1966, simulated a psychotherapist's function, repeating the users' sentences in an interrogative form. Its ability to communicate was limited; however, it can be considered a source of inspiration for further evolutions [13,14]. In 1972, *PARRY* was introduced. It acted as a patient with schizophrenia and defined its responses based on a system of assumptions and "emotional responses" [15]. AI was firstly used in the domain of the chatbots with the introduction of *Jabberwacky* in 1988. CleverScript was used in this system. It was a language based on spreadsheets, which was useful in the development of chatbots. This system was able to respond based on previous answers. It was limited in speed and number of users [16]. The term *CHATTERBOT* was introduced *in 1991*. It was an artificial player with a primary function of chatting [17]. *Dr. Sbaitso* appeared in 1992 [18]. It played the role of a psychologist seemingly without showing complications in its interactions with users [18]. *ALICE* was a further step forward in the world of chatbots. It used the artificial intelligence markup language. It performed better compared to ELIZA [19]. *SmarterChild* was introduced in *2001*. It was integrated with Messengers. This chatbot, for the first time, could help people with useful daily tasks using large databases with information related to movie times, sports, weather, and other information [20]. The chatbot using AI made another important step forward with the introduction of the smart personal voice assistants between 2010 and 2020. They were capable of understanding vocal commands and informative tasks. *Apple Siri, IBM Watson, Google Assistant, Microsoft Cortana, and Amazon Alexa* are the most popular voice assistants [21–27]. Early in 2016, a further evolution in AI technology radically changed the communicative interaction between users and manufactures. Social media platforms allowed developers to create chatbots that allowed the clients to complete specific tasks using their own messaging applications. At the end of 2016, the chatbots covered a wide range of applications ranging from entertainment to healthcare, including marketing, education, generalized support, cultural heritage, and much more. Moreover, the Internet of Things allowed new fields of application for chatbots; they played the role of connectors and mediators of "smart objects" [28]. See for example *Microsoft XiaoIce*, an AI-based chatbot with the role of satisfying the human need for sociability [29]. At the end of this process of evolution, the way of engaging in discussion with a chatbot was completely different from the *ELIZA* chatbot. Today, a chatbot is capable of sharing personal thoughts along with family drama events.

3.2. *Exploring the Wide Range of Applications*

Chatbots have many desirable characteristics that potentially make them an ideal interlocutor to interact and work with [30–32]. In fact, chatbots are potentially able to work efficiently 24/7. They can be customized with user data to try to create a better collaboration, compliance, and, ultimately, a *real virtual symbiosis* with the user. They also have the desirable IT property of scalability, being able to handle a large volume of data and requests/interactions simultaneously. Finally, they allow for savings in resources, being able, through their automatisms, to autonomously carry out tasks, resulting in economic savings in the management of the processes.

Today, chatbots have several applications.

For example, they can be applied, for example, to the following sectors [30–32]:

- *Customer service*: Chatbots can be used as virtual assistants to provide 24/7 customer support, answer frequently asked questions, resolve issues and handle customer complaints;
- *Sales and marketing*: Chatbots can assist customers in making purchases, provide product recommendations and cross-sell or upsell other products and services;
- *Healthcare:* Chatbots can provide health-related information, answer medical questions, assist patients in scheduling appointments and in other activities, such as refilling prescriptions, virtual triages, and self-monitoring.
- *Education:* Chatbots can help students with homework assignments, providing study documents and answering, for example, to specific questions.

- *Finance:* Chatbots can assist customers with banking tasks such as checking account balances, transferring money and paying invoices;
- *Travel and tourism*: Chatbots can help travelers plan their trips, book flights and hotels and provide information on tourist attractions;
- *Entertainment:* A chatbot can provide games or other tools to give fun time.
- *Industry:* Chatbots can provide suggestions, give general and specific assistance, support customers on specific products, checking inventory levels and providing assistance with returns and exchanges.

By analyzing the applications under a broader and cross-domain perspective independent of the categories, it can be noted that chatbots today can find up to 37 applications [33]. Population surveys have shown that chatbots are most expected to [33]: quickly respond to an emergency (37%), solve a complex problem (35%) and Respond quickly (35%). As far as the expected benefits are concerned, the following are mostly expected [34]: having access to a 24/7 service (64%), obtaining quick answers (55%) and answering elementary questions (55%). As can be seen, these "expectations" and "expected benefits" of the different categories identified seem to overlap, and touch many sectors, including the very important one of *healthcare*. In fact, in the healthcare domain, having a rapid response in the event of an emergency, in a 24 h mode, is strategic *in emergency medicine and intensive hospitalization medicine*.

3.3. The Chatbots in the Health Domain: Applications, Opportunities, Open Challenges, and Problems

The chatbots are increasingly showing several applications in the *health domain*. Diverse keywords [35–55] are associated with the concept of the chatbot in healthcare; among the most frequent we find: *patient engagement, clinical support, mental health, health monitoring, patient education, appointment scheduling, symptom checking, chronic disease management, triage, remote monitoring, telemedicine, health coaching and emergency response*.

Chatbots can be used in several fields in the health domain [35–55]: (a) As a tool for answering frequently asked questions [39]; (b) For the collection of data and patient details [39,42,43]; (c) To support patients finding a doctor or a specific service, on managing appointments, and on the medication dispensing procedure [42]; (d) As an interactive guide to the management of self-assessment and symptom control [39,42]; (e) As a tool to guide an interactive triage, applicable in the case of an emergency as well [36]; (f) In telehealth, digital health applications, and remote monitoring [37,39,40,42,43]. (g) In the learning process, in the construction of scientific knowledge and in supporting scientific dissemination [35,38]; (h) In mental health applications [37,38]; (i) For physical wellness and health coaching [40].

From a general point of view, the use of these tools has the potential to lighten the hospital and care facility load, decentralizing many of the activities, allowing them to be carried out in a remote mode, something that during a situation such as the COVID-19 pandemic better protects all the actors involved. The patients can be more responsible, self-diagnose independently, and invited and supported to take better care of themselves also in relation to the wellness and psychological aspects.

A search was performed on PubMed with the following composite key

(chatbot[Title/Abstract]) AND ((health [Title/Abstract]) OR (healthcare[Title/Abstract]) OR (health domain[Title/Abstract])).

The key showed the evolution of scientific dissemination in this area. Since 2010, *370* papers have been published, including 19 systematic reviews.

The research highlights a terrific growth in the volume of publications in the last three years, coinciding with the outbreak of the pandemic, with 340 of the papers were published from 2020 to 2022, which is *91.9%* of the total papers published, and the number of papers published in *2022* is *117*, which is *31.62%* of the total papers published.

We decided to analyze the systematic reviews [37,40,45,49,51,52,54,56–67] to detect the principal patterns of interest and highlight areas where additional research is needed or where current research is insufficient to support clinical decisions to introduce these tools in the clinical routine.

The analysis of the systematic review allowed us to detect five areas of interest, which are as follows:

- *Application of chatbots in mental health;*
- *Application of chatbots in the domain of the addiction;*
- *Application of chatbots in the domain of the chronic disease;*
- *Application of chatbots in the domain of the wellness and fitness;*
- *Heterogeneous applications in the health domain;*
- *Technology assessment.*

3.3.1. Application of Chatbots in Mental Health

Five studies have investigated the application of chatbots in *mental health* [45,61,65–67]. Lim et al. [61] reviewed the effectiveness of chatbot-delivered psychotherapy in improving depressive symptoms in adults with depression or anxiety. The review highlighted that chatbot-delivered psychotherapy significantly improved depressive symptoms. The preferred features for the design of chatbots include embodiment, a combination of input and output formats, less than 10 sessions, problem-solving therapy, offline platforms, and different regions of the United States. The study concluded that chatbot-delivered psychotherapy could be an alternative treatment for depression and anxiety, and further high-quality trials were needed to confirm its effectiveness.

Ruggiano et al. [65] identified current commercially available chatbots that were designed for use by people with dementia and their caregivers, and assessed their quality in terms of features and content. Although the chatbots were generally found to be easy to use, limitations were noted regarding their performance and programmed content for dialog. The authors concluded that evidence-based chatbots were needed to adequately educate and support people with dementia and their caregivers.

Vaidyam et al. [66] reviewed the use of conversational agents (chatbots or voice assistants) in the assessment and treatment of serious mental illnesses, such as depression, anxiety, schizophrenia, and bipolar disorder. The study highlighted positive outcomes for diagnostic quality, therapeutic efficacy, and acceptability. However, certain populations, such as pediatric patients and those with schizophrenia or bipolar disorder, were un-der-represented in the research. The authors recommended the standardization of studies to include measures of patient adherence and engagement, therapeutic efficacy, and clinician perspectives.

Gaffney et al. [67] investigated the use of conversational agent interventions in mental health. The interventions were diverse and targeted a range of mental health problems using various therapeutic orientations. All included studies reported reductions in psychological distress post-intervention, and the controlled studies demonstrated significant reductions in psychological distress compared to inactive control groups. However, the authors concluded that a more robust experimental design was required to demonstrate efficacy and efficiency.

Hoermann et al. [45] analyzed the feasibility and effectiveness of one-on-one mental health interventions that used chatbots. The interventions showed significant improvements compared to waitlist conditions, but were not superior to the usual treatment. The study also found substantial innovation in the use of trained volunteers and chatbot technologies. However, further research was needed to determine the feasibility of this mode of intervention in clinical practice.

3.3.2. Application of Chatbots in the Domain of the Addiction

The field of addiction was dealt with in three studies [49,57,62].

Aggarwal et al. [49] evaluated the feasibility, efficacy, and characteristics of AI chatbots for promoting health behavior change. The review found that AI chatbots have shown high efficacy in promoting healthy lifestyles, smoking cessation, treatment or medication adherence, and reduction in substance misuse. However, there were mixed results regarding feasibility, acceptability, and usability. Furthermore, the authors concluded that the reported results needed to be interpreted with caution due to limitations in internal validity, insufficient description of AI techniques, and limited generalizability. Future studies should adopt robust randomized control trials to establish definitive conclusions.

He et al. [57] also investigated conversational agents for smoking cessation. The systematic review and meta-analysis found that all studies reported positive effects on cessation-related outcomes. Meta-analyses of randomized controlled trials showed that conversational agents were more effective in promoting abstinence compared to control groups. However, the included studies were diverse in design, and evidence of publication bias was identified. The review also highlighted a lack of theoretical foundations and a need for relational communication in future designs. The standardization of reporting on and designing conversational agents was warranted for a more comprehensive evaluation. Overall, this review provided insights into the potential of conversational agents for smoking cessation and the need for further research and development to improve their effectiveness and acceptability.

Ogilvie et al. [62] researched the use of chatbots in the field of addiction, specifically as supportive agents for those with a substance use disorder. The findings suggested that the corpus of the research in this field is limited, and more research was needed to confidently report on the usefulness of chatbots in this area. While some papers reported a reduction in substance use in participants, caution was advised as expert input was needed to safely leverage existing data and avoid potential harm to the intended audience.

3.3.3. Application of Chatbots in the Domain of the Chronic Disease

Two studies focused on the domain of chronic disease [58,60].

Pernencar et al. [58] studied the field of e-Therapy and mobile apps integrated into healthcare systems. The study reviewed the connection between chatbots with inflammatory bowel disease patients' healthcare, with the goal of supporting the development of digital products for chronic diseases. The study highlighted that the chatbot technology for chronic disease self-management had high acceptance and usability levels. However, the chatbot ontology still needed strong guidelines for personalizing communication.

Sawad et al. [60] explored different conversational agents used in healthcare for chronic conditions, analyzing their communication technology, evaluation measures, and AI methods. They found that users provided positive feedback about the usefulness, satisfaction, and ease of use of conversational agents. However, there was still insufficient evidence to determine the efficacy of AI-enabled conversational agents for chronic health conditions due to the lack of reporting of technical implementation details.

3.3.4. Application of Chatbots in the Domain of the Wellness and Fitness

Two studies explored the application of chatbots in the domain of wellness and fitness [40,64].

Luo et al. [64] examined the use of conversational agents in promoting physical activity (PA). Conversational agents were found to have moderate usability and feasibility. The authors reported that conversational agents were effective in promoting PA. However, they highlighted the need for further research on the long-term effectiveness and safety of conversational agents in promoting PA, as well as the importance of using evidence-informed theories and addressing user preferences for variety and natural language processing.

Oh et al. [40] looked at studies evaluating the use of AI chatbots in changing physical activity, healthy eating, weight management behaviors, and other related health out-comes. The study found that chatbot interventions were promising in increasing physical activity but limited in changing diet and weight status. The review reported that the studies had inconsistent outcome assessments on chatbot characteristics. The study recommended standardization of designing and reporting chatbot interventions in the future. Overall, the authors concluded that chatbots might improve physical activity, but more research is needed on their efficacy for diet and weight management/loss.

3.3.5. Heterogeneous Applications in the Health Domain

Five studies analysed multiple applications simultaneously in the *health domain* [37,51, 52,56,63].

Milne-Ives et al. [52] discussed the increasing use of conversational agents in healthcare to support a variety of activities, such as behavior change, treatment support, health monitoring, training, triage and screening support. In particular, the review evaluated the effectiveness and usability of these agents and identified the elements that users liked and disliked. The evidence generally reported positive or mixed results for effective-ness, usability and satisfactoriness. However, qualitative feedback highlighted limitations of the agents, and the study design quality was limited. Further research was needed to evaluate the cost-effectiveness, privacy and security of these agents.

Geoghegan et al.'s [51] study reviewed the use of chatbots in the follow-up care of patients who underwent physical healthcare interventions. The included studies analyzed chatbots that were used for monitoring after cancer management, hypertension and asthma, orthopedic intervention, ureteroscopy, and intervention for varicose veins. All chatbots were deployed on mobile devices, and a range of metrics were identified. Importantly, no study examined patient safety. The authors suggested that further investigation was needed to evaluate the acceptability, efficacy and mechanistic evaluation of chatbots in routine clinical care.

Xu et al. [63] reviewed the recent advancements of and current trends in the use of chatbot technology in medicine, particularly in cancer therapy. The article provided a brief historical overview and discussed the design characteristics and the potential uses of chatbots in diagnosis, treatment, monitoring, patient support, workflow efficiency and health promotion. The article also addresses limitations and areas of concern, including ethical, moral, security, technical and regulatory standards. The authors concluded that chatbots have the potential to be integrated into clinical practice by working alongside health practitioners to reduce costs, refine workflow efficiencies and improve patient outcomes. However, they called for further research and interdisciplinary collaboration to advance this technology and improve the quality of care for patients.

Huq et al. [37] investigated the potential benefits of chatbots and conversational agents in improving the quality of life for aged and impaired individuals. The study emphasized the need for further research and development to fill knowledge gaps in remote healthcare and rehabilitation, which could ultimately lead to improved outcomes for patients.

Sallam [56] proposed a review on ChatGPT, an artificial intelligence (AI) chatbot that uses large language models. The review examined the potential benefits and limitations of using ChatGPT in healthcare education, research and practice. The article found that ChatGPT has several potential benefits, including improving scientific writing, enhancing research equity and versatility and improving personalized learning. However, there were also significant concerns surrounding ChatGPT's use, including ethical, copyright, transparency and legal issues, the risk of bias, plagiarism, lack of originality, cybersecurity issues and inaccurate content. Despite the potential benefits, the review recommended caution when using ChatGPT and other similar tools in healthcare and academia, calling for a code of ethics to guide their responsible use.

3.3.6. Technology Assessment

Technology assessment was investigated in two studies [54,59].

Denecke and May [59] discussed the use of conversational agents (CAs) in healthcare and the lack of a standard procedure to study their usability. The authors conducted a systematic literature review and found that a variety of tools and metrics were used to assess usability, but there was little consistency in the study designs. As a result, they found that it was difficult to compare usability among different CAs. The authors recommended the development of a standardized procedure for evaluating CA usability that can be applied consistently and can be tailored to specific features of individual CAs.

Chattopadhyay et al. [54] investigated the effectiveness of virtual humans (VH) in patient-facing systems. The study also identified two design categories—simple VH and VH augmented with health sensors and trackers. The intervention was mainly delivered using personal computers, and more focused analysis to identify what features of VH interventions contributed toward their effectiveness is needed in the future. Overall, the study offered evidence for the efficacy of VH in patient-facing systems, but further research is required to fully understand their potential benefits.

4. Discussion

Chatbots have undergone a terrific evolution through decades of technological innovation. The latest and most impactful advancement is the one we are experiencing, which is represented by artificial intelligence [8,9]. Today, the chatbots can find up to 37 applications [33]. Population expectations of chatbots are many and important. The major expectation of the population with respect to these systems is chatbots' rapid response in emergency situations [33]. The greatest expected benefit is that of being able to receive 24 h service from these systems. The main "expectations" and the main "expected benefits" of different domains seem to overlap, including the very important one of healthcare. In fact, specifically in healthcare, having a rapid response in the event of an emergency, in a 24 h mode, is strategic in emergency medicine and intensive hospitalization medicine. Additionally, *healthcare* is one of the domains affected by the introduction of these systems. After having retraced, through our study, the evolution of technology, and after having addressed the topic of applications, from which expectations of use and cross-benefits between categories (including the health domain) have emerged, we have focused on the *health domain*.

An overview on PubMed confirmed a rapid increase in scientific interest in the health domain [68], in line with the perceived general interest [69]. In particular, PubMed showed that in the last three years, a volume of publications equal to 91.2% of total publications on these systems has been published, which, as has been discussed, have existed since 1966 [13]. A substantial contribution to this growth was also made both by the COVID-19 pandemic, which has brought 91 publications since 2020 [70] (see the composite key in the first position of the Box 1 reported below), and by artificial intelligence, which was of interest in 123 studies [71] (see the composite key in the second position of the Box 1 reported below).

The topics of greatest interest emerged from a search on PubMed based on systematic reviews. These studies concerned two specific systematic reviews on *technology assessment* and systematic reviews analyzing chatbot applications in *mental health, addiction, chronic diseases, wellness and fitness, and, finally, the applications on heterogeneous applications in the health domain and technology assessment.*

The two studies [54,59] on *technology assessment*, even if limited to a few domains, gave us important indications on chatbot technology.

The first one [59] highlighted the lack of a standard procedure to study usability, with a variety of different tools and metrics used to assess usability with little consistency in the study designs. The authors recommend the development of a standardized procedure for evaluating usability.

The second [54], while acknowledging the potential effectiveness of these systems, deemed it necessary to further investigate the real requirements that make chatbots effective and the potential benefits.

The other studies have unanimously highlighted the potential opportunities of these systems in specific applications but have highlighted various *critical issues concerning different single domains.*

The need for more attention on the *domain of ethics* has been recommended, for example in [56], with regard to the applications in healthcare education, research and practice.

The need to deepen the *domain of safety*, to avoid the potentially harmful impact of these systems has been highlighted in [62], with a focus on the domain of drug addiction.

The need for more attention on the *domain of standardization* has been highlighted by three studies [40,49,58]. In [58], the development of specific guidelines was recommended in a *chronicity study*. Reference [49] suggested the standardization of procedures and protocols in the domain of *addiction*. Even in *applications dedicated to fitness and wellness*, the need for standardization has been recalled both in design and reporting [40].

Concerns regarding the *domain of efficacy* have been raised in several studies [40,45,51,57,60,61,64,65,67]. An in-depth study of this domain has been suggested in various applications of mental health [45,61,65,67] in the domain of addiction [57], in the field of chronicity [60], in post-intervention medicine [51] and in wellness and fitness [40,64].

The *domain of cybersecurity* has been touched upon to a certain extent by all the studies, but particularly in two studies that have addressed multiple applications simultaneously in the health domain [52,56]. The *domain of interdisciplinarity* was recalled as important in cancer therapy, where this aspect, as is known, is a key factor [63].

The *need to address some domains* together was highlighted in [56]. In this study, dedicated to AI-based chatbots used in healthcare education, research, and practice, concerns were expressed on *ethical aspects, copyright, transparency, legal issues, the risk of bias, plagiarism, lack of originality, cybersecurity issues and inaccurate content.'*

Box 1. The proposed composite keys.

("chatbot"[Title/Abstract] AND ("health"[Title/Abstract] OR "healthcare"[Title/Abstract] OR "health domain"[Title/Abstract]) AND ("COVID-19"[All Fields] OR "COVID-19"[MeSH Terms] OR "COVID-19 vaccines"[All Fields] OR "COVID-19 vac-cines"[MeSH Terms] OR "COVID-19 serotherapy"[All Fields] OR "COVID-19 nucleic acid test-ing"[All Fields] OR "COVID-19 nucleic acid testing"[MeSH Terms] OR "COVID-19 serological testing"[All Fields] OR "COVID-19 serological testing"[MeSH Terms] OR "COVID-19 test-ing"[All Fields] OR "COVID-19 testing"[MeSH Terms] OR "SARS-CoV-2"[All Fields] OR "SARS-CoV-2"[MeSH Terms] OR "severe acute respiratory syndrome coronavirus 2"[All Fields] OR "ncov"[All Fields] OR "2019 ncov"[All Fields] OR (("coronavirus"[MeSH Terms] OR "coro-navirus"[All Fields] OR "cov"[All Fields]) AND 2019/11/01:3000/12/31[Date–Publica-tion]))) AND (2020/1/1:2023/4/14[pdat])
("chatbot"[Title/Abstract] AND ("health"[Title/Abstract] OR "healthcare"[Title/Abstract] OR "health domain"[Title/Abstract]) AND ("artificial intelligence"[MeSH Terms] OR ("artificial"[All Fields] AND "intelligence"[All Fields]) OR "artificial intelligence"[All Fields])) AND (2020/1/1:2023/4/13[pdat])

5. Recommendation

A statement by Henry Ford reported that "*real progress happens only when the advantages of a new technology become available to everybody*". The consolidation of technologies based on chatbots is intended to bring benefits to everyone in several areas.

Among these areas we find the *health domain*, which is strategic since it has to do with the health of citizens.

The overview highlighted a particular increase in scientific interest in this area, which is accompanied, as for all other sectors of employment, by important expectations on the part of the citizen.

The overview also showed, through an analysis of the sectors most addressed by scholars in the *health domain*, that the need to deepen individual domains, such as effectiveness, legal aspects, and standardization, just to name a few, emerged from time to time.

What is necessary at this point in the evolution of these tools is to develop studies that simultaneously evaluate multiple domains all together in a synergistic way.

To do this, it is important that scholars, experts, politicians, and stakeholders stimulate and initiate large-scale consensus initiatives that address these issues by considering the different multiple domains of intervention. Concerted actions involving experts, international scientific societies and stakeholders could be useful for tackling these strategic issues more decisively. Initiatives such as studies on health technology assessment or the Consensus Conference are strongly recommended.

These initiatives could provide shared documents, including applications, organization models, training, regulations, ethics, and other domains [72,73].

This overview also highlights the cross-domain character of the topic of chatbots; 37 chatbot applications and some expectations have been identified [33,34].

There has always been a process of osmosis of technology between various areas and this is applicable to chatbots as well.

What is important, and this is where the stakeholders come into play, is the accurate monitoring of this process, when the process concerns the *health domain*, given that we are dealing with the health of citizens.

Another important aspect is the impact on the *health domain* that a distorted use of these tools as used in other contexts could generate.

There has recently been a discussion on addiction and the psychological impact (and therefore, on the consequences on the *health domain*) that some applications in use in the world of consumption could generate [74–76]. For example, *Replika* [74], which allows you to interact with virtual friends, has been banned in some countries, such as Italy, where the guarantor of privacy has banned its use after having identified the risk of behavioral and psychological problems, especially for young people.

Some chatbots allow you to talk to celebrities and others even to the dead, the so-called deadbots [75]. The latter are fed with memories, letters, messages from our loved ones and simulate interaction with the deceased. With these deadbots, important limits are being crossed, and we are entering a world where the implications are psychological, behavioral, and ethical. With these applications, one can enter delicate and special worlds, whose implications that can impact a person in unpredictable ways. In the religious sphere, there are chatbots created in Italy that address the sacred and the afterlife, as they simulate conversations with saints [76]. Other chatbots are venturing into very particular and specific sectors, with the implications that have been highlighted. Making a list would be unthinkable. It is precisely this difficulty that creates the need for activating serious monitoring actions in this field.

6. Conclusions

In conclusion, this study highlights the opportunity and potential of chatbots in the *health domain*. However, the studies carried out have highlighted from time to time the need to investigate issues relating to individual domains. Given the increase in the interest in this area, also driven by the introduction of artificial intelligence, concerted actions that address all related intervention domains simultaneously are recommended. It is also necessary to monitor the osmosis of technologies from other sectors in the *health domain*, which have to do with the health of citizens. The use of some chatbots used in other sectors could affect the mental health of citizens, and therefore, affect the health domain.

7. Limitations

This *study*, in relation to its objective, used all the available systematic reviews on PubMed on the topic of chatbots related to the *health domain*. More specific insights on narrow and very particular topics are suggested in future studies using other components of this biomedical database, as well as other national and international databases.

Funding: The APC was funded by Daniele Giansanti.

Conflicts of Interest: The authors declare no conflict of interest.

References

1. Available online: https://dictionary.cambridge.org/dictionary/english/chatbot (accessed on 15 April 2023).
2. Available online: https://www.dictionary.com/browse/chatbot (accessed on 15 April 2023).
3. Available online: https://www.collinsdictionary.com/dictionary/english/chatbot (accessed on 15 April 2023).
4. Available online: https://www.merriam-webster.com/dictionary/chatbot (accessed on 15 April 2023).

5. Kooli, C.; Al Muftah, H. Artificial intelligence in healthcare: A comprehensive review of its ethical concerns. *Technol. Sustain.* **2022**, *1*, 121–131. [CrossRef]
6. Available online: https://openai.com/blog/chatgpt (accessed on 15 April 2023).
7. Jungwirth, D.; Haluza, D. Artificial Intelligence and Public Health: An Exploratory Study. *Int. J. Environ. Res. Public Health* **2023**, *20*, 4541. [CrossRef]
8. Haluza, D.; Jungwirth, D. Artificial Intelligence and Ten Societal Megatrends: An Exploratory Study Using GPT-3. *Systems* **2023**, *11*, 120. [CrossRef]
9. Available online: https://www.mdpi.com/journal/ijerph/special_issues/C52Z967WA3 (accessed on 15 April 2023).
10. Giansanti, D. Artificial Intelligence in Public Health: Current Trends and Future Possibilities. *Int. J. Environ. Res. Public Health* **2022**, *19*, 11907. [CrossRef]
11. Kooli, C. Chatbots in Education and Research: A Critical Examination of Ethical Implications and Solutions. *Sustainability* **2023**, *15*, 5614. [CrossRef]
12. Adamopoulou, E.; Moussiades, L. Chatbots: History, technology, and applications. Machine Learning with Applications. *Mach. Learn. Appl.* **2020**, *2*, 100006.
13. Weizenbaum, J. ELIZA—A computer program for the study of natural language communication between man and machine. *Commun. ACM* **1966**, *9*, 36–45. [CrossRef]
14. Brandtzaeg, P.B.; Følstad, A. Why people use chatbots. In *International Conference on Internet Science*; Kompatsiaris, I., Cave, J., Satsiou, A., Carle, G., Passani, A., Kontopoulos, E., Diplaris, S., McMillan, D., Eds.; Springer: Cham, Switzerland, 2017; Volume 10673. [CrossRef]
15. Heiser, J.F.; Colby, K.M.; Faught, W.S.; Parkison, R.C. Can psychiatrists distinguish a computer simulation of paranoia from the real thing? The limitations of turing-like tests as measures of the adequacy of simulations. *J. Psychiatr.* **1979**, *15*, 149–162.
16. Jwala, K. Developing a Chatbot using Machine Learning. *Int. J. Recent Technol. Eng. (IJRTE)* **2019**, *8*. Available online: https://www.ijrte.org/wp-content/uploads/papers/v8i1S3/A10170681S319.pdf (accessed on 15 April 2023).
17. Mauldin. Chatterbots, Tinymuds, and the Turing Test: Entering the Loebner Prize Competition. *AAAI-94 Proceedings*. Available online: https://aaai.org/conference/Press/Proceedings/aaai94.php (accessed on 15 April 2023).
18. Zemčík, T. A brief history of chatbots. *DEStech Trans. Comput. Sci. Eng.* **2019**, *5*, 14–18. [CrossRef]
19. Bradeško, L.; Mladenić, D. A survey of chatbot systems through a loebner prize competition. In Proceedings of the Slovenian Language Technologies Society, Eighth Conference of Language Technologies, Ljubljana, Slovenia, 8–9 October 2012; Volume 34. Available online: http://nl.ijs.si/isjt12/JezikovneTehnologije2012.pdf (accessed on 15 April 2023).
20. Molnár, G.; Zoltán, S. The role of chatbots in formal education. In Proceedings of the IEEE 16th International Symposium on Intelligent Systems and Informatics, Subotica, Serbia, 13–15 September 2018. [CrossRef]
21. Siri Siri, Apple Website. Available online: https://www.apple.com/siri/ (accessed on 15 April 2023).
22. Watson Assistant | IBM Cloud. 2020. Available online: https://www.ibm.com/cloud/watson-assistant/ (accessed on 15 April 2023).
23. Google Assistant, Your Own Personal Google. 2019. Available online: https://assistant.google.com/ (accessed on 15 April 2023).
24. Microsoft, What Is Cortana? Available online: https://support.microsoft.com/en-us/topic/what-is-cortana-953e648d-5668-e017-1341-7f26f7d0f825 (accessed on 15 April 2023).
25. Alexa. Available online: https://developer.amazon.com/en-GB/alexa (accessed on 15 April 2023).
26. Microsoft Cortana, Your Intelligent Assistant. Available online: https://www.microsoft.com/en-us/cortana (accessed on 30 August 2019).
27. Digital Trends. Available online: https://www.digitaltrends.com/home/what-is-amazons-alexa-and-what-can-it-do/ (accessed on 30 August 2019).
28. Wizu. A Visual History of Chatbots. Medium. 2018. Available online: https://chatbotsmagazine.com/a-visual-history-of-chatbots-8bf3b31dbfb2 (accessed on 24 February 2020).
29. Zhou, L.; Gao, J.; Li, D.; Shum, H.-Y. The design and implementation of xiaoice, an empathetic social chatbot. *arXiv* **2019**, arXiv:1812.08989. [CrossRef]
30. Rapp, A.; Curti, L.; Boldi, A. The human side of human-chatbot interaction: A systematic literature review of ten years of research on text-based chatbots. *Int. J. Hum. Comput. Stud.* **2021**, *151*, 102630. [CrossRef]
31. Skjuve, M.; Følstad, A.; Fostervold, K.I.; Brandtzaeg, P.B. A longitudinal study of human–chatbot relationships. *Int. J. Hum. Comput. Stud.* **2022**, *168*, 102903. [CrossRef]
32. Park, M.; Aiken, M.; Salvador, L. How do Humans Interact with Chatbots? An Analysis of Transcript. *Int. J. Manag. Inf. Technol.* **2018**, *14*, 3338–3350. [CrossRef]
33. Available online: https://research.aimultiple.com/chatbot-applications/ (accessed on 15 April 2023).
34. Available online: https://research.aimultiple.com/conversational-ai-platforms/ (accessed on 15 April 2023).
35. Salvagno, M.; Taccone, F.S.; Gerli, A.G. Can artificial intelligence help for scientific writing? *Crit. Care* **2023**, *27*, 75. [CrossRef]
36. Zhai, C.; Wibowo, S. A systematic review on cross-culture, humor and empathy dimensions in conversational chatbots: The case of second language acquisition. *Heliyon* **2022**, *8*, e12056. [CrossRef] [PubMed]
37. Huq, S.M.; Maskeliūnas, R.; Damaševičius, R. Dialogue agents for artificial intelligence-based conversational systems for cognitively disabled: A systematic review. *Disabil. Rehabil. Assist. Technol.* 2022. *online ahead of print*. [CrossRef]

38. Cao, X.J.; Liu, X.Q. Artificial intelligence-assisted psychosis risk screening in adolescents: Practices and challenges. *World J. Psychiatry* **2022**, *12*, 1287–1297. [CrossRef]
39. Wilson, L.; Marasoiu, M. The Development and Use of Chatbots in Public Health: Scoping Review. *JMIR Hum. Factors* **2022**, *9*, e35882. [CrossRef]
40. Oh, Y.J.; Zhang, J.; Fang, M.L.; Fukuoka, Y. A systematic review of artificial intelligence chatbots for promoting physical activity, healthy diet, and weightloss. *Int. J. Behav. Nutr. Phys. Act.* **2021**, *18*, 160. [CrossRef]
41. Wollny, S.; Schneider, J.; Di Mitri, D.; Weidlich, J.; Rittberger, M.; Drachsler, H. Are We There Yet?—A Systematic Literature Review on Chatbots in Education. *Front. Artif. Intell.* **2021**, *4*, 654924. [CrossRef] [PubMed]
42. Tudor Car, L.; Dhinagaran, D.A.; Kyaw, B.M.; Kowatsch, T.; Joty, S.; Theng, Y.L.; Atun, R. Conversational Agents in Health Care: Scoping Review and Conceptual Analysis. *J. Med. Internet Res.* **2020**, *22*, e17158. [CrossRef]
43. Gabarron, E.; Larbi, D.; Denecke, K.; Årsand, E. What Do We Know About the Use of Chatbots for Public Health? *Stud. Health Technol. Inform.* **2020**, *270*, 796–800. [CrossRef]
44. Abd-Alrazaq, A.; Safi, Z.; Alajlani, M.; Warren, J.; Househ, M.; Denecke, K. Technical Metrics Used to Evaluate Health Care Chatbots: Scoping Review. *J. Med. Internet Res.* **2020**, *22*, e18301. [CrossRef] [PubMed]
45. Hoermann, S.; McCabe, K.L.; Milne, D.N.; Calvo, R.A. Application of Synchronous Text-Based Dialogue Systems in Mental Health Interventions: Systematic Review. *J. Med. Internet Res.* **2017**, *19*, e267. [CrossRef]
46. Chin, H.; Lima, G.; Shin, M.; Zhunis, A.; Cha, C.; Choi, J.; Cha, M. User-Chatbot Conversations During the COVID-19 Pandemic: Study Based on Topic Modeling and Sentiment Analysis. *J. Med. Internet Res.* **2023**, *25*, e40922. [CrossRef]
47. White, B.K.; Martin, A.; White, J.A. User Experience of COVID-19 Chatbots: Scoping Review. *J. Med. Internet Res.* **2022**, *24*, e35903. [CrossRef] [PubMed]
48. Chrimes, D. Using Decision Trees as an Expert System for Clinical Decision Support for COVID-19. *Interact. J. Med. Res.* **2023**, *12*, e42540. [CrossRef]
49. Aggarwal, A.; Tam, C.C.; Wu, D.; Li, X.; Qiao, S. Artificial Intelligence-Based Chatbots for Promoting Health Behavioral Changes: Systematic Review. *J. Med. Internet Res.* **2023**, *25*, e40789. [CrossRef]
50. Bowmans, R.; van de Sande, Y.; Thill, S.; Bosse, T. Voice-Enabled Intelligent Virtual Agents for People With Amnesia: Systematic Review. *JMIR Aging* **2022**, *5*, e32473. [CrossRef]
51. Geoghegan, L.; Scarborough, A.; Wormald, J.C.R.; Harrison, C.J.; Collins, D.; Gardiner, M.; Bruce, J.; Rodrigues, J.N. Automated conversational agents for post-intervention follow-up: A systematic review. *BJS Open* **2021**, *5*, zrab070. [CrossRef] [PubMed]
52. Milne-Ives, M.; de Cock, C.; Lim, E.; Shehadeh, M.H.; de Pennington, N.; Mole, G.; Normando, E.; Meinert, E. The Effectiveness of Artificial Intelligence Conversational Agents in Health Care: Systematic Review. *J. Med. Internet Res.* **2020**, *22*, e20346. [CrossRef] [PubMed]
53. Federici, S.; de Filippis, M.L.; Mele, M.L.; Borsci, S.; Bracalenti, M.; Gaudino, G.; Cocco, A.; Amendola, M.; Simonetti, E. Inside pandora's box: A systematic review of the assessment of the perceived quality of chatbots for people with disabilities or special needs. *Disabil. Rehabil. Assist. Technol.* **2020**, *15*, 832–837. [CrossRef]
54. Chattopadhyay, D.; Ma, T.; Sharifi, H.; Martyn-Nemeth, P. Computer-Controlled Virtual Humans in Patient-Facing Systems: Systematic Review and Meta-Analysis. *J. Med. Internet Res.* **2020**, *22*, e18839. [CrossRef] [PubMed]
55. Vaidyam, A.N.; Wisniewski, H.; Halamka, J.D.; Kashavan, M.S.; Torous, J.B. Chatbots and Conversational Agents in Mental Health: A Review of the Psychiatric Landscape. *Can. J. Psychiatry* **2019**, *64*, 456–464. [CrossRef] [PubMed]
56. Sallam, M. ChatGPT Utility in Healthcare Education, Research, and Practice: Systematic Review on the Promising Perspectives and Valid Concerns. *Healthcare* **2023**, *11*, 887. [CrossRef] [PubMed]
57. He, L.; Balaji, D.; Wiers, R.W.; Antheunis, M.L.; Krahmer, E. Effectiveness and acceptability of conversational agents for smoking cessation: A systematic review and meta-analysis. *Nicotine Tob. Res.* 2022. *online ahead of print*. [CrossRef] [PubMed]
58. Pernencar, C.; Saboia, I.; Dias, J.C. How Far Can Conversational Agents Contribute to IBD Patient Health Care—A Review of the Literature. *Front. Public Health* **2022**, *10*, 862432. [CrossRef]
59. Denecke, K.; May, R. Usability Assessment of Conversational Agents in Healthcare: A Literature Review. *Stud. Health Technol. Inform.* **2022**, *294*, 169–173. [CrossRef]
60. Bin Sawad, A.; Narayan, B.; Alnefaie, A.; Maqbool, A.; Mckie, I.; Smith, J.; Yuksel, B.; Puthal, D.; Prasad, M.; Kocaballi, A.B. A Systematic Review on Healthcare Artificial Intelligent Conversational Agents for Chronic Conditions. *Sensors* **2022**, *22*, 2625. [CrossRef]
61. Lim, S.M.; Shiau, C.W.C.; Cheng, L.J.; Lau, Y. Chatbot-Delivered Psychotherapy for Adults With Depressive and Anxiety Symptoms: A Systematic Review and Meta-Regression. *Behav. Ther.* **2022**, *53*, 334–347. [CrossRef] [PubMed]
62. Ogilvie, L.; Prescott, J.; Carson, J. The Use of Chatbots as Supportive Agents for People Seeking Help with Substance Use Disorder: A Systematic Review. *Eur. Addict. Res.* **2022**, *28*, 405–418. [CrossRef] [PubMed]
63. Xu, L.; Sanders, L.; Li, K.; Chow, J.C.L. Chatbot for Health Care and Oncology Applications Using Artificial Intelligence and Machine Learning: Systematic Review. *JMIR Cancer* **2021**, *7*, e27850. [CrossRef]
64. Luo, T.C.; Aguilera, A.; Lyles, C.R.; Figueroa, C.A. Promoting Physical Activity through Conversational Agents: Mixed Methods Systematic Review. *J. Med. Internet Res.* **2021**, *23*, e25486. [CrossRef]
65. Ruggiano, N.; Brown, E.L.; Roberts, L.; Framil Suarez, C.V.; Luo, Y.; Hao, Z.; Hristidis, V. Chatbots to Support People With Dementia and Their Caregivers: Systematic Review of Functions and Quality. *J. Med. Internet Res.* **2021**, *23*, e25006. [CrossRef]

66. Vaidyam, A.N.; Linggonegoro, D.; Torous, J. Changes to the Psychiatric Chatbot Landscape: A Systematic Review of Conversational Agents in Serious Mental Illness: Changements du paysage psychiatrique des chatbots: Une revue systématique des agents conversationnels dans la maladie mentale sérieuse. *Can. J. Psychiatry* **2021**, *66*, 339–348. [CrossRef]
67. Gaffney, H.; Mansell, W.; Tai, S. Conversational Agents in the Treatment of Mental Health Problems: Mixed-Method Systematic Review. *JMIR Ment. Health* **2019**, *6*, e14166. [CrossRef] [PubMed]
68. Pubmed Search. Available online: https://pubmed.ncbi.nlm.nih.gov/?term=%28chatbot%5BTitle%2FAbstract%5D%29+AND+%28%28health+%5BTitle%2FAbstract%5D%29+OR+%28healthcare%5BTitle%2FAbstract%5D%29+OR+%28health+domain%5BTitle%2FAbstract%5D%29%29&sort=pubdate&size=200 (accessed on 15 April 2023).
69. Available online: https://www.chatbot.com/blog/chatbot-statistics/ (accessed on 15 April 2023).
70. Pubmed Search. Available online: https://pubmed.ncbi.nlm.nih.gov/?term=%28chatbot%5BTitle%2FAbstract%5D%29+AND+%28%28health+%5BTitle%2FAbstract%5D%29+OR+%28healthcare%5BTitle%2FAbstract%5D%29+OR+%28health+domain%5BTitle%2FAbstract%5D%29%29+AND+%28Covid-19%29&filter=dates.2020%2F1%2F1-2023%2F4%2F14&sort=pubdate&size=200 (accessed on 15 April 2023).
71. Pubmed Search. Available online: https://pubmed.ncbi.nlm.nih.gov/?term=%28chatbot%5BTitle%2FAbstract%5D%29+AND+%28%28health+%5BTitle%2FAbstract%5D%29+OR+%28healthcare%5BTitle%2FAbstract%5D%29+OR+%28health+domain%5BTitle%2FAbstract%5D%29%29+AND+%28artificial+intelligence%29&filter=dates.2020%2F1%2F1-2023%2F4%2F13&sort=pubdate&size=200 (accessed on 15 April 2023).
72. Maccioni, G.; Ruscitto, S.; Gulino, R.A.; Giansanti, D. Opportunities and Problems of the Consensus Conferences in the Care Robotics. *Healthcare* **2021**, *9*, 1624. [CrossRef]
73. Consensus Conference Cicerone, Document Finale. Available online: https://www.simfer.it/consensusconference-ciceronedocumento-finale-conclusivo/ (accessed on 7 November 2022).
74. Available online: https://apps.apple.com/ch/app/replika-virtual-ai-friend/id1158555867?l=it (accessed on 15 April 2023).
75. Available online: https://www.cnet.com/culture/hereafter-ai-lets-you-talk-with-your-dead-loved-ones-through-a-chatbot/ (accessed on 15 April 2023).
76. Available online: https://www.prega.org/ (accessed on 15 April 2023).

life

Article

Robust Classification and Detection of Big Medical Data Using Advanced Parallel *K*-Means Clustering, YOLOv4, and Logistic Regression

Fouad H. Awad [1,*], Murtadha M. Hamad [1] and Laith Alzubaidi [2,3,4,*]

1 College of Computer Science and Information Technology, University of Anbar, Ramadi 31001, Iraq
2 Faculty of Science and Engineering, Queensland University of Technology, Brisbane, QLD 4000, Australia
3 ARC Industrial Transformation Training Centre-Joint Biomechanics, Queensland University of Technology, Brisbane, QLD 4000, Australia
4 Centre for Data Science, Queensland University of Technology, Brisbane, QLD 4000, Australia
* Correspondence: fouad.hammadi@uoanbar.edu.iq (F.H.A.); l.alzubaidi@qut.edu.au (L.A.)

Abstract: Big-medical-data classification and image detection are crucial tasks in the field of healthcare, as they can assist with diagnosis, treatment planning, and disease monitoring. Logistic regression and YOLOv4 are popular algorithms that can be used for these tasks. However, these techniques have limitations and performance issue with big medical data. In this study, we presented a robust approach for big-medical-data classification and image detection using logistic regression and YOLOv4, respectively. To improve the performance of these algorithms, we proposed the use of advanced parallel *k*-means pre-processing, a clustering technique that identified patterns and structures in the data. Additionally, we leveraged the acceleration capabilities of a neural engine processor to further enhance the speed and efficiency of our approach. We evaluated our approach on several large medical datasets and showed that it could accurately classify large amounts of medical data and detect medical images. Our results demonstrated that the combination of advanced parallel *k*-means pre-processing, and the neural engine processor resulted in a significant improvement in the performance of logistic regression and YOLOv4, making them more reliable for use in medical applications. This new approach offers a promising solution for medical data classification and image detection and may have significant implications for the field of healthcare.

Keywords: medical data; medical imaging; data classification; image detection; YOLOv4; logistic regression; machine learning; AI; deep learning

Citation: Awad, F.H.; Hamad, M.M.; Alzubaidi, L. Robust Classification and Detection of Big Medical Data Using Advanced Parallel *K*-Means Clustering, YOLOv4, and Logistic Regression. *Life* **2023**, *13*, 691. https://doi.org/10.3390/life13030691

Academic Editor: Daniele Giansanti

Received: 30 January 2023
Revised: 24 February 2023
Accepted: 28 February 2023
Published: 3 March 2023

1. Introduction

The advancement of digital medical technology, coupled with the exponential growth of medical data, has led to biomedical research becoming a data-intensive science, resulting in the emergence of the "big-data" phenomenon, as reported in the literature, such as in [1]. Data have become a strategic resource and a key driver of innovation in the era of big data, transforming not only the way biomedical research has been conducted, but also the ways in which people live and think, which has been highlighted in studies such as [2]. To capitalize on this, the relevant departments in the medical industry should focus on collecting and managing medical health data and use this information as a foundation for later developments through the integration, analysis, and application requirements required to employ big data in the medical field [3].

Big medical data and image detection is an essential element of healthcare that plays a critical role in the storage, organization, and analysis of medical information [4]. the effective classification of medical data enables the efficient retrieval and examination of patient records, which can aid in the diagnosis and treatment of illnesses. It can also assist in identifying trends and patterns in patient health data, enabling healthcare professionals to recognize potential risk factors and take preventative measures. Furthermore, medical data classification has facilitated the advancement of new treatments and therapies by allowing researchers to analyze large datasets and uncover potential correlations and trends [5].

COVID-19 data classification has involved organizing and labeling data related to the coronavirus pandemic, such as information about confirmed cases, deaths, and vaccination rates. These types of data have often been used to track the spread of the virus and inform public health decisions. Image detection techniques have been used to identify COVID-19-related images, such as X-ray scans showing lung abnormalities associated with the virus. These techniques have assisted healthcare professionals and researchers better understand and track the spread of the virus.

However, there have been several challenges and problems associated with COVID-19 data classification and image detection. One major challenge has been ensuring the accuracy and reliability of the data being used. There have been errors and biases in the data that affected the results. Additionally, there have been privacy concerns related to collecting and using personal health data. There have also been technical challenges in developing and implementing image detection algorithms, such as difficulties in obtaining a sufficiently large dataset for training. Overall, addressing these challenges is crucial in order to effectively use data and image detection techniques to understand and combat the COVID-19 pandemic.

In this study, an efficient and high-performance solution to enhance the accuracy of medical data classification and image detection was proposed. Advanced k-means clustering was merged with both classification and detection techniques to elevate the performance and accuracy of these techniques [6]. To evaluate the performance of medical data classification, a large medical dataset was used. Furthermore, to evaluate the effectiveness of the detection technique, a dataset comprising X-ray COVID-19 and CT images was utilized. The results indicated that the proposed models significantly improved the performances of classification and detection. The proposed model's contributions were the following:

1. The successful application of advanced parallel k-means clustering as a pre-processing step for both the images and the data to improve the accuracy of image feature extraction and detection, as well as the accuracy of data classification.
2. Both hardware and software improvements were employed to significantly accelerate the classification and detection processes. Hardware acceleration was achieved by utilizing the latest neural engine processor while the software optimization involved using parallel-processing mechanisms.

This paper is divided into seven sections. The introduction addresses the significance of medical data classification and medical image detection. Section 2 discusses various data classification and image detection algorithms, including their advantages and limitations. Section 3 addresses the current challenges and features of solutions for processing large amounts of medical data and images. Section 4 presents the proposed solution. Section 5 outlines the methodology and performance metrics used to evaluate the proposed solution. The implementation and results of the proposed solution are presented in Section 6. Section 7 concludes the paper.

2. Data Classification

Data classification is the process of organizing and categorizing data based on predetermined criteria [7]. It is a crucial aspect of many applications, including data management, data analysis, and information retrieval.

Logistic Regression Algorithm

Logistic regression is a type of binary classification algorithm that is used to predict the probability of an event occurring [8]. It has been commonly used in machine learning for applications such as spam detection, medical diagnosis, and sentiment analysis [9].

The logistic-regression model maps the input features $x_1, x_2, ..., x_n$ to a predicted output variable y that has a value between 0 and 1, representing the probability of the event occurring [10].

The logistic function, also known as the sigmoid function, is used to model the relationship between the input features and the predicted output variable. The sigmoid function is defined as [10]:

$$f(z) = \frac{1}{1+e^{-z}}$$

where $z = w_0 + w_1x_1 + w_2x_2 + ... + w_nx_n$ is the linear combination of the input features and their corresponding weights, with w_0 as the bias term.

The logistic regression algorithm aims to find the optimal values for the weights $w_0, w_1, w_2, ..., w_n$ that minimize the error between the predicted output variable and the true output variable [11]. This is achieved by maximizing the likelihood function, which is the probability of the observed data according to the model parameters [12]. The likelihood function for logistic regression is:

$$L(w) = \prod_{i=1}^{m} f(z_i)^{y_i}(1 - f(z_i))^{1-y_i}$$

where m is the number of training examples, y_i is the true output variable for the ith example, and z_i is the linear combination of the input features and weights for the ith example [13].

The optimal values for the weights can be found using gradient descent, which involves iteratively updating the weights in the direction of the negative gradient of the likelihood function [14]. The updated rule for the weights is:

$$w_j := w_j - \alpha\frac{\partial L(w)}{\partial w_j}$$

where α is the learning rate, and $\frac{\partial L(w)}{\partial w_j}$ is the partial derivative of the likelihood function with respect to the jth weight.

The logistic regression algorithm can be summarized in the following steps:

1. Initialize the weights $w_0, w_1, w_2, ..., w_n$ to random values.
2. Calculate the linear combination z_i for each training example using the current weights.
3. Calculate the predicted output variable y_i for each training example using the logistic function.
4. Calculate the error between the predicted output variable and the true output variable for each training example.
5. Calculate the gradient of the likelihood function with respect to each weight.
6. Update the weights using the gradient descent update rule.
7. Repeat steps 2–6 until the error converges or a maximum number of iterations is reached.

One of the main advantages of logistic regression is its simplicity and ease of implementation. It is a straightforward algorithm that can be easily implemented using standard statistical software [15]. Additionally, logistic regression is highly interpretable, allowing users to understand the contributions of each independent variable to the predicted probability. It is also robust regarding multicollinearity, meaning that it can handle correlated independent variables without producing biased estimates.

However, logistic regression is not without its challenges. One of the main limitations is that it is only suitable for binary classification problems, meaning that it can only predict the likelihood of an event occurring or not occurring [16].

3. Image Detection Technique

Image detection is a technique used to identify and locate specific objects, features, or patterns within an image. It is a crucial aspect of many applications, including object recognition, facial recognition, and scene comprehension [17]. In the field of healthcare, image detection is used to analyze and interpret medical images, such as X-rays, CT scans, and MRIs. These images provide important diagnostic information that can be used to identify and treat diseases.

YOLOv4 Algorithm

The You Only Look Once version 4 (YOLOv4) algorithm is a state-of-the-art object detection algorithm that processes an entire image and directly predicts the bounding boxes and class probabilities for all objects in the image. It uses a convolutional neural network (CNN) to extract features from the input image and then apply them a series of convolutional and fully connected layers in order to predict the class probabilities and bounding box coordinates for each object [18].

The YOLOv4 algorithm predicts object classes and bounding box coordinates by dividing the input image into a grid of cells and predicting the class probabilities and bounding box offsets for each cell [19]. Specifically, for each cell in the grid, the algorithm predicts:

- The probability of an object being present in that cell (denoted p_{obj}).
- The x and y coordinates of the center of the bounding box, relative to the coordinates of the cell (denoted by b_x and b_y, respectively).
- The width and height of the bounding box relative to the size of the cell (denoted by b_w and b_h, respectively).
- The class probabilities for each object class (denoted by $p_{c1}, p_{c2}, \ldots, p_{cn}$, where n is the number of classes).

These predictions are made using a series of convolutional and fully connected layers in the YOLOv4 network. The network architecture is based on a variant of the DarkNet architecture, which consists of multiple convolutional layers and followed by max-pooling layers, and ends with multiple fully connected layers [20]. The final layer of the network outputs a tensor that is the shape of (grid size) × (grid size) × (number of anchor boxes) × (5 + number of classes), where the 5 refers to the objectness score, b_x, b_y, b_w, and b_h [21].

The YOLOv4 algorithm then uses non-maximum suppression to remove redundant bounding boxes for the same object [22]. Specifically, for each class, it applies non-maximum suppression to the set of predicted bounding boxes with objectness scores above a certain threshold. This threshold is usually set to a value between 0.5 and 0.7, depending on the desired balance between precision and recall [23].

The YOLOv4 algorithm can be trained using a loss function that measures the errors between the predicted and ground-truth bounding boxes and class probabilities [24]. The loss function consists of two components: a localization loss that penalizes errors in the predicted bounding box coordinates, and a classification loss that penalizes errors in the predicted class probabilities. The localization loss is typically computed using the mean squared error (MSE) between the predicted and ground-truth bounding box coordinates, while the classification loss is typically computed using the cross-entropy loss between the predicted and ground-truth class probabilities [25].

Algorithm 1 shows the main steps of the YOLOv4 algorithm.

Algorithm 1 YOLOv4 object detection algorithm

Require: Input image I
Ensure: Bounding boxes B and class probabilities C
Pre-process I to obtain an input tensor X
Apply the backbone network to obtain feature maps $F_1, F_2, \ldots, F_n$
Apply the neck network to combine the feature maps and obtain a single feature map F
Apply the detection head to F to obtain a set of candidate boxes B_c and class probabilities C_c
Apply non-maximum suppression (NMS) to B_c and C_c to obtain the final set of bounding boxes B and class probabilities C, respectively
return B and C

Note that this algorithm assumes that the YOLOv4 architecture has already been trained on a large dataset of images with labeled objects and that the resulting model has been saved and can be loaded for inferences on new images. The backbone network, neck network, and detection head are all components of the YOLOv4 architecture, and their specific details are beyond the scope of this pseudo-coded algorithm [26].

One way that YOLOv4 has been used in medical image analysis has been in the detection of abnormalities and lesions in images [27]. For example, it was used to identify abnormalities in CT scans of the brain, which could then be used to diagnose and treat brain tumors. By analyzing CT scans with YOLOv4, healthcare professionals could more accurately identify abnormalities and determine the appropriate course of treatment.

During the COVID-19 pandemic, YOLOv4 has also been used to analyze chest X-rays, which have often been used to diagnose the virus [28]. By detecting characteristic patterns associated with COVID-19, such as lung abnormalities, YOLOv4 assisted healthcare professionals in making accurate diagnoses and providing timely treatments for patients [29]. In addition to its use in detecting abnormalities within images, YOLOv4 has also been used to detect objects in images, such as medical instruments and organs. This was particularly useful for identifying and tracking objects during surgical procedures, such as in the detection of brain tumors [30].

4. Medical Data Classification and Detection

Medical data classification and image detection are two critical areas in healthcare that could benefit from the latest advancements in machine-learning and computer-vision technologies. In recent years, there has been a significant increase in the amount of medical data generated due to the availability of electronic health records and medical imaging technologies. This growth in medical data has provided new opportunities for developing more accurate and efficient methods for classification and image detection, which could lead to improved diagnoses, treatments, and patient outcomes.

4.1. Medical Data and Image Classification

The classification of medical data refers to the process of assigning a label or category to a particular medical dataset. The classification of medical data could be used for various applications, such as disease diagnosis, drug discovery, and prognosis. The following are the state-of-the-art techniques used in medical data classification.

- Deep Learning: Deep learning has revolutionized the field of medical data classification and image detection, due to its ability to handle large and complex datasets with improved accuracy and efficiency [31]. Convolutional neural networks (CNNs) and recurrent neural networks (RNNs) are two widely used deep-learning techniques that have demonstrated exceptional performance in the medical field [32]. CNNs have been specifically designed to analyze visual imagery, making them a popular choice for medical image analysis. They consist of multiple layers that learn different features of an image, such as edges and textures, and then use these features to classify the image. The ability of CNNs to automatically extract relevant features from medical images has led to their use in a wide range of applications, such as mammogram analysis for breast cancer detection and brain tumor segmentation. Furthermore, RNNs have been designed to process sequential data and have been extensively used in various medical applications, such as medical signal processing, clinical event prediction, and ECG signal analysis [33]. They are able to analyze the temporal dependencies in sequential data by using a memory component that allows them to remember past inputs and use them to influence future predictions. RNNs have also been used in combination with CNNs to analyze both image and sequential data, such as in the case of electroencephalogram (EEG) signal analysis [34]. In addition to CNNs and RNNs, other deep-learning techniques, such as generative adversarial networks (GANs) and auto-encoders have also been explored in medical data classification and image detection [35]. GANs have been used to generate synthetic medical images,

which were then used to augment existing datasets and improve the performance of image classifiers. Auto-encoders, in contrast, have been used for feature extraction and dimensionality reduction, which improved the efficiency of classification algorithms.
- Support Vector Machines (SVMs): SVMs are a type of supervised learning algorithm that has been widely used for classification tasks in many areas, including in medical data classification. SVMs have been particularly useful for classification tasks in which the number of features was much greater than the number of samples [36]. SVMs find the optimal hyperplane that separates the different classes in a dataset. For medical data classification, SVMs have been used for tasks such as disease diagnosis, the classification of different types of cancer, and the identification of abnormal medical images [37]. SVMs have shown high accuracy and robustness in these tasks due to their ability to handle non-linear data and their resistance to over-fitting.
  One example of SVMs being used in medical data classification was for the identification of breast cancer using mammograms [38]. SVMs had a high accuracy in distinguishing between benign and malignant tumors, which is critical for the early detection and treatment of breast cancer. SVMs have also been used for the classification of brain tumors and the identification of Alzheimer's disease in medical imaging data.
- Random Forest: Random forest is a type of ensemble learning algorithm that combines multiple decision trees to improve its classification accuracy. The method is considered a supervised learning technique that operates by constructing several decision trees during training and then predicts the class label of an input data point by aggregating the predictions of all the decision trees [39]. Random forest has been effective in medical data classification due to its ability to handle high-dimensional data and its resistance to over-fitting. In medical applications, random forest has been used for various classification tasks, such as disease diagnosis, the prediction of treatment responses, and mortality risk assessments [40]. One advantage of random forest is its ability to handle missing data and noisy features. This is achieved by randomly selecting a subset of features at each node in the decision tree, which reduces the risk of over-fitting and improves the model's generalization performance. Additionally, the method allows for the calculation of feature importance, which can help identify the most important variables that contribute to the classification task.

However, the classification of medical data has not been without challenges. One of the main challenges has been the large volume of data that must be classified [41]. Medical data are typically generated at a rapid rate and can be difficult to manage due to size and complexity. Additionally, medical data classification often involves working with sensitive and personal information, which requires strict adherence to the privacy and security measures in place. Another challenge has been the lack of standardization in medical data classification, which has led to confusion and difficulties in data retrieval and analysis [42]. Finally, the constantly evolving nature of the healthcare field means that medical data classification systems must be regularly updated and adapted to meet changing needs.

4.2. Medical Image Detection

Image detection in healthcare refers to the process of detecting and identifying medical conditions or abnormalities in medical images such as X-rays, CT scans, and MRI scans. Image detection plays a vital role in the diagnosis and treatment of various medical conditions [43]. The following are the state-of-the-art techniques used in image detection in healthcare.

- Convolutional Neural Networks (CNNs): In medical imaging, CNNs have been used for a variety of applications, such as the detection of breast cancer, lung cancer, and brain tumors [44]. For example, in breast cancer detection, CNNs have been used to analyze mammograms and detect subtle changes that could indicate the presence of cancer. In lung cancer detection, CNNs have been used to analyze CT scans and identify nodules that could be indicative of cancer. In brain tumor detection,

CNNs have been used to analyze MRI scans and identify regions of abnormal tissue growth [45].

One of the advantages of using CNNs for medical image detection is their ability to learn and extract features automatically, without the need for manual feature extraction [46]. This makes them particularly useful for analyzing large and complex medical images, where manual feature extraction can be time-consuming and prone to error. Another advantage of CNNs is their ability to learn from large amounts of data. With the increasing availability of medical imaging data, CNNs can be trained on large datasets to improve their accuracy and generalization performance [47]. Additionally, CNNs can be fine-tuned and adapted for specific medical image detection tasks, which can further improve their performance.

- Transfer Learning: In the context of medical image detection, transfer learning was an effective method for improving the accuracy and efficiency of image classification tasks [48]. Pre-trained models, such as those based on CNNs, can learn generic image features that can be transferred to new medical imaging datasets, even when the size of the new dataset is relatively small [49]. This can be particularly useful in healthcare, where obtaining large labeled datasets can be challenging and time-consuming. By using transfer learning, researchers and clinicians leveraged the knowledge and expertise gained from pre-trained models to improve the accuracy and efficiency of image detection in healthcare [50]. For example, a pre-trained model that was trained on a large dataset of chest X-rays was then fine-tuned for a smaller dataset of lung cancer images, resulting in improved accuracy and faster training times.

5. Related Works

A literature review was conducted to examine the most recent approaches and techniques for medical data classification in this field.

The related works presented here were selected based on their technological similarity to the proposed solution and their focus on medical data. Furthermore, all papers were chosen based on their publication in high-quality journals. Furthermore, as COVID-19 has attracted the attention of researchers in the healthcare field, most of the papers selected in this review were related to the global COVID-19 pandemic.

In [51], the aim was to evaluate the performance of parallel computing and advanced *k*-means clustering as a pre-processing step for data classification and image detection in medical applications. To achieve this, the researchers utilized a parallel logistic regression algorithm and a mobile neural engine processor. The *k*-means clustering technique was used to pre-process both images and data, resulting in improvements in feature extraction, the removal of noise and outlier pixels, and classification accuracy. The results of this study showed that their proposed approach outperformed traditional methods both in terms of both accuracy and efficiency, making it a promising approach for medical data analysis and processing.

In 2021, the researchers in [52] proposed a new method for optimizing the performance of the *k*-means clustering algorithm on parallel and distributed computing systems. The study employed a hybrid approach that combined the traditional Lloyd's algorithm with a new partitioning technique. The proposed approach was evaluated using various datasets, and the results showed that the hybrid approach outperformed both the traditional Lloyd's algorithm and other state-of-the-art parallel *k*-means algorithms, in terms of both accuracy and efficiency. The study concluded that the proposed approach was a promising solution for large-scale clustering tasks on parallel and distributed computing systems. In [53], the authors proposed a new framework for automating the diagnosis of Alzheimer's disease (AD) using a machine-learning approach. The proposed framework utilized a combination of several machine-learning algorithms, including principal component analysis (PCA), support vector machine (SVM), and *k*-nearest neighbors (KNN) to classify brain images as normal or AD. The study used two different datasets, and the results showed that the proposed framework achieved high accuracy and specificity when classifying brain images as AD. The study concluded that the proposed framework could be a valuable tool for the early diagnosis and monitoring of AD.

In 2022, [54] investigated the potential use of deep learning algorithms for the detection of COVID-19 in chest X-ray images. The study proposed a deep-learning model based on convolutional neural networks (CNNs) that had been trained on a large dataset of chest X-ray images. The model was tested on a separate dataset of chest X-ray images, and the results showed that the proposed model achieved high accuracy, sensitivity, and specificity, in detecting COVID-19. The study concluded that the proposed deep-learning model could be a valuable tool for the rapid and accurate detection of COVID-19 in chest X-ray images, especially in regions with limited access to COVID-19 testing facilities.

A literature review was conducted in order to review the most recent approaches and techniques for medical image detection.

The authors of [55] developed a machine-learning algorithm that could accurately classify patients with severe COVID-19 and predict their risk of in-hospital mortality. The study collected data from electronic health records of patients with severe COVID-19, including demographics, vital signs, laboratory values, and comorbidities. A machine-learning algorithm based on a gradient-boosting machine (GBM) was developed and trained on the collected data. The results showed that the proposed GBM model achieved high accuracy in classifying patients with severe COVID-19 and predicting their risk of in-hospital mortality. The study concluded that the proposed machine-learning algorithm could be a valuable tool for clinicians to make more informed decisions about the management of patients with severe COVID-19.

In 2021, the authors of [56] proposed a classification solution using transfer learning to assess the suitability of 3 pre-trained CNN models (EfficientNetB0, VGG16, and InceptionV3) for mobile applications. These models were selected for their accuracy and efficiency with a relatively small number of parameters. The study used a dataset compiled from various publicly available sources and evaluated the models using performance measurements and deep-learning approaches, such as accuracy, recall, specificity, precision, and F1-scores. The results demonstrated that the proposed method produced a high-quality model with a COVID-19 sensitivity of 94.79% and an overall accuracy of 92.93%. The study suggested that computer-vision techniques could be utilized to improve the efficiency of detection and screening processes.

In 2021, the authors of [57] employed convolutional neural networks (ConvNets) to accurately identify COVID-19 in computed tomography (CT) images, enabling the early classification of chest CT images of COVID-19 by hospital staff. ConvNets automatically learned and extracted features from medical image datasets, including the COVID-CT dataset used in this study. The objective was to train the GoogleNet ConvNet architecture using 425 CT-coronavirus images from the COVID-CT dataset. The experimental results indicated that GoogleNet achieved a validation accuracy of 82.14% on the dataset in 74 min and 37 s. This study demonstrated the potential of ConvNets in improving the accuracy and efficiency of COVID-19 detection in medical imaging.

In 2022, the authors of [58] proposed a new method for improving the quality of CT scans using contrast limited histogram equalization (CLAHE) and developed a convolutional neural network (CNN) model to extract important features from a dataset of 2482 CT-scan images. These features were then used as input for machine-learning methods such as support vector machine (SVM), Gaussian naive Bayes (GNB), logistic regression (LR), random forest (RF), and decision tree (DT). The researchers recommended an ensemble method for classifying COVID-19 CT images and compared the performance of their model with other state-of-the-art methods. The proposed model outperformed existing models with an accuracy of 99.73%, a precision of 99.46%, and a recall of 100%.

In 2022, the authors of [59] described an approach that used a generative adversarial network (GAN) to improve the accuracy of a deep-learning model for classifying COVID-19 infections in chest X-ray images. To generate additional training data, the COVID-19 positive chest X-ray images were fed into a styleGAN2 model, which produced new images for training the deep-learning model. The resulting dataset was used to train a CNN binary classifier model that achieved a classification accuracy of 99.78%. This method could aid in the rapid and accurate diagnosis of COVID-19 infections from chest X-ray images.

6. Proposed Solution

The proposed solution was designed with two main objectives: medical data classification and medical image detection. Each model is described in detail in this section.

6.1. Advanced Parallel K-Means Clustering

In order to implement the modified parallel k-means clustering on the mobile execution unit and the SoC, the algorithm had to be modified to take advantage of a multi-core general-purpose processor and a multi-core neural engine. Each operating system offered a unique set of utilities for parallel operation.The iOS environment, due to its use of Objective-C programming, has an additional tool called dispatch queues, in addition to standard tools, such as processes and threads. Although iOS is a multi-tasking operating system, it did not allow multiple processes for a single program, resulting in only one procedure being available.

However, the Android OS had a limitation in its Java and Kotlin programming languages, which was the hardware-limited access and lack of pointer support, making it difficult to fully utilize the system hardware. A lightweight process is a thread of any type. Threads share memory with their parent process while processes themselves do not. This led to issues when two threads simultaneously modified the same resource, such as a variable, resulting in illogical outcomes. In the iOS environment, threads were a finite resource on any POSIX-compliant system. Only 64 threads could be active at once for a single process. While this is a large number, there were logical reasons to exceed this limit.

The overall processing, as shown in Figure 1, of the on-device parallel clustering consisted of two jobs: managing the dataset and clustering execution, and performing the parallel k-means clustering itself. The general-purpose processor cores were responsible for managing the clustering in the neural engine cores. After executing the k-means clustering on a sub-block of the data, each core sent the centroid point-value to the general-purpose cores. The general-purpose cores then evaluated whether the centroid value was less than the centroid threshold. If it was less, a signal was sent to the execution mechanism to process the clustering again.

Figure 2 shows a flowchart of advanced parallel k-means clustering on the neural engine and general-purpose cores.

6.2. Advanced Classification Solution

Pre-processing medical data with advanced parallel k-means clustering was a useful technique to improve the classification performance of logistic regression algorithms. K-means clustering is a machine-learning algorithm that is used to partition a dataset into a specified number of clusters. By using advanced parallel techniques, it is possible to process data more efficiently and quickly.

Pre-processing the medical data with k-means clustering improved the accuracy and precision of the logistic regression algorithms by ensuring the data were simpler to classify. The k-means algorithm divided the data into clusters based on similar characteristics, such as age or sex. This assisted in reducing the noise and the complexity of the data, making it simpler for the logistic regression algorithm to accurately classify the data.

In addition to improving the accuracy and precision of the classification process, pre-processing the medical data with k-means clustering also reduced the computational resources required to operate the logistic regression algorithm. By reducing the size and complexity of the dataset, it was possible to operate the logistic regression algorithm more efficiently and quickly.

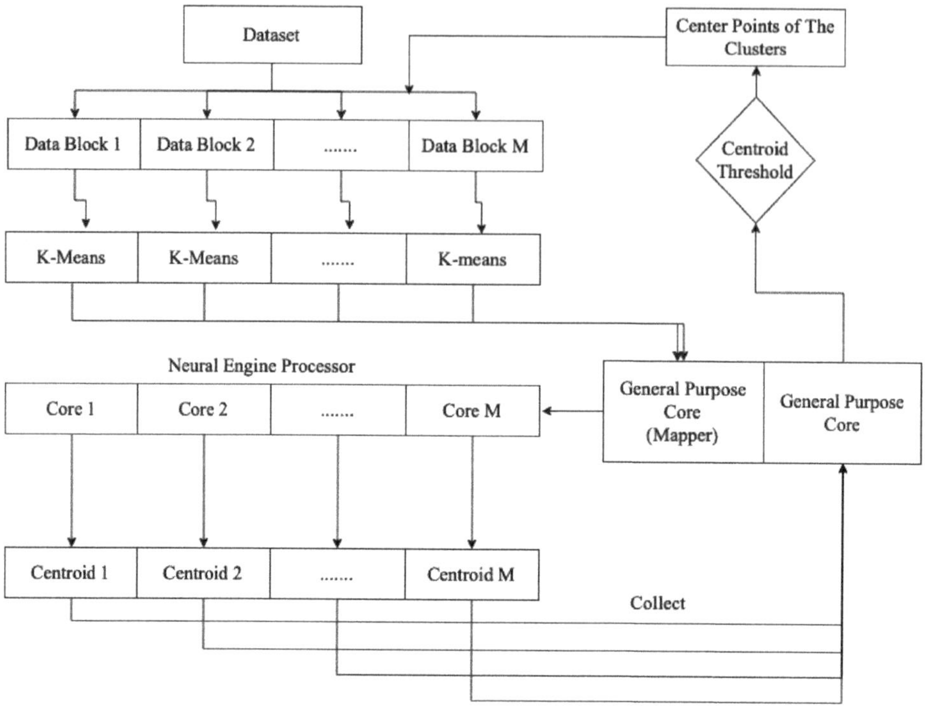

Figure 1. On-device parallel clustering processing.

After clustering the data using *k*-means clustering, the next step in the process was to perform the logistic-regression classification. The steps for performing parallel logistic-regression classification were the following:

- Pre-processing: As with non-parallel logistic regression, it was important to pre-process the data before applying the model. This included tasks such as missing-value imputation, scaling, and feature selection.
- Splitting the data: The data had to be split into training and testing sets in order to evaluate the model's performance on unfamiliar data.
- Choosing a parallelization method: We had to decide whether to use data parallelism, model parallelism, or a hybrid parallelism.
- Partitioning the data: Depending on the chosen parallelization method, the data had to be partitioned into smaller chunks and distributed across multiple processors or devices.
- Training the model: Each processor or device was responsible for training a separate logistic-regression model on its chunk of the data. The models were then combined to form the final model.
- Evaluating the model: The trained model was then evaluated on the testing data. This involved calculating evaluation metrics, such as accuracy, precision, and recall.
- Assessing the model's predictions: Once the model had been trained and evaluated, it was used to make predictions according to new data. To achieve this, the model's parameters were used to calculate the probability of an instance belonging to each class. The class with the highest probability was then predicted as the output.

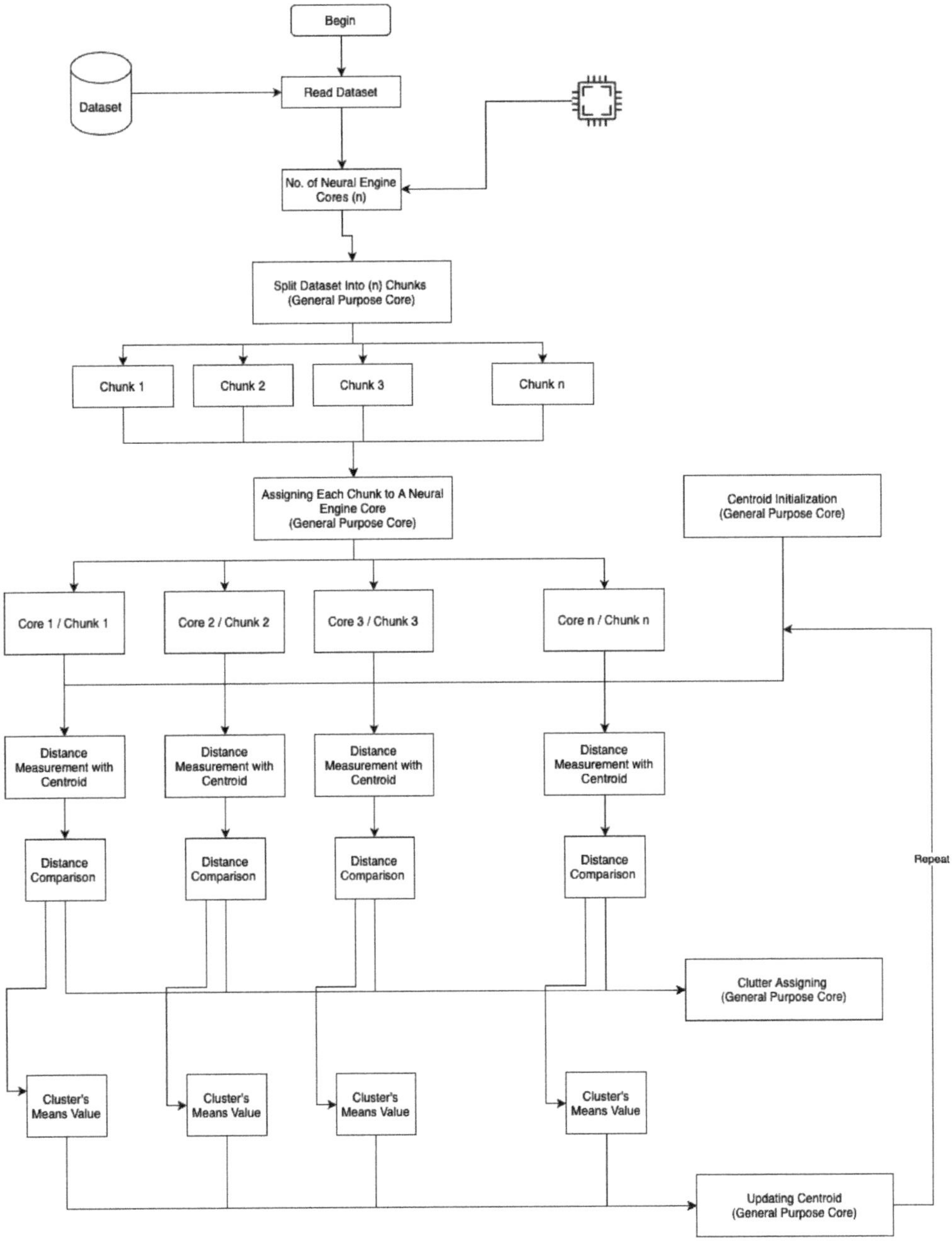

Figure 2. On-device parallel clustering flowchart.

In Algorithm 2, the input is the data D and the number of processors or devices n to be used for parallelization. The output is the trained logistic-regression model M. The data were pre-processed and split into training and testing sets. The parallelization method was chosen, and the training data were then partitioned into smaller chunks. A separate logistic-regression model was trained on each chunk of data, and the models were combined to form the final model. The model was then evaluated on the testing data and the returned results.

Algorithm 2 Parallel Logistic-Regression Classification

procedure PARALLELLOGISTICREGRESSIONCLASSIFICATION(D, n)
    Pre-process data D
    Split data into training and testing sets D_{train} and D_{test}
    Partition data D_{train} into n smaller chunks $D_{train,1}, D_{train,2}, \ldots, D_{train,n}$
    for $i \leftarrow 1$ **to** n **do**
        Train logistic-regression model M_i on chunk $D_{train,i}$
    end for
    Combine models $M_1, M_2, \ldots, M_n$ to form final model M
    Evaluate model M on testing data D_{test}
    return Model M
end procedure

The algorithm had two input parameters. The first was the clustered dataset, which included a new feature extracted by the clustering process. The second input was the number of chunks into which the dataset would be partitioned. The number of partitions depended on the number of neural engine cores available, with each chunk trained on a single core. The standard CPU cores handled general tasks, such as data partitioning; reading and writing data for the neural engine cores; combining models (M1, M2, ..., Mn); and evaluating models.

Classification Pre-Processing

Using k-means clustering as a pre-processing step could potentially improve the performance of the logistic-regression classification in several ways:

- Dimensionality reduction: K-means clustering was used to group similar data points together into clusters, which reduced the number of features in the dataset. By selecting the centroids of the clusters as the new features, we reduced the dimensionality of the data and removed the noise, which improved the performance of the logistic regression.
- Feature engineering: K-means clustering was used to create new features that captured the structure of the data. We added a new binary feature that indicated whether a data point belonged to a particular cluster or not. These new features enabled the logistic regression to capture complex relationships in the data that had not been apparent previously.
- Outlier detection: K-means clustering improved the identification and removal of outliers in the dataset. Outliers had a significant impact on the performance of the logistic regression, and removing them improved the accuracy of the model.
- Data normalization: K-means clustering was used to normalize the data by scaling it to a range from 0 to 1. Normalizing the data improved the performance of the logistic regression by reducing the impact of outliers and ensuring that all features were on a similar scale.

In the proposed parallel logistic regression, the weighted-combination method assisted in forming the final logistic-regression model from individual models that had been trained by each processor or device. An overview of the process is provided:

- Train individual models: The dataset was divided into subsets, and each subset was used to train a logistic-regression model on a separate processor or device.
- Obtain model weights: Once the individual models had been trained, each model was assigned a weight based on its performance on a validation set. The weights were determined using a variety of methods, such as the accuracy or the area under the receiver-operating characteristic curve (AUC-ROC).
- Combine the models: The predicted probabilities or coefficients from each individual model were multiplied by their corresponding weights, and the weighted sum was used as the final output. For example, if there were three individual models with weights of 0.3, 0.5, and 0.2, the predicted probabilities of each model were multiplied by 0.3, 0.5, and 0.2, respectively, and then summed to obtain their final predicted probabilities.
- Model selection: The performance of the final model was evaluated on a validation set, and the weights assigned to the individual models were adjusted to improve the performance of the final model. This process was repeated until the desired level of performance was achieved.
- Apply the final model: Once the final model was selected, it was implemented to make predictions on new data.

The weighted-combination method can be an effective way to leverage the power of multiple processors or devices to train logistic-regression models in parallel. By assigning weights to each individual model, the final model can benefit from the strengths of each model while mitigating their weaknesses.

6.3. Advanced Image Detection

In this study, we proposed a novel approach for pre-processing images using advanced parallel k-means clustering and then applying image detection using YOLOv4. The k-means clustering algorithm was used to divide the images into segments, which were then processed in parallel by multiple processors. The parallel-processing of the image segments resulted in a significant reduction in the overall processing time. The k-means algorithm is a popular method for clustering data based on similarity. It groups similar data points together and forms clusters. In the proposed approach, k-means was used to divide the images into segments, where each segment represented a cluster of similar pixels. The parallel-processing of these segments was achieved by distributing the segments across multiple processors. This allowed for a more efficient use of resources and resulted in a significant reduction in the overall processing time.

After the image had been segmented, the image detection algorithm YOLOv4 was applied to each segment. YOLOv4 is a state-of-the-art object detection algorithm that has been widely used for image-processing tasks. It can accurately detect and classify objects in an image, making it an ideal choice for this application. The proposed approach provided several advantages over traditional image-processing methods. The use of advanced parallel k-means clustering allowed for a more efficient use of resources, resulting in faster processing times. Additionally, the application of YOLOv4 to the image segments improved the accuracy of object detection. Overall, the proposed approach was a powerful tool for image processing on mobile devices.

Stage 1: Image Clustering and Pre-Processing

The intricate structure of the information in images makes the clustering of X-ray (radiographs) and CT-scan images challenging. A considerable visual resemblance exists between X-ray and CT images of the same class. Furthermore, because of the varied X-ray image types, orientation changes, alignments, and diseases, there was a significant variance within a class. The quality of the X-ray images also varied significantly, in addition to the contents. As illustrated in the accompanying diagram, the image clustering framework in this study was divided into two phases: image feature extraction and image clustering.

Then, the clustering process was carried out using the machine-learning engine-specific processors in contemporary mobile devices. Maintaining dataset characteristics while improving clustering efficiency was recommended [6].

Algorithm 3 outlined the primary steps for clustering the pixels in the input image, using the modified *k*-means clustering algorithm, as described earlier in this section.

Algorithm 3 *K*-Means Image Clustering

Require: Image Dataset
  Input: Random Centroid Points
  Start: Clustering Pixels
  while $pixels \neq end$ **do**
    Select: Neural Engine Core
    Assign: Processing to Core
    Calculate: Mean Value
    Set: Pixel-to-Cluster
  end while
  Output: Clustered Pixels

Initially, patient X-ray and CT-scan images of COVID-19 disease were segmented using the *k*-means clustering algorithm, which then split the image into a set of regions that could be processed and analyzed. Due to the high performance achieved through the modification of the aforementioned algorithm, this step resulted in a thorough scan of the images and the segmentation of their content at a high speed, in preparation for the next stage, which was the application of the YOLOv4 algorithm.

Second, incoming images were resized to 640 by 640 px and normalized using a normalize procedure. The improved *K*-means clustering algorithm, based on mobile neural engine processors [6], was then used to further match the training data with the *k*-mean YOLOv4 model. A suitable anchor size setting facilitated model convergence and provided useful prior information, and this sped up the model training process and resulted in more accurate values. The full implementation flowchart of anchor sizes is provided.

Figure 3 summarizes the main steps of the first stage of image clustering.

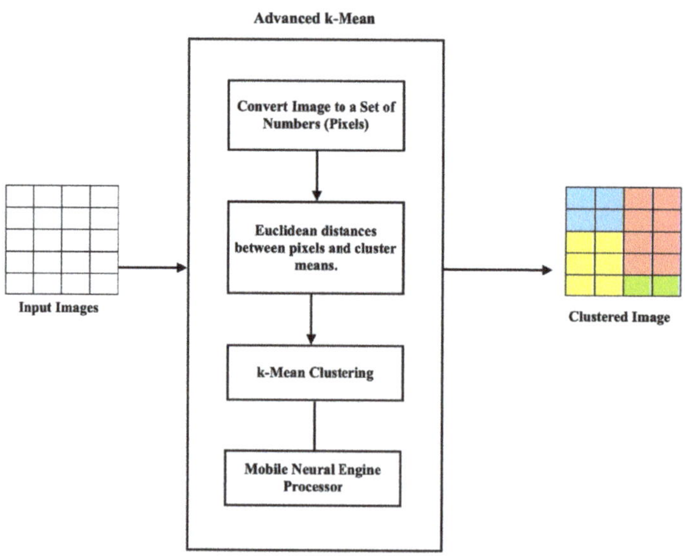

Figure 3. Stage 1 image clustering architecture.

By clustering pixels in an image, we simplified the image by reducing the number of colors and tones. This assisted in removing noise and unwanted details from the image, making it easier to extract relevant features.

Once the pixels were clustered, a new image was created where each pixel was assigned to its corresponding cluster, based on the map image. This new image was called a clustered image. The clustered image contained fewer colors and tones than the original image and could be used to extract features that were more representative of the image content.

For example, in the medical image analysis, *k*-means clustering was used to segment an X-ray or CT-scan image into regions based on the density of the tissue. By clustering the pixels in the image, we identified regions that corresponded to bones, organs, and other tissues, which were then evaluated for feature extraction. These features included the size, shape, and texture of the tissue, which was then used to detect abnormalities and other features that could be indicative of a disease or condition.

6.4. Stage 2: YOLOv4 Image Detection

After processing the images, the second stage of scanning the images commenced using the YOLOv4 algorithm, which could handle and detect objects in images at high speeds. Objects were easier to identify and detect in the pre-processed images due to the image content being segmented into consistent data aggregates.As shown in Figure 4, every object detector began by compressing and processing the images using a convolutional neural network backbone, which could then be used to make predictions at the endpoint of the image classification. To detect objects, several bounding boxes had to be constructed around images, requiring the concatenation of the convolutional feature layers of the backbone and the convergence of all the layers of features in the backbone at the neck.

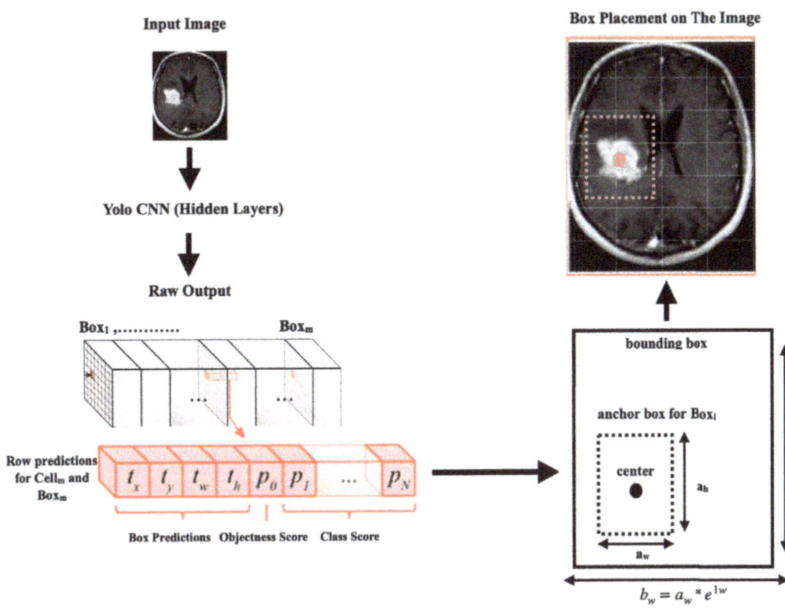

Figure 4. YOLOv4 image detection architecture.

The YOLOv4 system utilized image-resizing, non-maximal suppression, and a single convolutional neural network to identify objects. It generated multiple bounding boxes and class probabilities simultaneously. Although the system was efficient for detecting objects, it could have difficulty identifying the locations of smaller objects precisely.

The input images were divided into an S × S grid, with each grid cell responsible for identifying an object if the centroid of the object was within that grid cell. Using information from the entire image, each grid cell predicted the bounding boxes (B) and the confidence ratings for those boxes. These confidence scores represented the likelihood that an object was present in the box, as well as the accuracy of the object class prediction. The confidence score was defined as:

$$conf = Pr(class_i|obj) \times Pr(obj) \times IoU_{pred}^{truth} \tag{1}$$

where

$$Pre(obj) \in [0,1] \tag{2}$$

here, *Pr(object)* denotes the likelihood that there will be an object in the grid cell, and *Pr(classic | obj)* denotes the likelihood that a particular object will appear based on the presence of an item in the cell.

6.5. Stage 3: K-Means–YOLOv4 Clustering

YOLOv4 used Bag of Specials, which is a technique that adds minimal delays to inference times while significantly enhancing performance. The algorithm evaluated various activation functions. As features flowed through the network, the activation functions were altered, as depicted in Figure 5. Using conventional activation functions, such as ReLU, had not always been sufficient to push feature creation to its optimal limit, which has led to the development of novel techniques in the literature to slightly improve this method.

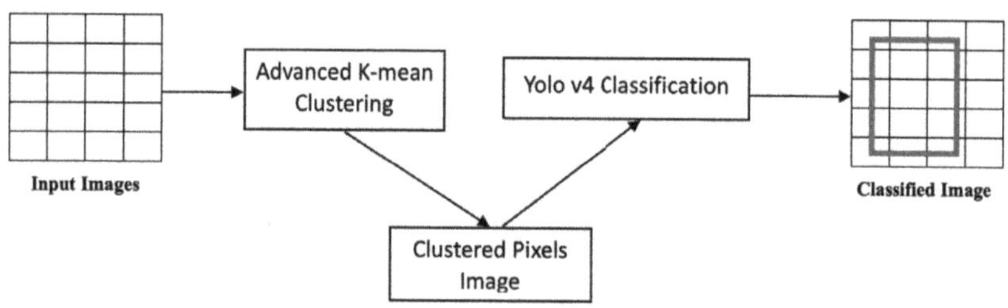

Figure 5. Proposed solution architecture.

To summarize Stage 3 as an algorithm, Algorithm 4 was written. As shown in the algorithm, the YOLOv4 detector received the clustered image before initializing the YOLOv4 layers on it. The clustered images had clustered pixels, which improved the performance of the layers in recognizing the objects, contents, and features of the images.

Algorithm 4 *K*-Means–YOLOv4 Classifier

```
Require: Image Dataset
  Input: Random Centroid Points
  Start: Clustering Pixels
  while pixels ≠ end do
      Select: Neural Engine Core
      Assign: Processing to Core
      Calculate: Mean Value
      Set: Pixel to Cluster
  end while
  Run: YOLO's Backbone on Clustered Image
  if Image Contains (COVID) then
      Flag: Image as Affected
  else
      Flag: Image as non-Affected
  end if
  Output: Classified Image
```

7. Performance Evaluation and Datasets

Performance metrics were crucial for evaluating both classification and detection techniques. In addition, the experiment environment, datasets, and data preparation used to assess these metrics were equally important. Therefore, this section provides a detailed explanation of the performance metrics, datasets, and environment, as they related to the obtained results and the implementation of the proposed solution.The dataset consisted of a diverse set of information, which was classified into four distinct categories. The dataset was split into a training set (70%) and a testing set (30%), with the training set being used to train the machine-learning algorithms and the testing set being used to evaluate their performance. The machine-learning algorithms were applied for classification, using features extracted through the feature-engineering process. The proposed algorithm was compared to various categorization approaches and was found to be highly effective on X-ray images in the experiments. The proposed solution was implemented using the Dart ARM-based programming language, which is suitable for resource-constrained mobile devices, along with specialized deep-learning code for machine-learning engines on mobile devices. For iOS devices, the Swift programming language was utilized, which is known for its ease-of-use and safety features, while Kotlin (the native Android language) was employed for Android devices. This approach allowed for the solution to be easily implemented on different mobile devices and platforms, providing a more versatile and widely accessible solution.

The *k*-means-YOLOv4 approach was evaluated on mobile devices equipped with machine-learning engines, including an iPhone 11 Pro Max with a dedicated 16-core machine-learning processor and the Samsung S22 with a system-on-a-chip, featuring a 16-bit floating-point neural processing unit (NPU). The testing dataset was divided into two categories: X-ray images and CT-scan images.

7.1. Performance Metrics

Recall (R), Precision (P), F1-score ($F1$), specificity (S), and accuracy were used as the performance criteria to examine deep-learning performance.

- Precision: This metric represented the fraction of genuine positives among the expected positives. As a result, true-positive (TP) and false-positive (FP) values were important.

$$P = TP/(TP + FP) \tag{3}$$

- Recall: The ratio of true positives accurately categorized by the model was the recall. The recall was calculated using *TP* and *FN* values.

$$R = TP/(TP + FN) \quad (4)$$

- Specificity: This was defined as the proportion of true negatives (those not caused by illness) correctly classified by the model. The *TN* and *FP* values were used to calculate specificity.

$$S = TN/(TN + FP) \quad (5)$$

- F1-Score: The F1-score measured the model's accuracy by combining precision and recall. Doubling the ratio of the total accuracy and recall values defined the F1-scores.

$$F1 = 2 \times (P \times R)/(P + R) \quad (6)$$

- Performance (Speed): This was an important performance metric in image detection and data classification and clustering, particularly when dealing with large datasets and real-time applications. It measured the time required to process and analyze the data and produce the desired output. In image detection, speed is important for applications such as autonomous vehicles, surveillance systems, and medical imaging, where the detection and analysis of images must be performed in real-time. The speed metric is usually measured in frames-per-second (FPS), which represents the number of images that can be processed in one second. In data classification and clustering, speed is important for applications such as recommendation systems, fraud detection, and customer segmentation, where large amounts of data must be analyzed and classified in a timely manner. The speed metric is usually measured in terms of processing time or throughput, which represents the number of data points that can be processed per unit of time.

7.2. Dataset

To validate the proposed solution in various scenarios and on varied dataset properties, experiments were conducted using a number of different datasets. The characteristics of all the datasets are summarized in Table 1. All datasets were downloaded from the Kaggle website.

A range of dataset sizes was used in this paper to evaluate the performance of the proposed solution with different dataset sizes ranging from a few thousand rows to millions of rows. Therefore, a dataset with 54 MB was used.

Table 1. Clustering datasets.

Dataset	Dataset Size
COVID-19 Dataset	54 MB
COVID-19	362 MB
COVID-19 Open Research Dataset Challenge	20 GB

For image detection and to confirm the model's robustness, two independent datasets were collected and tested. The dataset used in this paper was created using the analysis conducted by [60] and can be downloaded at https://github.com/muhammedtalo/COVID-19 (accessed on 20 February 2023). The dataset consisted of 500 pneumonia, 125 COVID-19, and 500 no-findings X-ray images. It was created using two separate resources: X-ray images obtained from multiple open-access sources of COVID-19 patients in the Cohen [61] database, and the chest X-ray database for normal and pneumonia X-ray images, provided by Wang et al. [62]. The COVID-19 dataset included 43 female patients and 82 male patients. Metadata were not available for all patients in this dataset. Positive COVID-19 patients were, on average, around 55 years old. This was a versatile dataset that could be used for multi-class and binary classification tasks.

The dataset from Harvard Lab [55] was also used in this study. The dataset consisted of non-enhanced chest CT scans of more than 1000 individuals diagnosed with COVID-19. The average age of the CT-scan patients was 47.18 years, with a standard deviation of 16.32 years and a range from 6 to 89 years. The population was composed of 60.9% males and 39.1% females. The most common self-reported co-morbidities among patients were coronary artery or hypertension disease, interstitial pneumonia or emphysema, and diabetes. The positive PTPCR patient images were obtained from in-patient treatment sites for COVID-19 and accompanying clinical symptoms, between March 2020 and January 2021. The scans were taken during end-inspiration with the subjects in a supine position.

The CT scans were conducted using a 16-slice helical mode on NeuViz equipment, without the use of intravenous contrast. The images were captured in DICOM format and were 16-bit gray-scale with 512 × 512 px. The slice thickness was determined by the operator and ranged from 1.5 to 3 mm, based on the clinical examination requirements. The CT scans were reviewed for the presence of COVID-19 infection by two board-certified radiologists. In cases where the first two radiologists were unable to reach a consensus, a third more-experienced radiologist provided the final judgment. The CT images showed a variety of patterns indicative of COVID-19-specific lung infections.

In the third phase of our comparison, two datasets were used. The specifics of the two major subsections of the sourced image graphs were as follows.

1. Radiography database for COVID-19 in [63]. The authors gathered chest X-ray images of COVID-19-positive individuals, along with healthy people and those with viral pneumonia, and made them accessible to the public on https://www.kaggle.com/ (accessed on 20 February 2023).
2. Actualmed, Pau Agust Ballester, and Jose Antonio Heredia from Universitat Jaume I (UJI) created the Actualmed COVID-19 Chest X-ray Dataset for study (https://github.com/agchung/Figure1-COVID-chestxray-dataset/tree/master/image (accessed on 20 February 2023)).

A total of 3106 images were utilized for model training, 16% of which were used for model validation. A total of 806 non-augmented images from various categories were used to test the proposed solution and assess the performance.

Furthermore, the large image datasets in Table 2 were used for the big-data evaluation. All the datasets were downloaded from the Kaggle website.

Table 2. Clustering datasets.

Dataset	Dataset Size	No. of Images/Slices	No. of Classes
Large COVID-19 CT-scan-slice dataset	2 GB	7593	9
COVIDx CT	65 GB	194,922	10
CT Low-Dose Reconstruction	20 GB	16,926	6

Data Preparation

The data clustering had to be prepared, and the primary parameters had to be selected before clustering, as follows:

- **Noise Removal:** The advanced parallel k-means clustering algorithm utilized the mean imputation as the method for handling missing data. In this approach, missing values were replaced with the mean value of the corresponding feature across all samples. This method is simple and computationally efficient, and it has been shown to be effective in practice. However, the mean imputation may introduce bias in the clustering results if the missing data were not missing completely-at-random (MCAR). If the missing data were missing-at-random (MAR) or missing not-at-random (MNAR), more sophisticated methods such as regression imputation and multiple imputation could be required to avoid bias.

- **Number of Clusters:** Selecting the optimal number of clusters in the advanced parallel *k*-means clustering was crucial for achieving effective cluster analysis. This is particularly true in the medical field, where the identification of meaningful clusters can lead to more accurate diagnoses and treatments. However, the traditional methods of finding *k*-value, such as the Elbow method or the Silhouette method, are not always sufficient in the medical field, where the data are often complex and high-dimensional. In such cases, expert knowledge could be required to identify clinically relevant subgroups, which could then be used to determine the optimal number of clusters. In this paper, the *k*-value set to 2 in the clustering of numeric and text data and set to 5 for image clustering, as there were 5 main gray-scale stages of colors in the X-ray and MRI images.

7.3. *Operating System Implementation*

Dispatch queues are a feature of the Grand Central Dispatch (GCD) system, which is a part of the iOS and macOS operating systems. GCD provides a high-level, asynchronous programming interface for managing concurrent tasks. Dispatch queues are lightweight and provide a simple interface for executing tasks concurrently without consuming an excessive amount of system resources. Dispatch queues are managed by the operating system and can be used to process tasks on a first-in, first-out (FIFO) basis. This makes it easy to manage task dependencies and avoid competitive conditions, and tasks submitted to a dispatch queue can be executed in parallel with other tasks in the queue. Dispatch queues can be created with different priorities to manage the order of execution of tasks and ensure that high-priority tasks are executed first.

Threads, in contrast, are a lower-level mechanism for achieving concurrency in a program. Threads achieve true parallelism, as multiple threads can execute simultaneously on different processor cores. Each thread has its own stack and program counter, and threads can share memory with other threads in the same process. Threads are managed by the operating system and can be used to process tasks concurrently in a more fine-grained way than dispatch queues. As compared to dispatch queues, threads have a higher overhead and require more system resources, making them less suitable for lightweight tasks. Threads can be used to implement more complex concurrency patterns, such as locking, synchronization, and message-passing.

The proposed solution for implementing the modified parallel *k*-means clustering algorithm on iOS leveraged the advantages of dispatch queues to achieve concurrency. The GCD framework provided several types of queues, including serial and concurrent dispatch queues. A serial dispatch queue executed tasks one at a time, while a concurrent dispatch queue executed tasks concurrently.

In the proposed solution, a concurrent dispatch queue was used to execute the *k*-means clustering algorithm on multiple cores simultaneously. Each task was scheduled on the dispatch queue, and the queue handled the scheduling of tasks across multiple cores. This allowed the algorithm to take advantage of the multi-core neural engine processor and general-purpose processor, leading to improved performance.

Furthermore, GCD provided mechanisms to ensure thread safety and avoid competitive conditions through the use of synchronization techniques, such as semaphores and barriers. By utilizing these features, the implementation of the parallel *k*-means clustering algorithm on dispatch queues was more efficient and reliable.

8. Results and Discussion

The proposed work was subjected to thorough testing and evaluation in multiple stages to ensure its effectiveness at various levels and within different contexts. The primary focus was on enhancing performance and leveraging the high speeds offered by the two integrated algorithms.

8.1. Operating System Performance

The proposed solution was designed to be operating system independent and hardware accelerated. This meant that the advanced parallel *k*-means clustering could be executed on any operating system that had two processors: a neural engine processor and a general-purpose processor. However, both iOS and Android operating systems were designed to manage and take advantage of hardware allocation and management that included their neural engine processor. These dedicated operating systems were able to send specific tasks to a particular processor core, enabling the implementation and execution of the advanced parallel *k*-means clustering.

Overall, this type hardware acceleration provides opportunities for future advancements of the operating systems, which is expected since the new M-family MacOS already supports dedicated neural-engine-core assignments.

In order to evaluate the performance of the advanced *k*-means clustering across different operating systems, Table 3 presents two large datasets, each with over 9 million records. These were clustered using the advanced parallel *k*-means clustering algorithm on Windows OS, Android, and iOS systems.

Table 3. Clustering datasets.

Dataset	Number of Records
Google Play Store	11,000,000
KDD99 [64]	9,000,000

The performance results, as presented in Table 4, showed that the processing of 11 million records from the Google Play Store dataset doubled in speed with a dedicated ML processor. The next experiment was conducted using the education-sector dataset, and the mobile processor exhibited a performance up to 10-times faster than the desktop OS (Windows 11). Additionally, the performance of iOS was twice as fast as that of the Android OS.

Table 4. Clustering performance of big-data sets in minutes.

Dataset	Windows OS	iOS	Android OS
Google Play Store	90 min	46.1	56.4
Education Sector	24.3 ms	2.4	6.3

The performance differences observed between the iOS and Android operating systems, within the context of advanced parallel *k*-means clustering, could be due to several factors. It could be related to the differences in the underlying architectures of the two operating systems. Specifically, iOS was designed to take full advantage of its hardware resources, including the dedicated neural engine cores, which could explain the observed faster performance, as compared to Android.

Additionally, the iOS architecture was based on the use of Objective-C and dispatch queues, which were designed to facilitate concurrent processing and task scheduling. These features provide a more efficient way to execute the parallel *k*-means clustering algorithm, potentially resulting in the observed faster performance.

However, the performance differences observed could have also been influenced by other factors, such as the differences in the hardware configurations of the devices used to test the algorithms, as well as the specific implementation of the parallel *k*-means clustering algorithm on the different operating systems.

8.2. Data Classification Model

The performance of logistic regression and naive Bayes algorithms could have been influenced by various factors, such as the size and complexity of the data, the hardware and software utilized, and the specific implementation of the algorithms. Typically, logistic regression has been faster than naive Bayes when working with large datasets, as the naive Bayes algorithm can become computationally demanding as the number of features increases. However, naive Bayes can be faster when working with smaller datasets or when the number of features is relatively limited. Table 5 illustrates the performance of both algorithms after classifying 10 million records of medical data. The table shows the speed and accuracy of both algorithms, which aided in determining which algorithm was more suitable for a specific application. While the speed of an algorithm was an important consideration, accuracy was also taken into account when a compromise between speed and accuracy may be necessary.

Table 5. Data classification performance.

Algorithm	Performance (m.)
Logistic Regression	23.1
Naive Bayes	31.4

Based on the datasets described in Section 7.2, the proposed solution was analyzed to evaluate its classification performance and accuracy.

The results presented in Tables 6 and 7 demonstrated the superior performance of the proposed solution, as compared to the logistic-regression and naive Bayes algorithms. The naive Bayes algorithm is known to be efficient for small datasets, but the proposed solution outperformed both algorithms, even when the dataset size increased. This highlighted the effectiveness of the proposed solution in handling larger datasets, which pose a significant challenge for traditional classification methods. Additionally, the strong performance of the proposed solution, as compared to the standard classification algorithms, such as logistic regression and naive Bayes, further emphasized its potential for practical applications. Overall, the results demonstrated the exceptional performance and potential of the proposed solution.

Table 6. Data classification performance (Dataset 1).

Dataset	Algorithm	Speed (m.)
1	Logistic Regression	12.2
	Naive Bayes	8.5
	K-Means–Logistic Regression	9.7
2	Logistic Regression	63.2
	Naive Bayes	83.1
	K-Means–Logistic Regression	45.1
3	Logistic Regression	2754
	Naive Bayes	3571
	K-Means–Logistic Regression	1693

Table 7. Data classification accuracy of Dataset 1.

Dataset	Algorithm	Accuracy (%)
1	Logistic Regression	93.4
	Naive Bayes	92.1
	K-Means–Logistic Regression	95.3
2	Logistic Regression	94.2
	Naive Bayes	93.5
	K-Means–Logistic Regression	97.2
3	Logistic Regression	93.1
	Naive Bayes	91.3
	K-Means–Logistic Regression	97.6

One possible reason for the higher accuracy of the proposed solution was that it had been specifically designed to handle larger datasets, which may have been more challenging for traditional classification algorithms. For example, logistic regression and naive Bayes algorithms could have struggled to effectively classify data when the number of features increased significantly, as they can become computationally demanding as the number of features increases. In contrast, the proposed solution used more advanced techniques, including the advanced parallel *k*-means, parallel logistic regression, and the neural engine processor, to effectively classify the large datasets. Additionally, the proposed solution incorporated additional factors and features that were relevant to the classification task, which further improved its accuracy. Overall, the results demonstrated the effectiveness of the proposed solution in handling large datasets and achieving high accuracy in classification tasks.

In order to examine the performance of the proposed solution with the recent advancements in medical data classification, the proposed solution was compared with the three most recent medical-data-classification approaches, which were: [65–67]. All solutions were compared with the proposed solution in terms of classification performance and classification accuracy, as shown in tables below.

As shown in Table 8, the proposed solution significantly outperformed the three compared solutions, while the naive Bayes-based algorithm tended to be slower, the proposed solution was more effective than both the binary logistic regression and the logistic regression. This suggested that the proposed solution was particularly well suited for handling larger datasets, which could be more challenging for traditional classification algorithms. The results demonstrated the strong performance of the proposed solution, as compared to the conventional classification algorithms, indicating that it was an effective and reliable method for classification tasks.

Table 8. Data classification speed (min.) when compared with recent approaches.

Algorithm	Performance (m.)
Logistic Regression [66]	34.1
Novel Binary Logistic Regression [65]	28.3
Correlated Naive Bayes [67]	37.4
K-Means–Logistic Regression	21.1

The accuracy of the proposed solution was compared with previous solutions, and the results, as shown in Table 9, demonstrated its high accuracy. Specifically, the proposed solution outperformed the comparable solutions, achieving an accuracy rate of 99.8% while a novel binary-logistic-regression solution only achieved 98% accuracy. The worst performance in terms of accuracy was observed in the logistic-regression solution designed for the prediction of myocardial infarction disease in [66]. These results suggested that the proposed solution was particularly effective at achieving high accuracy in classification tasks, and that it outperformed other approaches.

Table 9. Data classification accuracy when compared with recent approaches.

Algorithm	Accuracy (%)
Logistic Regression [66]	88
Novel Binary Logistic Regression [65]	98
Correlated Naive Bayes [67]	97
Shared Bayesian Variable Shrinkage [68]	93
Classification of Breast Cancer Metastasis Using Machine-Learning Algorithms [69]	92
K-Means–Logistic Regression	99.8

8.3. Training Proposed Model

Feature extraction and classification are the two crucial components of the proposed image detection system. The quality of the extracted features was critical to the success of the classification process. Therefore, the extracted features were used to train the model in order to demonstrate its effectiveness in feature extraction. Figure 6 shows an image after applying the *k*-means clustering technique with feature selection. The resulting image was divided into two main clusters, black and white, in the first stage of learning. This clustered image was then used as a map for pixel-based feature extraction, where each pixel was assigned to its corresponding cluster based on the mapped image.

In the next step, the pixel values were processed with their original values for the image detection process. This approach provided two benefits. Firstly, any outlier pixels due to the X-ray device or CT-scan process were removed. Secondly, a new feature was added to the image pixels, which was the pixel group. The cluster value associated with each pixel provided valuable information for image feature extraction and detection. By considering the cluster value, we could efficiently extract the relevant features from the image and ignore the noise and other irrelevant pixels. This approach significantly improved the accuracy of image detection and reduced false positives. When processing an X-ray image, the proposed solution began by extracting the lung features of the patient and then determined whether the lungs were normal or abnormal by classifying them as positive or negative, accordingly.

8.4. Object Detection Speed

During the initial phase, the proposed work was compared with a range of standard algorithms frequently used for image classification. The proposed solution demonstrated exceptional performance, outperforming the other algorithms by up to 15-fold. It also outperformed the YOLOv4 algorithm by approximately 60%, as shown in the comparison presented in Table 10.

Table 10. Detection speeds of object algorithms.

Algorithm	Speed (ms.)
SPP-net	1500
R-CNN	900
Fast R-CNN	750
Faster R-CNN	600
R-FCN	550
Mask R-CNN	400
YOLOv3	250
YOLOv4	150
Advanced Parallel *K*-means–YOLOv4 (APK-YOLO)	90

To assess the performance of the proposed solution under various scenarios and with varied device specifications, it was tested using both a standard computer CPU (Intel Core i5-3.5 GHz) and GPU (AMD Radeon R9 M290X 2 GB). The results of the experiments showed that the proposed detection solution maintained its high performance, as compared to the YOLOv4 algorithm, as demonstrated in Table 11.

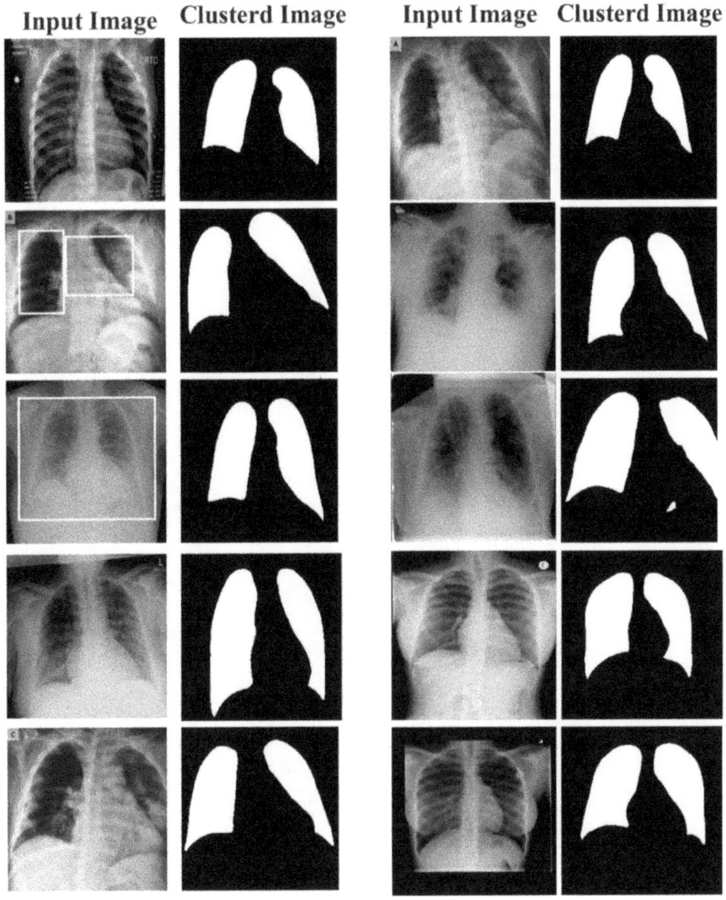

Figure 6. Learned features from the first layer.

Table 11. Classification performance on neural engine, CPU, and GPU.

Algorithm	CPU (ms.)	GPU (ms.)	Neural Engine (ms.)
YOLOv4	350	280	150
APK-YOLOv4	220	120	90

The results showed that the proposed solution exhibited a high performance, which was up to 2.3 times faster on the GPU and up to 1.5 times faster on the CPU, as compared to the standard YOLOv4. Additionally, the proposed algorithm demonstrated a significant speed advantage, achieving speeds that were up to 7 times faster due to the high speed of the proposed algorithm and the efficient use of the artificial intelligence processors in modern mobile devices, as compared to recent solutions, such as (VGGCOV19-NET [70] and CAD-based YOLOv4 [71]).

8.5. Object Detection Performance

In the second phase of the performance comparison, as shown in Table 12, the proposed solution was compared with two recent approaches that made adjustments to classification algorithms to handle X-ray images of COVID-19 patients.

Table 12. Image classification performance comparison.

Algorithm	Speed (ms.)
VGGCOV19-NET [70]	620
CAD-based YOLOv4 [71]	530
APK-YOLOv4	90

Due to the importance of the TN, TP, FN, and FP values [72], their values had been calculated first, as shown in Figure 7.

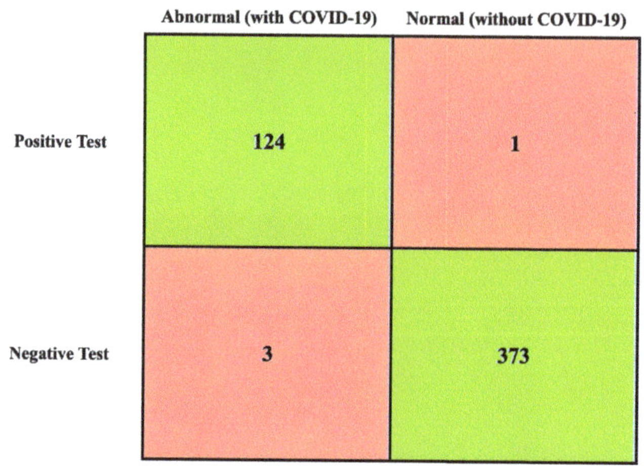

Figure 7. *TN*, *TP*, *FN*, and *FP* values of X-ray dataset.

In the last part of the comparison, the proposed work was compared to the benchmark examples, based on four performance measures, including recall, precision, F1-score, and accuracy. These represented the best testing factors for evaluating the performance of the classification algorithms and to ensure that the improvements achieved [73] by the proposed algorithm were accurate across all levels, which, in turn, would indicate its potential application in the medical field. The results, as shown in Table 13, illustrated the excellent performance of the proposed algorithm in the classification task of images when applied to the Fold 1–5 levels.

Table 13. Recall, precision, F1-score, and accuracy performance of Folds 1–5 of the chest X-ray images.

Fold	Algorithm	Recall	Precision	F1	Accuracy
1	VGGCOV 19-NET [70]	78.20	78.80	78.30	78.22
	CAD-based YOLOv4 [71]	75.90	75.6	75.8	75.4
	APK-YOLOv4	82.2	82.6	82.7	82.7
2	VGGCOV 19-NET [70]	91.10	91.10	91.10	91.11
	CAD-based YOLOv4 [71]	89.5	89.4	89.5	89.5
	APK-YOLOv4	93.4	93.4	93.3	93.4
3	VGGCOV 19-NET [70]	84.40	84.80	84.50	84.44
	CAD-based YOLOv4 [71]	90.2	90.1	90.2	90.1
	APK-YOLOv4	94.2	94.23	94.2	94.3
4	VGGCOV 19-NET [70]	95.10	95.20	95.10	95.11
	CAD-based YOLOv4 [71]	94.5	94.2	94.4	94.4
	APK-YOLOv4	96.7	96.4	96.5	96.5
5	VGGCOV 19-NET [70]	95.60	95.70	95.60	95.56
	CAD-based YOLOv4 [71]	94.2	94.1	94.2	94.2
	APK-YOLOv4	97.2	97.6	97.4	97.5

When the algorithm treated images classified as infected images, it also showed superior accuracy, and the performance measures of the rest of the results are shown in Tables 14 and 15.

Table 14. Recall, precision, F1-score, and accuracy performance on COVID-19 images.

Algorithm	Recall	Precision	F1	Accuracy
VGGCOV 19-NET [70]	92.80	99.15	95.87	87.89
CAD-based YOLOv4 [71]	91.5	95.7	85.7	90.67
APK-YOLOv4	93.8	99.7	97.44	96.21

An advanced *K*-means clustering [6] combined with YOLOv4 solution enabled the rapid and accurate detection of COVID-19 within milliseconds, making it a useful tool in regions with a shortage of experienced doctors and radiologists. Additionally, the model could be utilized to identify patients in settings with limited healthcare facilities, even when only X-ray technology is available, and it could ensure more timely treatments for positive COVID-19 patients. One practical benefit of the concept was that it allowed for the identification of patients who did not require PCR testing, thereby reducing the overcrowding in medical facilities.

Table 15. Recall, precision, F1-score, and accuracy performance on no-findings images.

Algorithm	Recall	Precision	F1	Accuracy
VGGCOV 19-NET [70]	90.20	86.40	88.26	85.80
CAD-based YOLOv4 [71]	89.1	82.3	80.4	89.7
APK-YOLOv4	92.9	95.4	92.6	91.82

In the second part of the performance comparison, as shown in Table 16, the proposed solution was compared with recent studies in which classification algorithms were modified to handle CT-scan images of COVID-19 patients. Due to the high speed of the suggested method and the extensive use of artificial intelligence processors prevalent in recent mobile devices, the proposed algorithm demonstrated superiority in its accuracy, recall, and other performance metrics.

Table 16. Recall, precision, F1-score, and accuracy performance of Folds (1–5) with CT-scan images.

Fold	Algorithm	Recall	Precision	F1	Accuracy
1	Compressed Chest CT Image through Deep Learning [74]	79.10	79.30	79.10	79.87
	APK-YOLOv4	83.5	83.4	83.4	83.3
2	Compressed Chest CT Image through Deep Learning [74]	93.10	92.80	92.60	92.8
	APK-YOLOv4	94.1	94.3	94.7	94.6
3	Compressed Chest CT Image through Deep Learning [74]	89.10	89.70	89.60	89.8
	APK-YOLOv4	93.8	94.1	93.9	93.8
4	Compressed Chest CT Image through Deep Learning [74]	96.20	96.30	96.18	96.20
	APK-YOLOv4	97.1	97.5	97.4	97.2
5	Compressed Chest CT Image through Deep Learning [74]	98.78	98.75	98.80	98.7
	APK-YOLOv4	99.4	99.7	99.3	99.2

Figure 8 shows the learning curve accuracy of the proposed solution in both the training and testing stages. The accuracy of the proposed solution had consistent improvement. Furthermore, the learning curve began with an accuracy near 32% and continued to improve, up to 99%.

Figure 8. Accuracy learning curve.

In the third part of the performance comparison, as shown in Table 17, the proposed solution was compared with recent studies that used a classification technique on brain MRI images to maximize the generalizability of the proposed solution. This comparison was conducted to show that the proposed solution could be adapted for various datasets and image types, as well as to classify other diseases, such as brain tumors. The results showed that the proposed solution had excellent performance across all four comparison parameters (recall, precision, F1-score, and accuracy). The dataset used in [75], which consisted of 280 samples of MRI images, was also used in this test. The dataset contained 100 images with normal tumors and 180 with abnormal tumors.

Table 17. Recall, precision, F1-score, and accuracy performance of Folds (1–5) on MRI images.

Fold	Algorithm	Recall	Precision	F1	Accuracy
1	hybrid deep CNN-Cov-19-Res-Net [75]	78.50	78.70	78.40	78.12
	APK-YOLOv4	82.4	82.3	82.4	82.3
2	hybrid deep CNN-Cov-19-Res-Net [75]	91.30	91.20	91.60	91.3
	APK-YOLOv4	93.5	93.4	93.2	93.3
3	hybrid deep CNN-Cov-19-Res-Net [75]	90.10	90.2	90.3	90.1
	APK-YOLOv4	94.5	94.5	94.3	94.3
4	hybrid deep CNN-Cov-19-Res-Net [75]	95.30	95.20	95.10	95.40
	APK-YOLOv4	96.3	96.5	96.3	96.2
5	hybrid deep CNN-Cov-19-Res-Net [75]	97.18	97.15	97.30	97.2
	APK-YOLOv4	98.1	98.2	98.1	98.1

Table 17 shows the proposed solution's performance, as compared to a recent solution [75]. The results showed that the proposed solution outperformed the comparable solution, with an accuracy of up to 98%.

The proposed solution was compared with a high-performance and highly accurate solution, which had been proposed in 2020 [76]. For datasets 1 and 2, the solution obtained 98.7%, 98.2%, 99.6%, and 99% for classification accuracy and F1-Score, respectively. However, as shown in Tables 18 and 19 with the best values across 7 folds, the proposed solution had excellent classification performance in terms of accuracy, recall, precision, and F1-score, as compared to the comparable 2020 approach.

Table 18. Data classification (recall, precision, accuracy, and F1-score) on dataset 1.

Algorithm	Recall	Precision	F1	Accuracy
Deep Features and Fractional-Order Marine Predators [76]	98.2	98.5	99.6	98.7
APK-YOLOv4	98.8	99.1	99.8	99.1

Table 19. Data classification (recall, precision, accuracy, and F1-score) on dataset 2.

Algorithm	Recall	Precision	F1	Accuracy
Deep Features and Fractional-Order Marine Predators [76]	97.7	98.1	99	98.2
APK-YOLOv4	98.5	99.3	98.1	99.6

The excellent performance and accuracy of the proposed solution could be attributed to the optimization of the *k*-means clustering, which enhanced the recognition of the image characteristics by the classifier. Additionally, the optimization of the YOLOv4 algorithm through modified layers improved the ability to detect and recognize features, resulting in an overall improvement in performance.

In order to evaluate the performance of the proposed solution on a vast amount of medical image detection, a set of big-medical-data was used, as described in Section 7.2 and (Table 3). Table 20 shows the performance of the proposed solution, as compared to recent approaches. The performance of the proposed solution in terms of recall, precision, F1-score and accuracy was up to 10% better than the comparable solutions.

The results of the proposed approach using advanced parallel *k*-means clustering, logistic regression, and YOLOv4 for medical data classification and image detection could have important implications for the field of healthcare. The accurate classification and detection of medical data could have a significant impact on patient outcomes by enabling earlier diagnoses and more effective treatment planning. The proposed approach has potential for improving the accuracy and efficiency of these tasks, which could ultimately lead to better patient outcomes and reduced healthcare costs.

Furthermore, the proposed approach has the potential to contribute to the development of new solutions in these areas by providing a more efficient and effective means of pre-processing medical data. The use of advanced parallel *k*-means clustering for pre-processing reduced the dimensionality of the data, which made it easier to classify and detect patterns. This could lead to the development of new algorithms that are more effective for identifying specific medical conditions and abnormalities and could, ultimately, lead to new treatments and therapies.

Table 20. Image detection recall, precision, F1-score, and accuracy performance of Folds (1–5) (big-medical-data image sets).

Fold	Algorithm	Recall	Precision	F1	Accuracy
1	VGGCOV 19-NET [70]	77.30	77.20	77.10	77.34
	CAD-based YOLOv4 [71]	78.80	78.9	78.5	78.4
	APK-YOLOv4	85.1	84.1	85.3	85.7
2	VGGCOV 19-NET [70]	90.30	90.40	90.60	90.4
	CAD-based YOLOv4 [71]	90.6	91.2	91.6	91.8
	APK-YOLOv4	94.3	95.1	95.2	95.1
3	VGGCOV 19-NET [70]	88.60	88.85	87.40	87.74
	CAD-based YOLOv4 [71]	91.3	91.1	91.5	91.4
	APK-YOLOv4	96.52	96.27	97.1	96.8
4	VGGCOV 19-NET [70]	95.30	95.24	95.34	95.61
	CAD-based YOLOv4 [71]	94.7	94.25	94.7	94.8
	APK-YOLOv4	97.7	98.5	97.8	97.9
5	VGGCOV 19-NET [70]	96.60	96.30	95.9	95.76
	CAD-based YOLOv4 [71]	95.2	95.13	95.22	95.12
	APK-YOLOv4	98.8	98.4	98.7	98.6

Additionally, the proposed approach could aid in the development of new medical imaging technologies. By improving the accuracy of image detection, the proposed approach could assist in identifying abnormalities that are difficult to detect using traditional imaging methods. This could lead to the development of new imaging technologies that are more accurate and effective and could, ultimately, improve patient outcomes.

In terms of the overall medical-data field, the proposed approach using advanced parallel *k*-means clustering for pre-processing medical data, combined with logistic regression and YOLOv4 for classification and image detection, respectively, could contribute to the development of new solutions for medical data classification and image detection.

Firstly, the use of advanced parallel *k*-means clustering for pre-processing medical data could significantly reduce the processing time and improve the accuracy of subsequent classification and detection tasks. This could be especially beneficial for large-scale medical datasets, where traditional clustering methods may not be feasible due to computational limitations.

Secondly, the combination of logistic regression and YOLOv4 for classification and image detection, respectively, could improve the accuracy of these tasks in medical applications. Logistic regression is a simple and efficient algorithm that could be used for both binary and multi-class classification, while YOLOv4 is a state-of-the-art object detection algorithm that can detect multiple objects in an image with high accuracy.

Thirdly, the proposed approach could potentially aid in the diagnosis, treatment planning, and disease monitoring in healthcare. The accurate classification and detection of medical data could provide clinicians with valuable insights into a patient's condition and assist them in making informed decisions regarding treatments.

Lastly, the proposed approach could also serve as a framework for the development of new solutions in medical data classification and image detection. The combination of advanced clustering methods, logistic regression, and object detection algorithms could be customized and optimized for specific medical applications and datasets. This could lead to the development of innovative solutions that address the unique challenges and complexities of medical data analysis.

9. Conclusions

The proposed approach using advanced parallel *k*-means clustering for pre-processing medical data, combined with logistic regression and YOLOv4 for classification and image detection, respectively, effectively improved the performance of these algorithms, particularly when applied to large medical datasets. The results of the classification task showed that the approach was able to accurately classify the medical data, and the results of the image detection task using X-ray and CT scan images showed that the approach was able to effectively detect and classify the medical images. The use of advanced parallel *k*-means pre-processing and acceleration of the neural engine processor contributed to the improved accuracy and efficiency of the approach. This approach has the potential to significantly impact the field of healthcare, as it can aid in diagnostics, treatment planning, and disease monitoring. Further research and evaluation on larger and more diverse medical datasets could reveal additional benefits and potential applications. While the proposed solution has shown promise in improving the accuracy and efficiency of these tasks on large medical datasets, there were still limitations that should be considered. One limitation was the hardware dependency, as the acceleration of the *k*-means clustering was highly dependent on the neural engine processor, multi-core processor, and the operating system's support for hardware management. Another limitation was the ability to improve 24-bit color images, which require a different number of *k*-values and could affect the clustering performance negatively.

Author Contributions: Conceptualization, F.H.A., M.M.H. and L.A.; methodology, F.H.A., M.M.H. and L.A.; software, F.H.A., M.M.H. and L.A.; validation, F.H.A., M.M.H. and L.A.; formal analysis, F.H.A., M.M.H. and L.A.; investigation, F.H.A., M.M.H. and L.A.; resources, F.H.A., M.M.H. and L.A.; data curation, F.H.A., M.M.H. and L.A.; writing—original draft preparation, F.H.A., M.M.H. and L.A.; writing—review and editing, F.H.A., M.M.H. and L.A.; visualization, F.H.A., M.M.H. and L.A.; supervision M.M.H.; project administration, F.H.A., M.M.H. and L.A. All authors have read and agreed to the published version of the manuscript.

Funding: Laith Alzubaidi would like to acknowledge the support received through the following funding schemes of Australian Government: Australian Research Council (ARC) Industrial Transformation Training Centre (ITTC) for Joint Biomechanics under grant IC190100020 and QUT ECR SCHEME 2022, The Queensland University of Technology.

Informed Consent Statement: Not applicable.

Data Availability Statement: COVID-19 Research Challenge: (https://www.kaggle.com/datasets/allen-institute-for-ai/CORD-19-research-challenge); Large COVID- 19 CT (https://www.kaggle.com/datasets/maedemaftouni/large-covid19-ct-slice-dataset); COVIDx- CT (https://www.kaggle.com/datasets/hgunraj/covidxct); CT- LOW- Dose (https://www.kaggle.com/datasets/andrewmvd/ct-low-dose-reconstruction); Google Play Store (https://www.kaggle.com/datasets/gauthamp10/google-playstore-apps); KDD99 (https://datahub.io/machine-learning/kddcup99).

Conflicts of Interest: The authors declare no conflict of interest.

References

1. Lawonn, K.; Smit, N.N.; Buhler, K.; Preim, B. A survey on multimodal medical data visualization. In *Computer Graphics Forum*; Wiley Online Library: Hoboken, NJ, USA, 2018; Volume 37, pp. 413–438.
2. Seo, H.; Badiei Khuzani, M.; Vasudevan, V.; Huang, C.; Ren, H.; Xiao, R.; Jia, X.; Xing, L. Machine learning techniques for biomedical image segmentation: An overview of technical aspects and introduction to state-of-art applications. *Med. Phys.* **2020**, *47*, e148–e167. [CrossRef]
3. Alzubaidi, L.; Fadhel, M.; Al-Shamma, O.; Zhang, J.; Santamaria, J.; Duan, Y. Robust application of new deep learning tools: An experimental study in medical imaging. *Multimed. Tools Appl.* **2022**, 1–29. [CrossRef]
4. Boyapati, S.; Swarna, S.R.; Dutt, V.; Vyas, N. Big Data Approach for Medical Data Classification: A Review Study. In Proceedings of the 2020 3rd International Conference on Intelligent Sustainable Systems (ICISS), Palladam, India, 3–5 December 2020; IEEE: New York, NY, USA, 2020; pp. 762–766.
5. Yadav, S.S.; Jadhav, S.M. Deep convolutional neural network based medical image classification for disease diagnosis. *J. Big Data* **2019**, *6*, 113. [CrossRef]
6. Awad, F.H.; Hamad, M.M. Improved k-Means Clustering Algorithm for Big Data Based on Distributed SmartphoneNeural Engine Processor. *Electronics* **2022**, *11*, 883. [CrossRef]
7. Patel, H.; Singh Rajput, D.; Thippa Reddy, G.; Iwendi, C.; Kashif Bashir, A.; Jo, O. A review on classification of imbalanced data for wireless sensor networks. *J. Distrib. Sens. Netw.* **2020**, *16*, 1550147720916404. [CrossRef]
8. De Menezes, F.S.; Liska, G.R.; Cirillo, M.A.; Vivanco, M.J. Data classification with binary response through the Boosting algorithm and logistic regression. *Expert Syst. Appl.* **2017**, *69*, 62–73. [CrossRef]
9. Karasoy, O.; Ballı, S. Spam SMS detection for Turkish language with deep text analysis and deep learning methods. *Arab. J. Sci. Eng.* **2022**, *47*, 9361–9377. [CrossRef]

10. Theodoridis, S. *Machine Learning: A Bayesian and Optimization Perspective*; Academic Press: Cambridge, MA, USA, 2015.
11. Tigga, N.P.; Garg, S. Predicting type 2 diabetes using logistic regression. In *Proceedings of the Fourth International Conference on Microelectronics, Computing and Communication Systems: MCCS 2019*; Springer: Berlin/Heidelberg, Germany, 2021; pp. 491–500.
12. Itoo, F.; Singh, S. Comparison and analysis of logistic regression, Naïve Bayes and KNN machine-learning algorithms for credit card fraud detection. *Int. J. Inf. Technol.* **2021**, *13*, 1503–1511. [CrossRef]
13. Sen, S.; Kundu, D.; Das, K. Variable selection for categorical response: A comparative study. *Comput. Stat.* **2022**, 1–18 . [CrossRef]
14. Sun, Y.; Zhang, Z.; Yang, Z.; Li, D. Application of logistic regression with fixed memory step gradient descent method in multi-class classification problem. In Proceedings of the 2019 6th International Conference on Systems and Informatics (ICSAI), Shanghai, China, 2–4 November 2019; IEEE: New York, NY, USA, 2019; pp. 516–521.
15. De Caigny, A.; Coussement, K.; De Bock, K.W. A new hybrid classification algorithm for customer churn prediction based on logistic regression and decision trees. *Eur. J. Oper. Res.* **2018**, *269*, 760–772. [CrossRef]
16. Gibert, D.; Mateu, C.; Planes, J. The rise of machine learning for detection and classification of malware: Research developments, trends and challenges. *J. Netw. Comput. Appl.* **2020**, *153*, 102526. [CrossRef]
17. Galvez, R.L.; Bandala, A.A.; Dadios, E.P.; Vicerra, R.R.P.; Maningo, J.M.Z. Object detection using convolutional neural networks. In *Proceedings of the 2018 IEEE Region 10 Conference (TENCON 2018), Jeju Island, Republic of Korea, 28–31 October 2018*; IEEE: New York, NY, USA, 2018; pp. 2023–2027.
18. Yu, J.; Zhang, W. Face mask wearing detection algorithm based on improved YOLO-v4. *Sensors* **2021**, *21*, 3263. [CrossRef]
19. Li, S.; Gu, X.; Xu, X.; Xu, D.; Zhang, T.; Liu, Z.; Dong, Q. Detection of concealed cracks from ground penetrating radar images based on deep learning algorithm. *Constr. Build. Mater.* **2021**, *273*, 121949. [CrossRef]
20. Jiang, P.; Ergu, D.; Liu, F.; Cai, Y.; Ma, B. A Review of YOLOv4 algorithm developments. *Procedia Comput. Sci.* **2022**, *199*, 1066–1073. [CrossRef]
21. Haggui, O.; Bayd, H.; Magnier, B. Centroid human tracking via oriented detection in overhead fisheye sequences. *Vis. Comput.* **2023**, 1–19. [CrossRef]
22. Fan, S.; Liang, X.; Huang, W.; Zhang, V.J.; Pang, Q.; He, X.; Li, L.; Zhang, C. Real-time defects detection for apple sorting using NIR cameras with pruning-based YOLOV4 network. *Comput. Electron. Agric.* **2022**, *193*, 106715. [CrossRef]
23. Bao, W.; Xu, B.; Chen, Z. Monofenet: Monocular 3d object detection with feature enhancement networks. *IEEE Trans. Image Process.* **2019**, *29*, 2753–2765. [CrossRef]
24. Saponara, S.; Elhanashi, A.; Gagliardi, A. Implementing a real-time, AI-based, people detection and social distancing measuring system for COVID-19. *J.-Real-Time Image Process.* **2021**, *18*, 1937–1947. [CrossRef]
25. Sun, J.; Ge, H.; Zhang, Z. AS-YOLO: An improved YOLOv4 based on attention mechanism and SqueezeNet for person detection. In Proceedings of the 2021 IEEE 5th Advanced Information Technology, Electronic and Automation Control Conference (IAEAC), Chongqing, China, 12–14 March 2021; IEEE: New York, NY, USA, 2021; Volume 5, pp. 1451–1456.
26. Singh, A.; Kalaichelvi, V.; DSouza, A.; Karthikeyan, R. GAN-Based Image Dehazing for Intelligent Weld Shape Classification and Tracing Using Deep Learning. *Appl. Sci.* **2022**, *12*, 6860. [CrossRef]
27. Singh, S.; Ahuja, U.; Kumar, M.; Kumar, K.; Sachdeva, M. Face mask detection using YOLOv3 and faster R-CNN models: COVID-19 environment. *Multimed. Tools Appl.* **2021**, *80*, 19753–19768. [CrossRef]
28. Nair, R.; Vishwakarma, S.; Soni, M.; Patel, T.; Joshi, S. Detection of COVID-19 cases through X-ray images using hybrid deep neural network. *World J. Eng.* **2021**, *19*, 33–39. [CrossRef]
29. Yoshitsugu, K.; Nakamoto, Y. COVID-19 Diagnosis Using Chest X-ray Images via Classification and Object Detection. In Proceedings of the 2021 4th Artificial Intelligence and Cloud Computing Conference, Kyoto Japan, 17–19 December 2021; pp. 62–67.
30. Arunkumar, N.; Mohammed, M.A.; Abd Ghani, M.K.; Ibrahim, D.A.; Abdulhay, E.; Ramirez-Gonzalez, G.; de Albuquerque, V.H.C. *K*-means clustering and neural network for object detecting and identifying abnormality of brain tumor. *Soft Comput.* **2019**, *23*, 9083–9096. [CrossRef]
31. Razzak, M.I.; Naz, S.; Zaib, A. Deep learning for medical image processing: Overview, challenges and the future. In *Classification in BioApps: Automation of Decision Making*; Springer: Berlin/Heidelberg, Germany, 2018; pp. 323–350.
32. Alzubaidi, L.; Fadhel, M.; Al-Shamma, O.; Zhang, J.; Duan, Y. Deep learning models for classification of red blood cells in microscopy images to aid in sickle cell anemia diagnosis. *Electronics* 2020, *9*, 427. [CrossRef]
33. Khalifa, Y.; Mandic, D.; Sejdić, E. A review of Hidden Markov models and Recurrent Neural Networks for event detection and localization in biomedical signals. *Inf. Fusion* **2021**, *69*, 52–72. [CrossRef]
34. Altaheri, H.; Muhammad, G.; Alsulaiman, M.; Amin, S.U.; Altuwaijri, G.A.; Abdul, W.; Bencherif, M.A.; Faisal, M. Deep learning techniques for classification of electroencephalogram (EEG) motor imagery (MI) signals: A review. *Neural Comput. Appl.* **2021**, 1–42. [CrossRef]
35. Heidari, A.; Navimipour, N.J.; Unal, M.; Toumaj, S. The COVID-19 epidemic analysis and diagnosis using deep learning: A systematic literature review and future directions. *Comput. Biol. Med.* **2022**, *141*, 105141. [CrossRef] [PubMed]
36. Battineni, G.; Chintalapudi, N.; Amenta, F. Machine learning in medicine: Performance calculation of dementia prediction by support vector machines (SVM). *Inform. Med. Unlocked* **2019**, *16*, 100200. [CrossRef]
37. Houssein, E.H.; Emam, M.M.; Ali, A.A.; Suganthan, P.N. Deep and machine learning techniques for medical imaging-based breast cancer: A comprehensive review. *Expert Syst. Appl.* **2021**, *167*, 114161. [CrossRef]

38. Kaur, P.; Singh, G.; Kaur, P. Intellectual detection and validation of automated mammogram breast cancer images by multi-class SVM using deep learning classification. *Inform. Med. Unlocked* **2019**, *16*, 100151. [CrossRef]
39. Charbuty, B.; Abdulazeez, A. Classification based on decision tree algorithm for machine learning. *J. Appl. Sci. Technol. Trends* **2021**, *2*, 20–28. [CrossRef]
40. Shakhovska, N.; Yakovyna, V.; Chopyak, V. A new hybrid ensemble machine-learning model for severity risk assessment and post-COVID prediction system. *Math. Biosci. Eng.* **2022**, *19*, 6102–6123. [CrossRef] [PubMed]
41. Ma, J.J.; Nakarmi, U.; Kin, C.Y.S.; Sandino, C.M.; Cheng, J.Y.; Syed, A.B.; Wei, P.; Pauly, J.M.; Vasanawala, S.S. Diagnostic image quality assessment and classification in medical imaging: Opportunities and challenges. In Proceedings of the 2020 IEEE 17th International Symposium on Biomedical Imaging (ISBI), Iowa City, IA, USA, 3–7 April 2020; IEEE: New York, NY, USA, 2020; pp. 337–340.
42. Sarvamangala, D.; Kulkarni, R.V. Convolutional neural networks in medical image understanding: A survey. *Evol. Intell.* **2022**, *15*, 1–22. [CrossRef] [PubMed]
43. Alshamma, O.; Awad, F.; Alzubaidi, L.; Fadhel, M.; Arkah, Z.; Farhan, L. Employment of multi-classifier and multi-domain features for PCG recognition. In Proceedings of the 2019 12th International Conference On Developments In ESystems Engineering (DeSE), Kazan, Russia, 7–10 October 2019; pp. 321–325.
44. Kattenborn, T.; Leitloff, J.; Schiefer, F.; Hinz, S. Review on Convolutional Neural Networks (CNN) in vegetation remote sensing. *ISPRS J. Photogramm. Remote Sens.* **2021**, *173*, 24–49. [CrossRef]
45. Alzubaidi, L.; Al-Shamma, O.; Fadhel, M.; Arkah, Z.; Awad, F. A deep convolutional neural network model for multi-class fruits classification. In Proceedings of the Intelligent Systems Design And Applications: 19th International Conference On Intelligent Systems Design And Applications (ISDA 2019), Auburn, WA, USA, 3–5 December 2019; pp. 90–99.
46. Anwar, S.M.; Majid, M.; Qayyum, A.; Awais, M.; Alnowami, M.; Khan, M.K. Medical image analysis using convolutional neural networks: A review. *J. Med. Syst.* **2018**, *42*, 226. [CrossRef]
47. Alzubaidi, L.; Fadhel, M.; Oleiwi, S.; Al-Shamma, O.; Zhang, J. DFU QUTNet: Diabetic foot ulcer classification using novel deep convolutional neural network. *Multimed. Tools Appl.* **2020**, *79*, 15655–15677. [CrossRef]
48. Kora, P.; Ooi, C.P.; Faust, O.; Raghavendra, U.; Gudigar, A.; Chan, W.Y.; Meenakshi, K.; Swaraja, K.; Plawiak, P.; Acharya, U.R. Transfer learning techniques for medical image analysis: A review. *Biocybern. Biomed. Eng.* **2022**, *42*, 79–107. [CrossRef]
49. Alzubaidi, L.; Al-Shamma, O.; Fadhel, M.; Farhan, L.; Zhang, J.; Duan, Y. Optimizing the performance of breast cancer classification by employing the same domain transfer learning from hybrid deep convolutional neural network model. *Electronics* **2020**, *9*, 445. [CrossRef]
50. Chen, W.; Li, X.; Gao, L.; Shen, W. Improving computer-aided cervical cells classification using transfer learning based snapshot ensemble. *Appl. Sci.* **2020**, *10*, 7292. [CrossRef]
51. Khanday, A.M.U.D.; Rabani, S.T.; Khan, Q.R.; Rouf, N.; Mohi Ud Din, M. Machine learning based approaches for detecting COVID-19 using clinical text data. *Int. J. Inf. Technol.* **2020**, *12*, 731–739. [CrossRef]
52. Deepa, N.; Prabadevi, B.; Maddikunta, P.K.; Gadekallu, T.R.; Baker, T.; Khan, M.A.; Tariq, U. An AI-based intelligent system for healthcare analysis using Ridge-Adaline Stochastic Gradient Descent Classifier. *J. Supercomput.* **2021**, *77*, 1998–2017. [CrossRef]
53. Wu, J.; Hicks, C. Breast cancer type classification using machine learning. *J. Pers. Med.* **2021**, *11*, 61.
54. Krishnamoorthi, R.; Joshi, S.; Almarzouki, H.Z.; Shukla, P.K.; Rizwan, A.; Kalpana, C.; Tiwari, B. A novel diabetes healthcare disease prediction framework using machine learning techniques. *J. Healthc. Eng.* **2022**, *2022*, 1684017. [CrossRef]
55. Shakouri, S.; Bakhshali, M.A.; Layegh, P.; Kiani, B.; Masoumi, F.; Ataei Nakhaei, S.; Mostafavi, S.M. COVID19-CT-dataset: An open-access chest CT image repository of 1000+ patients with confirmed COVID-19 diagnosis. *BMC Res. Notes* **2021**, *14*, 178. [CrossRef]
56. Gaur, L.; Bhatia, U.; Jhanjhi, N.; Muhammad, G.; Masud, M. Medical image-based detection of COVID-19 using deep convolution neural networks. *Multimed. Syst.* **2021**, 1–10. [CrossRef]
57. Mijwil, M.M.; Al-Zubaidi, E.A. Medical Image Classification for Coronavirus Disease (COVID-19) Using Convolutional Neural Networks. *Iraqi J. Sci.* **2021**, *62*, 2740–2747.
58. Islam, M.R.; Nahiduzzaman, M. Complex features extraction with deep-learning model for the detection of COVID19 from CT scan images using ensemble based machine learning approach. *Expert Syst. Appl.* **2022**, *195*, 116554. [CrossRef]
59. Abirami, R.N.; Vincent, P.; Rajinikanth, V.; Kadry, S. COVID-19 Classification Using Medical Image Synthesis by Generative Adversarial Networks. *Int. J. Uncertain. Fuzziness-Knowl.-Based Syst.* **2022**, *30*, 385–401. [CrossRef]
60. Ozturk, T.; Talo, M.; Yildirim, E.A.; Baloglu, U.B.; Yildirim, O.; Acharya, U.R. Automated detection of COVID-19 cases using deep neural networks with X-ray images. *Comput. Biol. Med.* **2020**, *121*, 103792. [CrossRef]
61. Cohen, J.P.; Morrison, P.; Dao, L. COVID-19 image data collection. *arXiv* **2020**, arXiv:2003.11597.
62. Wang, X.; Peng, Y.; Lu, L.; Lu, Z.; Bagheri, M.; Summers, R.M. Chestx-ray8: Hospital-scale chest x-ray database and benchmarks on weakly-supervised classification and localization of common thorax diseases. In Proceedings of the IEEE Conference on Computer Vision and Pattern Recognition, Honolulu, HI, USA, 21–26 July 2017; pp. 2097–2106.
63. Chowdhury, M.E.; Rahman, T.; Khandakar, A.; Mazhar, R.; Kadir, M.A.; Mahbub, Z.B.; Islam, K.R.; Khan, M.S.; Iqbal, A.; Al Emadi, N.; et al. Can AI help in screening viral and COVID-19 pneumonia? *IEEE Access* **2020**, *8*, 132665–132676.
64. Lu, W. Improved *K*-means clustering algorithm for big data mining under Hadoop parallel framework. *J. Grid Comput.* **2020**, *18*, 239–250. [CrossRef]

65. Wu, Y.; Zhang, Q.; Hu, Y.; Sun-Woo, K.; Zhang, X.; Zhu, H.; Li, S. Novel binary logistic-regression model based on feature transformation of XGBoost for type 2 Diabetes Mellitus prediction in healthcare systems. *Future Gener. Comput. Syst.* **2022**, *129*, 1–12. [CrossRef]
66. Dubey, P.K.; Naryani, U.; Malik, M. Logistic Regression Based Myocardial Infarction Disease Prediction. In *Intelligent System Algorithms and Applications in Science and Technology*; Apple Academic Press: Palm Bay, FL, USA, 2022; pp. 39–51.
67. Mansour, N.A.; Saleh, A.I.; Badawy, M.; Ali, H.A. Accurate detection of COVID-19 patients based on Feature Correlated Naive Bayes (FCNB) classification strategy. *J. Ambient Intell. Humaniz. Comput.* **2022**, *13*, 41–73. [CrossRef]
68. Uddin, M.N.; Gaskins, J.T. Shared Bayesian variable shrinkage in multinomial logistic regression. *Comput. Stat. Data Anal.* **2023**, *177*, 107568. [CrossRef]
69. Botlagunta, M.; Botlagunta, M.D.; Myneni, M.B.; Lakshmi, D.; Nayyar, A.; Gullapalli, J.S.; Shah, M.A. Classification and diagnostic prediction of breast cancer metastasis on clinical data using machine-learning algorithms. *Sci. Rep.* **2023**, *13*, 485. [CrossRef]
70. Karacı, A. VGGCOV19-NET: Automatic detection of COVID-19 cases from X-ray images using modified VGG19 CNN architecture and YOLOv4 algorithm. *Neural Comput. Appl.* **2022**, *34*, 8253–8274. [CrossRef]
71. Al-Antari, M.A.; Hua, C.H.; Bang, J.; Lee, S. Fast deep learning computer-aided diagnosis of COVID-19 based on digital chest x-ray images. *Appl. Intell.* **2021**, *51*, 2890–2907. [CrossRef]
72. Alzubaidi, L.; Duan, Y.; Al-Dujaili, A.; Ibraheem, I.K.; Alkenani, A.H.; Santamaría, J.; Fadhel, M.A.; Al-Shamma, O.; Zhang, J. Deepening into the suitability of using pre-trained models of ImageNet against a lightweight convolutional neural network in medical imaging: An experimental study. *PeerJ Comput. Sci.* **2021**, *7*, e715. [CrossRef]
73. Alzubaidi, L.; Hasan, R.I.; Awad, F.H.; Fadhel, M.A.; Alshamma, O.; Zhang, J. Multi-class breast cancer classification by a novel two-branch deep convolutional neural network architecture. In Proceedings of the 2019 12th International Conference on Developments in eSystems Engineering (DeSE), Kazan, Russia, 7–10 October 2019; IEEE: New York, NY, USA, 2019; pp. 268–273.
74. Zhu, Z.; Xingming, Z.; Tao, G.; Dan, T.; Li, J.; Chen, X.; Li, Y.; Zhou, Z.; Zhang, X.; Zhou, J.; et al. Classification of COVID-19 by compressed chest CT image through deep learning on a large patients cohort. *Interdiscip. Sci. Comput. Life Sci.* **2021**, *13*, 73–82. [CrossRef]
75. Kumar, K.A.; Prasad, A.; Metan, J. *A Hybrid Deep CNN-Cov-19-Res-Net Transfer Learning Architype for an Enhanced Brain Tumor Detection and Classification Scheme in Medical Image Processing*; Elsevier: Amsterdam, The Netherlands, 2022; Volume 76, p. 103631.
76. Sahlol, A.T.; Yousri, D.; Ewees, A.A.; Al-Qaness, M.A.; Damasevicius, R.; Elaziz, M.A. COVID-19 image classification using deep features and fractional-order marine predators algorithm. *Sci. Rep.* **2020**, *10*, 15364. [CrossRef]

life

MDPI

Article

Understanding COVID: Collaborative Government Campaign for Citizen Digital Health Literacy in the COVID-19 Pandemic

Mónica López-Ventoso [1], Marta Pisano González [1,2,*], Cristina Fernández García [1,2], Isabel Diez Valcarce [1], Inés Rey Hidalgo [3], María Jesús Rodríguez Nachón [1], Ana María Menéndez García [1], Michelle Perello [4], Beatrice Avagnina [4], Oscar Zanutto [5] and Alberto Lana [2,6]

1 General Directorate of Care, Humanization and Socio-Health Care, 33005 Oviedo, Asturias, Spain
2 Health Research Institute of the Principality of Asturias (ISPA), 33006 Oviedo, Asturias, Spain
3 Foundation for the Promotion in Asturias of Applied Scientific Research and Technology (FICYT), 33007 Oviedo, Asturias, Spain
4 Consult Europa Projects and Innovation, 35006 Las Palmas de Gran Canaria, Las Palmas, Spain
5 Institute for Hospitalization and Assistance Services for the Elderly (ISRAA), 31100 Treviso, Veneto, Italy
6 Department of Preventive Medicine and Public Health, School of Medicine and Health Sciences, University of Oviedo, 33003 Oviedo, Asturias, Spain
* Correspondence: martamaria.pisanogonzalez@asturias.org

Abstract: The strategy "Understanding COVID" was a Public Health campaign designed in 2020 and launched in 2021 in Asturias-Spain to provide reliable and comprehensive information oriented to vulnerable populations. The campaign involved groups considered socially vulnerable and/or highly exposed to COVID-19 infection: shopkeepers and hoteliers, worship and religious event participants, school children and their families, and scattered rural populations exposed to the digital divide. The purpose of this article was to describe the design of the "Understanding COVID" strategy and the evaluation of the implementation process. The strategy included the design and use of several educational resources and communication strategies, including some hundred online training sessions based on the published studies and adapted to the language and dissemination approaches, that reached 1056 people of different ages and target groups, an accessible website, an informative video channel, posters and other pedagogical actions in education centers. It required a great coordination effort involving different public and third-sector entities to provide the intended pandemic protection and prevention information at that difficult time. A communication strategy was implemented to achieve different goals: reaching a diverse population and adapting the published studies to different ages and groups, focusing on making it comprehensible and accessible for them. In conclusion, given there is a common and sufficiently important goal, it is possible to achieve effective collaboration between different governmental bodies to develop a coordinated strategy to reach the most vulnerable populations while taking into consideration their different interests and needs.

Keywords: eHealth; (d)health literacy; health literacy; health intervention; health strategy; digital health; pandemic; COVID-19

Citation: López-Ventoso, M.; Pisano González, M.; Fernández García, C.; Diez Valcarce, I.; Rey Hidalgo, I.; Rodríguez Nachón, M.J.; Menéndez García, A.M.; Perello, M.; Avagnina, B.; Zanutto, O.; et al. Understanding COVID: Collaborative Government Campaign for Citizen Digital Health Literacy in the COVID-19 Pandemic. *Life* **2023**, *13*, 589. https://doi.org/10.3390/life13020589

Academic Editor: Daniele Giansanti

Received: 2 January 2023
Revised: 14 February 2023
Accepted: 17 February 2023
Published: 20 February 2023

1. Introduction

1.1. COVID-19 Pandemic and Vulnerable Populations

A viral outbreak of an unknown coronavirus (SARS-CoV-2) was declared a pandemic by the World Health Organization (WHO) in March 2020. The increasing rate of incidence and mortality from the associated disease (COVID-19) challenged and stressed healthcare institutions and the global economy and had an impact on the physical and mental health of people around the world. The effects of this pandemic forced the adoption of drastic collective prevention measures throughout the world. In Spain, a state of alarm was decreed, followed by a series of agreements and resolutions on preventive measures and recommendations related to SARS-CoV-2 infection [1], which broke into the normal organization of administrations and the population. Collective confinement stands out among all measures adopted and how it profoundly transformed the general way of living and how to act and confront COVID-19.

As the pandemic evolved, the natural history of COVID-19 and its comprehension suffered great advances. From the beginning, it was suspected that the burden of disease between urban and rural areas could be different [2]. According to Lakhani and colleagues, once the virus entered a rural community, there was a higher relative rate of morbidity and mortality [3]. These higher rates would go unnoticed if specific epidemiological monitoring were not carried out in rural areas, given that the population impact is low because they represent a low percentage of the total population [4]. In addition, the rural environment is characterized by inequity in access to health infrastructure, health literacy, preparation and adaptation for the pandemic, greater difficulties in changing social and/or work habits, and demographic aging that conditions high rates of physical frailty, morbidity and dependency [3,5].

Although the characteristics of the Spanish rural environment are not comparable to those of the aforementioned studies, in Spain, there was also concern about the epidemiological vulnerability of the rural population. This situation, together with the analysis presented, justified the need to adapt intervention strategies in the prevention and control of transmission in rural and aged areas. The rural population is the most vulnerable, but other existing vulnerable populations must be included in the prevention strategies as there is strong evidence about the different impacts of COVID-19 on other groups. Making a general health information approach expecting that the resulting information will be acceptable for different groups is not sufficient to achieve equity, and a multifaceted approach that reaches every group is therefore needed.

1.2. Decision-Making in the COVID-19 Crisis and Public Health Strategies

During the early phases of the COVID-19 pandemic, healthcare professionals worked under high levels of uncertainty. Soon a pressing need emerged to translate knowledge into practice more efficiently, with rapid assessment and dissemination of scientific evidence to guide decision-making [6]. Some studies found that bringing together experts from academia, science and clinical practice to search for and summarize information of high scientific quality was effective for informed decision-making [7]. However, knowledge was not only needed by clinical and public health decision-makers as the general population also had a compelling information need to make the best choices for their health. In addition, an 'infodemic' led to confusion and distrust in health workers weakening public health responses [8].

Moreover, the continuous demand for efforts from the population to comply with the requirements and recommendations to improve the epidemiological and health situation (for example, confinements, social distancing, travel restrictions, cuts in benefits, vaccination, etc.) also required efforts to convey information clearly and understandably. However, with the passage of time, the transmission of information and the training of citizens from the public administrations became increasingly complex, a circumstance that especially affected the most vulnerable groups [9], and the first signs of "pandemic fatigue" in the general population began to show. This is why the WHO urged the inclusion of four recommendations in dissemination campaigns and actions summarized in Figure 1 [10].

1.3. The "Understanding COVID" Strategy

For all these reasons, in most regions, it became necessary to implement information strategies tailor-made for the general population, especially for the most vulnerable people. "Understanding COVID" was a strategy that began to be developed in March 2020 at the General Directorate of Care, Humanization and Socio-Health Care of Asturias (Spain). Asturias is the most aged region in Spain and one of the most aged in Europe. In addition, although a large part of its population lives in diffuse urban environments [11,12], a large number of small and dispersed rural centers that are difficult to access also exist. Therefore, the rural area of Asturias presents some particular problems leading to a risk of poverty or social exclusion: the demographic situation (shortage of population, exodus of inhabitants and aging of the population in rural areas); the difficulties for the mobility of the population (lack of infrastructures and basic services, lack of adequate transport connections); or problems related to the labor market (lower employment rates and long-term unemployment).

Figure 1. WHO: Pandemic fatigue: proposal of four key strategies for governments to maintain and reinvigorate public support for protective behaviors.

The general objective of the "Understanding COVID" strategy was to increase the offer of digital information on prevention and protection against COVID-19, individual care and emotional approach to both the largest number of citizens, prioritizing vulnerable groups and the rural environment of Asturias. The specific objectives were to:

1. Listen to citizens' voices to redesign training actions, keeping in mind the suggestions from the community and acknowledging the difficulties and successes in carrying out the recommended protection measures against COVID-19.
2. Search, simplify, and adapt in a more comprehensive way to the community all the information and evidence available to increase people's protection against COVID-19.
3. Promote accessible information on protection measures for citizens in general and for people with hearing or visual disabilities.
4. Adapt digital health literacy (or (d)HL) to the particular needs of the population groups to which it is directed (adult population, young people, fathers and mothers, etc.).
5. Design specific campaigns for sectors of activity that are particularly exposed, such as workers in poorly ventilated places and/or environments with a large influx of people.
6. Work with children and adolescents to increase safety in the school environment.

The "Understanding COVID" strategy had various target population groups, which included citizens of rural areas, citizens of urban areas, municipal technical professionals, people of Roma ethnic origin, immigrants, and citizens with impaired vision and/or hearing. Another relevant focus to work with was the job sectors most exposed to COVID-19: local commercial activities carried out in indoor areas like hairdressers, beauty salons, and places of worship. In particular, tourism outlets and hotels developed training to protect their own activity, increasing their own security measures and "how to use the facilities" materials for their customers. Finally, in the design of the "Understanding COVID" strategy, two groups were also taken into account in a differentiated way, schoolchildren and also their families, since children and adolescents were left out of the pedagogical approach used for the adult population, and other participation and information methodologies were incorporated.

Given that the "Understanding COVID" strategy was a population-based public health campaign, no sampling method was used, but it was disseminated throughout the target population in order to reach the largest possible number of vulnerable people and later study the scope and effectiveness of the strategy under real conditions, within the broad framework of the implementation science.

1.4. Purpose of this Article

Currently, a large number of national and subnational governments are conducting retrospective evaluations of COVID-19 health policy decisions and actions to reflect on their strengths and weaknesses and, thus, to find opportunities to reinforce public institutions for future crises [13]. Thereby, the main aim of this article was to describe the design of the "Understanding COVID" strategy of the Ministry of Health of the Principality of Asturias (Spain), which helped to offer and to adapt information on protection measures against COVID-19 for the entire population of Asturias, including how specific criteria were incorporated for the design of interventions in rural areas with the active participation of their inhabitants. Secondarily, the article also shows the preliminary evaluation of the implementation process of the strategy under real conditions. Ultimately, this work allows us to understand the Spanish public health sector's capacities to deal with crises and seeks to generate learning toward a more effective and equitable response in the future.

2. Design Phases and Preliminary Evaluation of the Implementation Process of the "Understanding COVID" Strategy

2.1. Hearing Citizen's Voice

The first step in designing the strategy was based on analysis through a survey to evaluate weaknesses, perceived strengths, and topics and contents of interest. This was sent by phone using the WhatsApp application, adapting to the technology available at the time of confinement. The survey met the requirements of Organic Law 3/2018 [14] in terms of data analysis and dissemination. Participants were asked to answer six questions on a 6-item Likert scale (0: Strongly disagree; 1: Disagree; 2: Slightly Disagree; 3: Slightly Agree; 4: Agree; 5: Strongly agree) and an additional open-ended question to capture proposals on COVID-19 prevention and control training ("What else would you like us to include in that training? Write down what you deem important in these times of pandemic").

Table 1 shows the responses to this first online survey incorporating the opinion of citizens between 9 March 2020 and 13 March 2020 (106 responses in one week). The questions were not mandatory, so not all of them were answered by all respondents, and the average response rate was 93%. The percentage of agreement was greater than or equal to 80% in the answer "strongly agree" in questions 1, 3, 5, and 6 and greater than or equal to 65% in the same answer in questions 2 and 4.

In the open-ended question, respondents answered that their greatest interest was the formations focused on the use of "personal protective equipment", information on face masks and on dealing with emotions in times of pandemics.

Table 1. Initial questionnaire for collecting information from citizens.

Question	Likert Scale *					
	0	1	2	3	4	5
1. Do you think it is necessary to receive training on how to carry out your work safely in times of coronavirus?	0%	2%	2%	2%	9%	84%
2. Do you think that receiving training from the Ministry of Health in the format of short videos that would reach your phone could be a good way to do so?	5%	3%	7%	9%	11%	65%
3. Do you think it is appropriate that this training includes information on how to act safely in the activities of your daily work (cleaning, cleaning, shopping, …)?	0%	2%	2%	3%	13%	80%
4. Do you think it appropriate that this training includes information on safe behavior rules during transfers between one home and another in your daily work?	0%	1%	4%	4%	12%	79%
5. Do you think it appropriate that this training includes information on how to disinfect your private car or that of your company that you use in your work transfers?	1%	2%	2%	4%	11%	80%
6. Does it seem appropriate to you that this training includes information on how to act when you feel a lot of stress or are overwhelmed at work?	0%	2%	1%	5%	10%	82%

* 0: "Strongly disagree" → 5: "Strongly agree".

The continuous collection of information from participants was considered a priority due to the relevance of incorporating real doubts and needs, as well as adapting the strategy to all target audiences. Thereby, in a later phase, a survey was designed and sent to all participants who took part in the training so that they could anonymously and voluntarily evaluate the usefulness, accessibility, contents, teaching methodology and satisfaction, as shown in Table 2. Along with the previous results, information was also collected from the participants in the training sessions (e.g., opinions or testimonials), thanks to which new content was al-so developed and work was done with groups or sectors that were valued as relevant at different times.

A total of 472 responses were collected from the continuous information and satisfaction survey. Of these, 65 were from participants in the activity aimed at the school families' associations, 111 from those aimed at secondary schools, 256 from the "Drop by Drop" action training designed for individual citizens, and 40 from the "First Quality Air", the specific training offered to the hotel industry.

The percentages of agreement for usefulness, time and accessibility based on the Likert scale are shown in Table 3. In summary, when asked about the usefulness of the training, the activity with the best result was the training for catering "First Quality Air" with 88% "strongly agree" followed by the training of secondary schools with 86%, family associations with 78% and "Gota a Gota" with 77%. In relation to the duration of the training, all four pieces of training have high scores of "yes", with the secondary schools' training reaching 100% agreement, followed by the Family Associations' training with 97% and "Gota a Gota" and "First Quality Air" with 95% each. In the section related to accessibility, the "strongly agree" scores ranged from 86% for training in secondary schools, 83% for Family Associations and "Gota a Gota," and finally, "First Quality Air" with 80%.

Table 2. Questionnaire for continuous collection of citizen information.

	Questions	Options
1.	Do you think that the training received is useful for you? (Usefulness)	6-item Likert scale 0 = not useful 5 = very useful
2.	Is the training time adequate? (Time)	yes/no/don't know
3.	Has the training format been accessible to you? (Accessibility)	6-item Likert scale 0 = little accessible 5 = very accessible
4.	General satisfaction with the training received. (Satisfaction)	11-item Likert scale 0 = no satisfaction 10 = very satisfied
5.	Would you add any content to this training?	Open-ended question
6.	Would you remove any content from this formation?	Open-ended question
7.	Write any comment that you want to send us.	Open-ended question

Table 3. Percentage of agreement with usefulness, time and accessibility from the questionnaire for the continuous collection of information for citizens.

Type of Training	Usefulness (Likert)						Time			Accessibility (Likert)					
	0	1	2	3	4	5	Yes	No	Don't Know	0	1	2	3	4	5
Families associations	0%	0%	0%	0%	22%	78%	97%	0%	3%	0%	0%	0%	2%	15%	83%
Secondary schools	0%	0%	0%	0%	14%	86%	100%	0%	0%	0%	0%	0%	0%	14%	86%
Drop by Drop	0%	0%	0%	4%	18%	77%	95%	1%	4%	0%	0%	0%	2%	15%	83%
First Quality Air	0%	0%	0%	3%	10%	88%	95%	5%	0%	0%	0%	0%	0%	20%	80%

The percentages of agreement with the satisfaction-based question on the Likert scale are shown in Table 4. In this table, we can be seen that satisfaction was measured with a broader Likert scale (0–10), with the highest score in all cases being 10, with 49% in families' associations, 85% in "First Quality Air ", 86% in secondary schools and finally 89% in the "Drop by Drop".

Table 4. Percentage of agreement with satisfaction from the questionnaire for the continuous collection of information for citizens.

Type of Training	Satisfaction (Likert)										
	0	1	2	3	4	5	6	7	8	9	10
Families associations	0%	0%	0%	0%	0%	0%	2%	6%	9%	34%	49%
Secondary schools	0%	0%	0%	0%	0%	0%	0%	0%	4%	11%	86%
Drop by Drop	0%	0%	0%	0%	0%	0%	1%	2%	2%	5%	89%
First Quality Air	0%	0%	0%	0%	0%	0%	0%	5%	5%	5%	85%

Along with the previous results, information was also collected from the participants in the training sessions (e.g., opinions or testimonials), thanks to which new content was also developed and work was done with groups or sectors that were valued as relevant at different times.

2.2. Informative Content

In order to prepare the contents of the "Understanding COVID" strategy, the needs expressed by the citizens and identified in the previous phase were taken into consideration. These needs were articulated around four core themes:

- Self-protection and collective protection measures against COVID-19: including frequent hand washing, interpersonal distance and coughing into the cubital fossa, use of masks, cleaning of domestic environments, collective protection measures and specific environments, and the disinfection of physical spaces.
- Identification and containment of the sources of contagion: early diagnosis of people with symptoms, isolation of cases and tracing and quarantine of close contacts. Therefore, it was important to publicize the symptoms and the protocols for reporting them (e.g., health personnel in the area).
- Content related to emotional management: assertiveness, managing emotions in difficult times, such as facing fear, leaving home after confinement, love, learning to trust, positive thinking, guided visualization, etc.
- Contents related to maintaining healthy habits: healthy diet, physical exercise, maintenance of routines, communication and sleep.

Next, personnel trained in documentation and communication conducted a search for all the information available at that time in different sources: scientific literature (PubMed and Web of Science), gray literature, expert information (documents and explanatory videos), documentation of official health agencies and web resources. Finally, the relevant information was analyzed and synthesized by a group of experts and distributed in the communication channels of the "Understanding COVID" strategy: one created the informative content on the web (in text, in video or in infographics) and others for online training. The contents were continuously reviewed to ensure the maximum topicality of the information that was so changing at that time, to adapt it and adapt it to an understandable language for the different target audiences, as well as to align the messages with the policies that were required in terms of prevention and protection by the authority at any given time.

As a result, more than 150 documentary sources were consulted. A total of 10 sections containing the most relevant information were grouped together, and 50 subsections of key information (Figure 2), with 55 infographics and videos.

2.3. Digital Health Literacy

The central methodological pillar of the "Understanding COVID" strategy was the live training sessions. The sessions were virtual, using the office tool "Microsoft Teams", which allows access via smartphone (preferred), PC or tablet without the need to install any type of software. The sessions lasted 60–90 min and were structured into two well-differentiated parts: a first part where updated information on the pandemic was presented (30–45 min), and a second part where there was free time for questions from attendees to resolve doubts and explore their needs and barriers to implement protection measures (30–45 min). During the design phase of the training sessions, special emphasis was placed on adapting the content, images and language to vulnerable populations (e.g., residents of rural areas, Roma, caregivers, etc.). In addition, attention was paid to adapting the temporary programming of the sessions to the working hours of the professional groups.

For the development of the online training sessions and the recruitment of attendees, there was a collaboration from 54 municipalities of Asturias (out of a total of 78, or 69.2%) that adhered to the strategy. Likewise, the Ministry of Education, Ministry of Tourism and Sports, associations of the third sector (hotels, patients, neighbors, etc.), associations of families, as well as associations of Roma ethnic. Thanks to this collaborative work, a total of 100 interventions were carried out with an overall attendance of 1056 people. A summary of the population groups reached in the training sessions can be seen in Table 5.

Figure 2. Sections and web content subsections at www.entendercovid.es accessed on 1 February 2023.

Figure 3 depicts the main training actions developed by the strategy, among which the following clearly stand out: "Drop by drop", that is, general training for individually enrolled citizens, with 32 actions; training for vulnerable populations, with 19 actions; and training for parents of children in confinement and/or isolation, as well as specific training for parents of children with special needs, with 14 actions.

Lastly, in the open-ended questions of the questionnaire, the most repeated messages correspond to the following codes: "gratitude", "appreciation of the live session for questions and needs", "request to repeat the training to update knowledge", and "verification of the need for training for the entire population".

Table 5. Target population reached in the "Understanding COVID" training.

Group /Environment	Specific Populations
Citizenship	Citizens in rural areas Citizens in urban areas Municipal technical professionals Ethnic minorities: Roma, Immigrants Citizens with impaired vision Citizens with impaired hearing Associations of people with mental health problems or addictions Non-professional caregivers
School environment	School-age students (6–12 years old) Secondary education and vocational training students (12–18 years) Families with school-age children ages 6–12 years Families Association of students aged 3 to 16 years according to educational levels of public and private centers and special education
Professional sector	Professionals from shelters Home caregiving professionals Non-professional caregivers Risk prevention services Hostelry Tourism Small business Religious groups

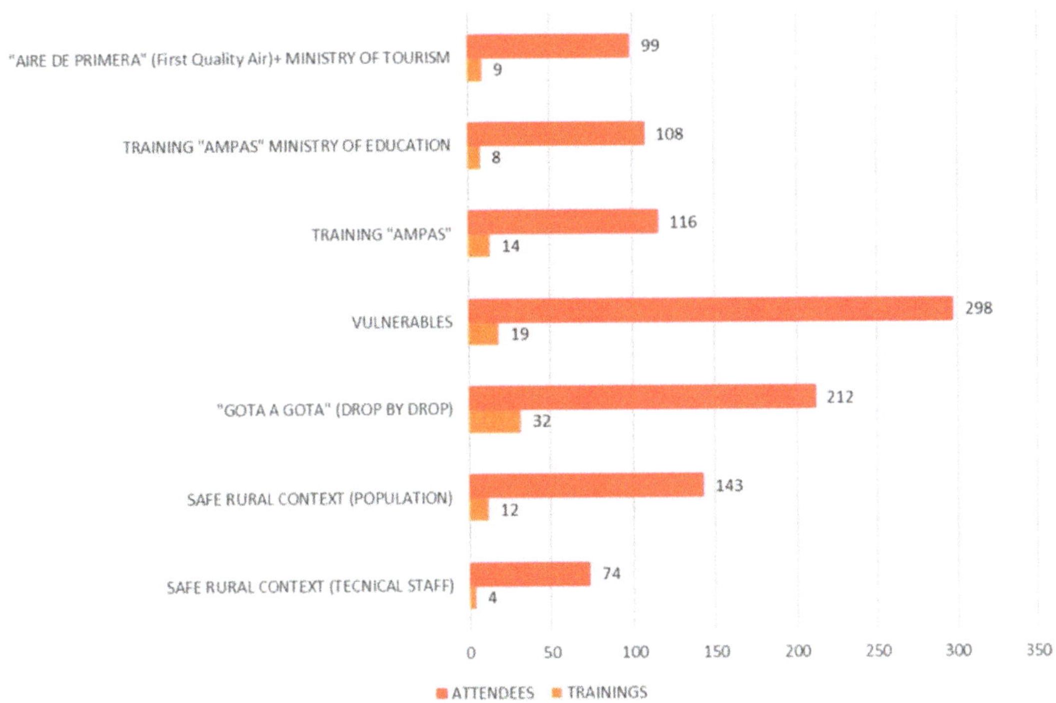

Figure 3. Number of training sessions and participants. AMPAS: Spanish acronym for Families Associations.

2.4. Communicative Materials and Accessibility of Information

In order to achieve the objectives related to reaching the maximum number of citizens, making adaptations for different groups, achieving accessibility of language and content of materials, breaking the digital divide and making information accessible to citizens with accessibility and equity, the strategy "Understanding COVID" designed, coordinated and produced the following communication materials, which were made available in March 2021 to the target population.

2.4.1. Logo and Graphic Identity

The design of the logo and the graphic identity of the campaign were part of the methodology of the "Understanding COVID" strategy (Figure 4). Reaching the public, involving them in their health decisions and reflecting on the available evidence were the core of the starting elements for the design of the logo and the graphic identity of the strategy.

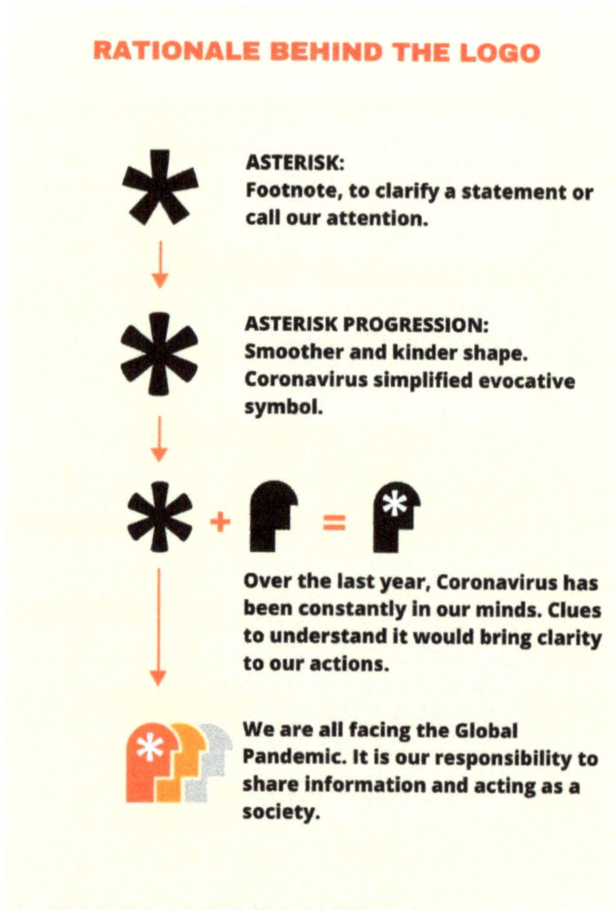

Figure 4. Understanding COVID: logo development.

2.4.2. Web Page

An independent web page (www.entendercovid.es, accessed on 1 February 2023) with free and open access was developed, which provided access to all the informative material of the strategy, including documents, ad hoc infographics and videos. The web page also had an interactive virtual space for solving doubts, as well as suggestions and actions for the campaign. The accessibility of the page was reviewed from its design by the Spanish National Organization for the Blind to guarantee inclusiveness. In addition, the videos were designed to include sign language for the deaf community. The following principles were taken into account when designing the website:

- Didactic vocation: Present the information in an orderly, clear and attractive way. Carry out positive communication avoiding contributing to general pandemic fatigue.
- Usability: Simple and intuitive navigation. Increase click efficiency (relevant information in the minimum number of clicks). Prioritize information in plain text.
- Accessibility: Information intended for the whole of society. Visual codes are understandable by all. Its simple structure and adaptation for people with visual disabilities aim to increase the friendliness of its reading, as well as its possible use from mobile phones. Respect for the Accessibility Guidelines for WEB Content (WCAG).

The web page is made up of:

- Training content: grouped information, frequently asked questions document, self-assessment questionnaire, training videos and registration.
- Sections for special groups: people with chronic illness or special situations, pregnancy and lactation, children and caregivers.
- Press room: included a press kit to download according to the communication objectives of each campaign, press releases, press clipping, elements of future campaigns, etc.
- Frequently Asked Questions: Frequently asked questions document prepared by the General Directorate of Care, Humanization and Socio-Health Care in collaboration with the General Directorate of Public Health and the Agency for Food Safety, Environmental Health and Consumption.
- How much do you know? Self-assessment form with 16 questions to verify the level of basic knowledge about SARS-CoV2.
- Specific campaigns: images about the different campaigns that were carried out.
- Secondary Education Institutes of Asturias: This section was created to house the campaign and the contest that was carried out for school adolescents in Asturias.

According to Google Analytics web service, during the study period (March 2021–January 2022), the web page received 7080 visits from 5842 users (1.20 sessions per user). On each visit, users viewed an average of 1.70 pages from the main web page. The bounce rate, that is, the percentage of visitors who left the web page without taking any action, was 76%. Accesses to the web page were highest immediately after its creation, with a maximum of 500 weekly visitors between April and May 2021, and especially in December 2021, the week before the Christmas period, with more than 1000 weekly entries.

Of all the users of the web page, the age group most represented was that of 25–34 years (33.5%), followed by the group of 18–24 (24.5%), the group of 35–44 (15.5%) and that of 45–54 years (12.5%). Finally, those over 55 years of age accounted for 11% of accesses. Regarding sex, the percentage of men (54.2%) was slightly higher than that of women (45.8%).

In the analysis by country, accesses from Spain stood out (82.0%). Overall, Spanish-speaking countries accounted for more than 89.9% of accesses. Of the remaining percentage, the United States stood out with 2.08% and China with 1.90%. Finally, the devices used to access the website are shown in Figure 5.

2.4.3. Actions for the Child and Adolescent Population

Following the WHO recommendations in times of pandemic fatigue, co-creation and participatory actions for the underage population were designed.

First, a creative contest for adolescents (from 12 to 18 years old) was run. In addition to the previously described adapted training sessions, a creative contest was held in collaboration with the Asturian Ministry of Education for the involvement of adolescents. First, an email was sent in April 2021 to all educational centers that provide secondary education, high school, and vocational training with an invitation and instructions for participation. The email contained a training video to be viewed in class with the students, which was definitively projected in 954 classrooms. The teachers encouraged debate and reflection on its content, and later, the students voluntarily created a creative product to compete in one of the following modalities: audiovisual, written, poster, and free creative. Campaign promoters received 111 creations of the four modalities, each one from a classroom of students between 12 and 16 years of age. In May 2021, the awards were delivered in a collaborative virtual ceremony organized by the Departments of Health and Education.

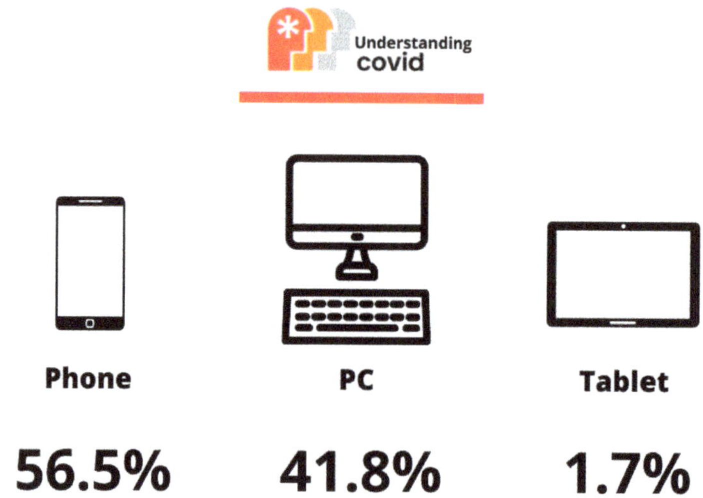

Figure 5. Device distribution.

Second, a handwashing campaign called *"Bichos fuera"* (Bugs out!) was carried out in the Early Childhood and Primary Education Centers (from 3 to 11 years old). The teachers wrote the lyrics of a song in the Asturian language, designed the music, and devised the choreography and staging of the song "Bugs out!". A video was recorded where the protagonists were the children (https://www.youtube.com/watch?v=vU1Kphiukdw, accessed on 1 February 2023). Classroom materials were also designed, such as cards and games. All the material was summarized in a guide for teachers (Supplementary Figure S1).

2.4.4. Complementary Actions to Highly Exposed Workers/People

Some complementary actions were also carried out for specific activity sectors with high exposure to COVID-19 infection. For instance, a protocol for safe actions against COVID-19 was carried out during the celebration of Catholic masses indoors, with its corresponding posters ("Safe churches help us stop the COVID and reduce risks"). Moreover, a campaign called "First Quality Air" (Aire de Primera) was specifically designed for the hotel industry. Taking advantage of the Asturias tourist campaign "Asturias, Natural Paradise", which promotes its pure and clean natural environment, posters of "First Quality Air" were prepared with COVID-19 protection measures to be used in hospitality and tourism establishments (Supplementary Figure S2).

2.4.5. Other Dissemination Actions

To reinforce as well as to make room for consultation and reminders of the above elements, we also designed: (1) visual presentations with educational material adapted to different groups; (2) co-design of materials and recruitment mailings; (3) dissemination through social networks (Facebook, Twitter and WhatsApp) and mailing; and (4) a YouTube channel to host educational videos with simultaneous recording in sign language. The channel definitively hosted 10 training videos that had a total of 3,866 views.

3. Discussion

3.1. The "Understanding COVID" Strategy

The government of Asturias launched a public health campaign to improve the population response to the COVID-19 crisis and to fight against pandemic fatigue and infodemia. The main novelty of the "Understanding COVID" strategy consisted of specifically targeting a population selected on the basis of vulnerability criteria and identifying the topics on which they should be trained in. In addition, the training actions were delivered through a wide variety of methodologies tailor-made for recipients. For example, online training for all interested vulnerable citizens, posters for the productive sectors with the highest risk of transmission, pedagogical contests and educational games for schools, presence in social networks, etc. Definitively, 100 training actions were carried out for 21 subgroups of the vulnerable population, reaching more than 1000 individuals, but also students from almost 1000 classrooms and countless users from the hostelry industry and church sector.

A large number of graphic and audiovisual materials were developed that supported a positive and preventive discourse in the face of the COVID-19 pandemic. In addition, some of the materials disseminated in the school environment were co-created by children and adolescents since including them in the design was considered to increase acceptability.

One of the main challenges of the strategy was to reach as many vulnerable people of different ages as possible, as has been done in similar studies and interventions [15–25], but at the same time minimizing the technological gap that could leave someone behind, which is a common problem when trying to reach a vulnerable population using information technologies [9,26,27]. To do this, everyday technology tools already existing in homes, such as tablets and smartphones, were used, with no need for additional installation of complex programs. The design of the communication and dissemination strategy through digital technologies was in line with similar studies [28]. The good reception of the strategy "Understanding COVID" reinforced the choice of the method of dissemination and implementation, bringing this information accessible also to the population with hearing disabilities (with the support of professional translators in sign language) and visual (adaptation of audiovisual media).

Of all the actions of the "Understanding COVID" strategy, the ones that generated the most participation were those carried out in schools and in the catering sector since educational activities were added to the online training sessions in the centers and schools, and the distribution of posters occurred in restaurants. Gray et al. described the need to develop protection strategies within the school community and responded to an important need to provide information and support both to the teaching community and to families and students [29]. In our strategy, creativity and horizontal and ascending training were encouraged: from some students to others and from students to their parents. The information strategy in the hospitality sector through the "First Quality Air" campaign allowed commercial establishments to display posters with recommendations for the population, as well as to have a certificate accrediting the training received, thus promoting confidence and security among customers. As the restaurant industry is particularly sensitive to disasters, specific campaigns were run in some countries to encourage people to go out for lunch or dinner. Campaigns such as "Go to Eat" in Japan or "Eat Out to Help Out" in the United Kingdom applied discounts for dining in restaurants and simultaneously achieved an increase in sales and a rebound in cases [30,31].

The "Understanding COVID" strategy focused more on security and less on the economy because it was understood that by pursuing the first goal, it would achieve the second one. Finally, although the results referring to visits to the website are difficult to measure, it was relevant that the highest volume of unique visitors occurred two weeks before Christmas 2021, a time when the restrictions had been modified, and the population was looking for information to safely carry out trips, family reunions and other recreational activities. This increase in the number of hits to the page may reflect the confidence that the population has in seeking accurate, verified, accessible and adapted information, as was the objective of this strategy.

The "Understanding COVID" strategy contemplated some key elements that the scientific literature identifies for a campaign to be successful [32]. These include (1) messages that focus on the identity of the population, (2) the use of visual aids, and (3) the use of social networking features to encourage interaction. In addition, although it used online resources (web pages, webinars, social networks, etc.) to be consistent with the message of limiting physical and social contacts, other more appropriate resources were also used to reach the vulnerable population (posters, songs, etc.), which was somewhat less common in campaigns from other countries. In addition, it has been shown that the high penetration of mobile devices and technology in the younger population [33] opened a very interesting door to their inclusion in schools as a means to achieve early health literacy [34]. Additionally, parents of students can benefit from health literacy strategies from schools in collaboration with government health policies [17,35–37], as has been appreciated throughout this strategy.

3.2. Other Strategies and Campaigns

Most countries in the world disseminated information and prevention campaigns for COVID-19 through official statements and other mass media. In Spain, the national government developed four population campaigns exclusively on the internet in order to fight against the spread of the pandemic, reinforcing individual security measures and community action [38,39]. The campaigns were disseminated via Twitter, and the analysis of their design and implementation allowed some interesting conclusions to be drawn. Although the campaigns promoted the dissemination of health security measures, they did not serve to encourage debate and interaction between governments/public institutions and citizens [39]. In addition, the campaigns generated polar responses, with very positive visions that were faced with other very negative ones, which did not help to improve union and community action [38]. However, a similar campaign carried out in Italy through Facebook, the #I-am-engaged campaign, was built around a community perspective, with a participatory process that favored co-creation among peers. In addition, the campaign adopted a positive tone of voice by focusing on the promotion of good practices [40]. In these respects, the Italian campaign was similar to the "Understanding COVID" campaign, although the latter included a wide range of actions to be carried out beyond the digital world on the basis of trying to reach as many vulnerable people as possible.

Other campaigns carried out in various countries also tried to address the vulnerable population. For example, in the USA, an alliance of institutions launched a multifaceted national campaign whose objective was to increase confidence in vaccines and decrease misinformation within Hispanic communities. They successfully used social networks, webinars, radio and newsletters, with the participation of volunteers, key people for the Hispanic community and influencers [41]. In Maryland (USA), another regional campaign was developed through social networks and a web page to promote testing for COVID-19 and acceptance of the vaccine among Latinos with limited English proficiency [42]. Also, in the USA, campaigns were created on social networks to promote scientific information on the risks of COVID-19 in pregnancy and the benefits of vaccination, such as the "One Vax Two Lives" campaign in Seattle [43]. In Sydney, Australia, there were also efforts to engage culturally and linguistically diverse communities in the effective and appropriate public health response to COVID-19 [44]. A novel and rapid inter-agency campaign was established that included tailored public education and testing, the establishment of a local clinic, and inspections of local businesses to achieve a safe environment.

3.3. Lessons Learned and Limitations

An important lesson learned from the "Understand COVID" strategy was the importance of various public institutions working in a coordinated manner in pursuit of a common goal, something common to other similar campaigns [43,44]. It was also learned that in vulnerable populations, the public health response in crises must be adapted and react to their needs since, in these population groups, the information channels and conventional health messages are often insufficient. It was particularly interesting to see the acceptance of the campaign in the education sector, perhaps because teachers are very used to introducing transversal content into the academic curriculum, especially when the topic is linked to a problem in the real environment.

Another lesson provided by the implementation of this strategy is that in order to achieve successful health communication, the adoption of a participatory approach is essential where the stakeholders participate in the training and change process. In general, health communication based on evidence, culturally relevant and acceptable to the recipients is essential to educate and involve the population in situations that require a rapid and forceful response, either to educate about practical aspects or to combat the infodemic. The lessons learned in this strategy can be applied to other public health programs that seek to engage vulnerable communities.

The "Understanding COVID" strategy also presented some weaknesses.

First, the campaign was implemented in 2021, when pandemic fatigue was already becoming chronic. Bringing its launch back a few months might have been more successful in preventing fatigue. In addition, the execution deadlines for some activities to adapt to the environment where they were carried out (for example, actions in schools) and the evolution of the pandemic itself forced decisions to be made with little time for reflection.

Second, although most of the activities were always evidence-based and oriented towards infection prevention and management in a pandemic setting [45], other activities and groups, such as the promotion of physical activity [46], college students [47], and the 'emotional well-being' intervention [48], could have been taken more into account. On the contrary, it was decided not to focus solely on encouraging vaccination, as was done in many countries [49–56], since in Spain, the public response to the vaccine was very favorable, probably due to high confidence in the vaccination and in the health system [57].

Third, no data on the effectiveness of the campaign was obtained. This is a common limitation of public health campaigns, especially if they are launched under the pressure of an emergency. Evaluating the impact of public health strategies disseminated in an uncontrolled environment is a methodological challenge due to the many factors involved that can influence the results. In any case, at least one study based on surveys could have been carried out. It would have allowed us to know the impressions of people about the strategy. Although several opinion surveys were conducted, these were only used to tailor the strategy and not to explore the satisfaction of the participants in detail.

4. Conclusions

The "Understanding COVID" strategy was a public health campaign launched by the government of Asturias to improve the population's response and adaptation to the COVID-19 crisis and to combat pandemic fatigue and infodemia. The main innovation of the campaign was to target a population selected on the basis of vulnerability criteria, whose voices were taken into account to identify training topics. Capacity building was achieved through a variety of tailor-made methodologies, such as online activities, posters for hotels and catering establishments, educational quizzes and games for schools, social media presence, etc. More than 100 training activities were conducted for 21 subgroups of the vulnerable population, reaching more than 1000 people, as well as students from almost 1000 classrooms and users of various hospitality establishments and vulnerable populations. The strategy faced the challenge of reaching as many vulnerable people of different ages as possible while minimizing the technological gap, which was addressed by using technologies accessible to the population, such as tablets and smartphones, that did not require large technological features. The "Understanding COVID" strategy was well-received and reinforced the choice of dissemination and implementation method, making the information inclusive for the deaf and visually impaired population (with the support of professional sign language translators and adaptations of audiovisual materials). The most participatory actions were those carried out in the school environment and in the hospitality sector, where educational activities were added to the online training sessions and posters were displayed in restaurants. The information campaign in the hospitality sector, "First Air Quality", allowed commercial establishments to display posters with recommendations for the public and to have a certificate of the training received, promoting confidence and safety among customers.

Overall, the collaboration between different government agencies with the ultimate goal of reaching the population most vulnerable to the COVID-19 pandemic is possible if a coordinated strategy is developed that takes into account the citizens and their interests and adapts to their different needs.

Supplementary Materials: The following supporting information can be downloaded at: https://www.mdpi.com/article/10.3390/life13020589/s1, Figure S1: School material from the *"Bichos fuera"* (Bugs out!) campaign; Figure S2: *"Aire de Primera"*(First Quality Air) poster.

Author Contributions: Conceptualization, M.L.-V. and M.P.G.; methodology, M.P.G.; formal analysis, M.L.-V., M.P.G. and C.F.G.; writing—original draft preparation, M.L.-V., M.P.G., C.F.G. and A.L.; writing—review and editing M.L.-V., M.P.G., C.F.G., A.L., I.D.V., I.R.H., M.J.R.N., A.M.M.G., M.P., B.A. and O.Z.; supervision, M.P.G. and A.L. All authors have read and agreed to the published version of the manuscript.

Funding: This research received no external funding.

Institutional Review Board Statement: Not applicable.

Informed Consent Statement: Not applicable.

Data Availability Statement: Not applicable.

Acknowledgments: We thank Raquel Vázquez Álvarez and Laura González Lozano for their work in the design and implementation of the "Understanding COVID" strategy. We thank the Management Department of the Ministry of Health of Asturias (Sergio Vallés, Lidia Clara, Elena Llorente and Jose Antonio Altolaguierre) for supporting the "Understanding COVID" strategy. We thank Francisco Sánchez Refusta for his help in translating this article. We thank Tierra Voz Communication S.L. for its graphic support.

Conflicts of Interest: The authors declare no conflict of interest.

References

1. Boletín Oficial del Principado de Asturias (BOPA) Acuerdo de 12 de marzo de 2020, del Consejo de Gobierno, sobre medidas preventivas y recomendaciones relacionadas con la infección del SARS CoV-2 (COVID-19). Available online: https://sede.asturias.es/bopa/2020/03/13/2020-02687.pdf (accessed on 18 February 2023).
2. Khose, S.; Moore, J.X.; Wang, H.E. Epidemiology of the 2020 Pandemic of COVID-19 in the State of Texas: The First Month of Community Spread. *J. Community Health* **2020**, *45*, 696–701. [CrossRef] [PubMed]
3. Lakhani, H.V.; Pillai, S.S.; Zehra, M.; Sharma, I.; Sodhi, K. Systematic Review of Clinical Insights into Novel Coronavirus (COVID-19) Pandemic: Persisting Challenges in U.S. Rural Population. *Int. J. Environ. Res. Public Health* **2020**, *17*, 4279. [CrossRef] [PubMed]
4. Probst, J.C.; Crouch, E.L.; Eberth, J.M. COVID-19 risk mitigation behaviors among rural and urban community-dwelling older adults in summer, 2020. *J. Rural. Health* **2021**, *37*, 473–478. [CrossRef]
5. Greer, M.L.; Sample, S.; Jensen, H.K.; McBain, S.; Lipschitz, R.; Sexton, K.W. COVID-19 Is Connected with Lower Health Literacy in Rural Areas. *Stud. Health Technol. Inform.* **2021**, *281*, 804–808. [CrossRef] [PubMed]
6. Schippers, M.C.; Rus, D.C. Optimizing Decision-Making Processes in Times of COVID-19: Using Reflexivity to Counteract Information-Processing Failures. *Front. Psychol.* **2021**, *12*, 650525. [CrossRef]
7. Groot, G.; Witham, S.; Badea, A.; Baer, S.; Dalidowicz, M.; Reeder, B.; Froh, J.; Carr, T. Evaluating a learning health system initiative: Lessons learned during COVID-19 in Saskatchewan, Canada. *Learn. Health Syst.* **2022**, e10350. [CrossRef]
8. Ghoushchi, S.J.; Bonab, S.R.; Ghiaci, A.M. A decision-making framework for COVID-19 infodemic management strategies evaluation in spherical fuzzy environment. *Stoch. Environ. Res. Risk Assess.* **2023**. *online early publication*. [CrossRef]
9. Mistry, S.K.; Shaw, M.; Raffan, F.; Johnson, G.; Perren, K.; Shoko, S.; Harris-Roxas, B.; Haigh, F. Inequity in Access and Delivery of Virtual Care Interventions: A Scoping Review. *Int. J. Environ. Res. Public Health* **2022**, *19*, 9411. [CrossRef]

10. World Health Organization. *Regional Office for Europe Pandemic Fatigue: Reinvigorating the Public to Prevent COVID-19: Policy Framework for Supporting Pandemic Prevention and Management: Revised Version November 2020*; World Health Organization. Regional Office for Europe: Copenhagen, Denmark, 2020.
11. Consejería de Presidencia y Participación Ciudadana Plan demográfico del Principado de Asturias 2017–2027. Available online: https://www.age-geografia.es/site/wp-content/uploads/2017/10/Plan-Demogr%C3%A1fico-Asturias-2017-2027.pdf (accessed on 10 December 2022).
12. Sudhira, H.; Ramachandra, T.; Jagadish, K. Urban sprawl: Metrics, dynamics and modelling using GIS. *Int. J. Appl. Earth Obs. Geoinf.* **2004**, *5*, 29–39. [CrossRef]
13. Brubacher, L.J.; Hasan, Z.; Sriram, V.; Keidar, S.; Wu, A.; Cheng, M.; Lovato, C.Y.; Berman, P.; U. B. C. Working Group on Health Systems. Response to COVID-19 Investigating the influence of institutions, politics, organizations, and governance on the COVID-19 response in British Columbia, Canada: A jurisdictional case study protocol. *Health Res. Policy Syst.* **2022**, *20*, 74. [CrossRef]
14. Jefatura del Estado Ley Orgánica 3/2018, de 5 de diciembre, de Protección de Datos Personales y garantía de los derechos digitales. Available online: https://www.boe.es/eli/es/lo/2018/12/05/3 (accessed on 11 December 2022).
15. Woodrow, N.; Fairbrother, H.; D'Apice, K.; Breheny, K.; Albers, P.; Mills, C.; Tebbett, S.; Campbell, R.; De Vocht, F. Exploring the Potential of a School-Based Online Health and Wellbeing Screening Tool: Young People's Perspectives. *Int. J. Environ. Res. Public Health* **2022**, *19*, 4062. [CrossRef] [PubMed]
16. Spitzer, M. Open schools! Weighing the effects of viruses and lockdowns on children. *Trends Neurosci. Educ.* **2021**, *22*, 100151. [CrossRef] [PubMed]
17. Odone, A.; Bricchi, L.; Signorelli, C. COVID-19 Control School-Based Interventions: Characteristics and Impact of a Nation-al-Level Educational Programme in Italy. *Acta Biomed.* **2022**, *92*, e2021495. [CrossRef] [PubMed]
18. Barcenilla-Guitard, M.; Espart, A. Influence of Gender, Age and Field of Study on Hand Hygiene in Young Adults: A Cross-Sectional Study in the COVID-19 Pandemic Context. *Int. J. Environ. Res. Public Health* **2021**, *18*, 13016. [CrossRef] [PubMed]
19. Hefferon, C.; Taylor, C.; Bennett, D.; Falconer, C.; Campbell, M.; Williams, J.G.; Schwartz, D.; Kipping, R.; Taylor-Robinson, D. Priorities for the child public health response to the COVID-19 pandemic recovery in England. *Arch. Dis. Child.* **2020**, *106*, 533–538. [CrossRef] [PubMed]
20. Amran, M.S.; Jamaludin, K.A. The Impact of Unplanned School Closures on Adolescent Behavioral Health During the COVID-19 Pandemic in Malaysia. *Front. Public Health* **2021**, *9*, 639041. [CrossRef]
21. Almoslem, M.M.; Alshehri, T.A.; Althumairi, A.A.; Aljassim, M.T.; Hassan, M.E.; Berekaa, M.M. Handwashing Knowledge, Attitudes, and Practices among Students in Eastern Province Schools, Saudi Arabia. *J. Environ. Public Health* **2021**, *2021*, 6638443. [CrossRef]
22. Sugita, E.W. Water, Sanitation and Hygiene (WASH) in Japanese elementary schools: Current conditions and practices. *Pediatr. Int.* **2021**, *64*, e15062. [CrossRef]
23. Rohwer, E.; Mojtahedzadeh, N.; Neumann, F.A.; Nienhaus, A.; Augustin, M.; Harth, V.; Zyriax, B.-C.; Mache, S. The Role of Health Literacy among Outpatient Caregivers during the COVID-19 Pandemic. *Int. J. Environ. Res. Public Health* **2021**, *18*, 11743. [CrossRef]
24. Pham, Q.; El-Dassouki, N.; Lohani, R.; Jebanesan, A.; Young, K. The Future of Virtual Care for Older Ethnic Adults Beyond the COVID-19 Pandemic. *J. Med. Internet Res.* **2022**, *24*, e29876. [CrossRef]
25. Zenone, M.A.; Cianfrone, M.; Sharma, R.; Majid, S.; Rakhra, J.; Cruz, K.; Costales, S.; Sekhon, M.; Mathias, S.; Tugwell, A.; et al. Supporting youth 12–24 during the COVID-19 pandemic: How Foundry is mobilizing to provide information, resources and hope across the province of British Columbia. *Glob. Health Promot.* **2021**, *28*, 51–59. [CrossRef] [PubMed]
26. Leader, A.E.; Capparella, L.M.; Waldman, L.B.; Cammy, R.B.; Petok, A.R.; Dean, R.; Shimada, A.; Yocavitch, L.; Rising, K.L.; Garber, G.D.; et al. Digital Literacy at an Urban Cancer Center: Implications for Technology Use and Vulnerable Patients. *Kimmel Cancer Center Faculty Papers JCO Clin. Cancer Inform.* **2021**, *5*, 872–880. [CrossRef]
27. Alford-Teaster, J.; Wang, F.; Tosteson, A.N.A.; Onega, T. Incorporating broadband durability in measuring geographic access to health care in the era of telehealth: A case example of the 2-step virtual catchment area (2SVCA) Method. *J. Am. Med. Inform. Assoc.* **2021**, *28*, 2526–2530. [CrossRef] [PubMed]
28. Malkin, M.; Mickler, A.K.; Ajibade, T.O.; Coppola, A.; Demise, E.; Derera, E.; Ede, J.O.; Gallagher, M.; Gumbo, L.; Jakopo, Z.; et al. Adapting High Impact Practices in Family Planning During the COVID-19 Pandemic: Experiences From Kenya, Nigeria, and Zimbabwe. *Glob. Health Sci. Pract.* **2022**, *10*, e2200064. [CrossRef]
29. Gray, D.J.; Kurscheid, J.; Mationg, M.L.; Williams, G.M.; Gordon, C.; Kelly, M.; Wangdi, K.; McManus, D.P. Health-education to prevent COVID-19 in schoolchildren: A call to action. *Infect. Dis. Poverty* **2020**, *9*, 142–144. [CrossRef]
30. Fetzer, T. Subsidising the spread of COVID-19: Evidence from the UK'S Eat-Out-to-Help-Out Scheme*. *Econ. J.* **2021**, *132*, 1200–1217. [CrossRef]
31. Tamura, M.; Suzuki, S.; Yamaguchi, Y. Effects of tourism promotion on COVID-19 spread: The case of the "Go To Travel" campaign in Japan. *J. Transp. Health* **2022**, *26*, 101407. [CrossRef]

32. Argyris, Y.A.; Nelson, V.R.; Wiseley, K.; Shen, R.; Roscizewski, A. Do social media campaigns foster vaccination adherence? A systematic review of prior intervention-based campaigns on social media. *Telemat. Inform.* **2023**, *76*, 101918. [CrossRef]
33. Puzio, D.; Makowska, I.; Rymarczyk, K. Raising the Child—Do Screen Media Help or Hinder? The Quality over Quantity Hypothesis. *Int. J. Environ. Res. Public Health* **2022**, *19*, 9880. [CrossRef] [PubMed]
34. Bhyat, R.; Hagens, S.; Bryski, K.; Kohlmaier, J.F. Digital Health Value Realization through Active Change Efforts. *Front. Public Health* **2021**, *9*, 741424. [CrossRef] [PubMed]
35. Linardon, J.; Westrupp, E.M.; Macdonald, J.A.; Mikocka-Walus, A.; Stokes, M.A.; Greenwood, C.J.; Youssef, G.J.; Teague, S.; Hutchinson, D.; Sciberras, E.; et al. Monitoring Australian parents' shifting receptiveness to digital mental health interventions during the COVID-19 pandemic. *Aust. N. Z. J. Psychiatry* **2021**, *56*, 1503–1514. [CrossRef] [PubMed]
36. Vamos, S.D.; McDermott, R.J. Rebranding School Health: The Power of Education for Health Literacy. *J. Sch. Health* **2021**, *91*, 670–676. [CrossRef] [PubMed]
37. Mörelius, E.; Robinson, S.; Arabiat, D.; Whitehead, L. Digital Interventions to Improve Health Literacy among Parents of Children Aged 0 to 12 Years with a Health Condition: Systematic Review. *J. Med. Internet Res.* **2021**, *23*, e31665. [CrossRef] [PubMed]
38. Santoveña-Casal, S.; Gil-Quintana, J.; Ramos, L. Digital citizens' feelings in national #COVID 19 campaigns in Spain. *Heliyon* **2021**, *7*, e08112. [CrossRef] [PubMed]
39. Santoveña-Casal, S.; Pérez, M.D.F. Relevance of E-Participation in the state health campaign in Spain: #EstoNoEsUnJuego/#ThisIsNotAGame. *Technol. Soc.* **2022**, *68*, 101877. [CrossRef] [PubMed]
40. Graffigna, G.; Bosio, C.; Savarese, M.; Barello, M.; Barello, S. "#I-Am-Engaged": Conceptualization and First Implementation of a Multi-Actor Participatory, Co-designed Social Media Campaign to Raise Italians Citizens' Engagement in Preventing the Spread of COVID-19 Virus. *Front. Psychol.* **2020**, *11*, 567101. [CrossRef] [PubMed]
41. Silesky, M.D.; Panchal, D.; Fields, M.; Peña, A.S.; Diez, M.; Magdaleno, A.; Frausto-Rodriguez, P.; Bonnevie, E. A Multifaceted Campaign to Combat COVID-19 Misinformation in the Hispanic Community. *J. Community Health* **2022**, *Online early publication*. [CrossRef]
42. Miller, A.F.; Yang, C.; Grieb, S.M.; Lipke, M.; Bigelow, B.F.; Phillips, K.H.; Palomino, P.; Page, K.R. A Community-Engaged Social Marketing Campaign to Promote Equitable Access to COVID-19 Services among Latino Immigrants. *Am. J. Public Health* **2023**, *113*, e1–e4. [CrossRef]
43. Marcell, L.; Dokania, E.; Navia, I.; Baxter, C.; Crary, I.; Rutz, S.; Monteverde, M.J.S.; Simlai, S.; Hernandez, C.; Huebner, E.M.; et al. One Vax Two Lives: A social media campaign and research program to address COVID-19 vaccine hesitancy in pregnancy. *Am. J. Obstet. Gynecol.* **2022**, *227*, 685–695.e2. [CrossRef]
44. Ioannides, S.; Hess, I.; Lamberton, C.; Luisi, B. Engaging with culturally and linguistically diverse communities during a COVID-19 outbreak: A NSW Health interagency public health campaign. *Public Health Res. Pract.* **2022**. [CrossRef]
45. Park, S.; Oh, S. Factors associated with preventive behaviors for COVID-19 among adolescents in South Korea. *J. Pediatr. Nurs.* **2021**, *62*, e69–e76. [CrossRef] [PubMed]
46. Luo, L.; Zeng, X.; Wu, Y.; An, F.; Huang, J.; Yang, H.; Jiang, Q.; Ou, Q.; Du, J.; Song, N. Influencing Factors of Students Aged 10–20 Non-participating in Home Physical Exercise During the COVID-19 Isolation Policy Period: A Cross-Sectional Study From China. *Front. Public Health* **2022**, *10*, 787857. [CrossRef] [PubMed]
47. Chang, J.; Yuan, Y.; Wang, D. Analysis of mental health status and influencing factors of College Students under the epidemic of novel coronavirus pneumonia. *J. South. Med. Univ.* **2020**, *2*, 171–176. [CrossRef]
48. Shen, X.; Li, Y.; Feng, J.; Lu, Z.; Tian, K.; Gan, Y. Current status and associated factors of psychological resilience among the Chinese residents during the coronavirus disease 2019 pandemic. *Int. J. Soc. Psychiatry* **2020**, *68*, 34–43. [CrossRef]
49. Almansour, A.; Hussein, S.M.; Felemban, S.G.; Mahamid, A.W. Acceptance and hesitancy of parents to vaccinate children against coronavirus disease 2019 in Saudi Arabia. *PLoS ONE* **2022**, *17*, e0276183. [CrossRef]
50. Fedele, F.; Aria, M.; Esposito, V.; Micillo, M.; Cecere, G.; Spano, M.; De Marco, G. COVID-19 vaccine hesitancy: A survey in a population highly compliant to common vaccinations. *Hum. Vaccines Immunother.* **2021**, *17*, 3348–3354. [CrossRef]
51. Zhang, M.-X.; Lin, X.-Q.; Chen, Y.; Tung, T.-H.; Zhu, J.-S. Determinants of parental hesitancy to vaccinate their children against COVID-19 in China. *Expert Rev. Vaccines* **2021**, *20*, 1339–1349. [CrossRef]
52. Lubis, T.A.; Gunardi, H.; Soedjatmiko, S.; Satari, H.I.; Alatas, F.S.; Pulungan, A.B. Educational videos to address vaccine hesitancy in childhood immunization. *Vaccine* **2022**, *40*, 5965–5970. [CrossRef]
53. Ngandjon, J.K.; Ostermann, T.; Kenmoue, V.; Laengler, A. Insights into Predictors of Vaccine Hesitancy and Promoting Factors in Childhood Immunization Programs—A Cross-Sectional Survey in Cameroon. *Int. J. Environ. Res. Public Health* **2022**, *19*, 2721. [CrossRef]
54. Huang, L.-L.; Yang, Y.-P.; Mao, H.-P.; Hu, W.-W.; Jiang, Y.-H.; Jiesisibieke, Z.L.; Tung, T.-H. Parental hesitancy towards vaccinating their children with a booster dose against COVID-19: Real-world evidence from Taizhou, China. *J. Infect. Public Health* **2022**, *15*, 1006–1012. [CrossRef]
55. Xu, Y.; Zhang, R.; Zhou, Z.; Fan, J.; Liang, J.; Cai, L.; Peng, L.; Ren, F.; Lin, W. Parental psychological distress and attitudes towards COVID-19 vaccination: A cross-sectional survey in Shenzhen, China. *J. Affect. Disord.* **2021**, *292*, 552–558. [CrossRef] [PubMed]

56. Altulahi, N.; AlNujaim, S.; Alabdulqader, A.; Alkharashi, A.; AlMalki, A.; AlSiari, F.; Bashawri, Y.; Alsubaie, S.; AlShahrani, D.; AlGoraini, Y. Willingness, beliefs, and barriers regarding the COVID-19 vaccine in Saudi Arabia: A multiregional cross-sectional study. *BMC Fam. Pract.* **2021**, *22*, 247. [CrossRef] [PubMed]
57. Microsoft Power BI. Available online: https://app.powerbi.com/view?r=eyJrIjoiNTdhYzlhYjUtZmFjNi00NjBhLThiNTktMmNjNDY5NzYzNjBlIiwidCI6IjI4ZmI0NmYwLTU0OWYtNDI5Ny1iOTZmLWFjNjJhZTkxY2YwYyIsImMiOjl9&pageName=ReportSectionda82d8ffb60be1590dd8 (accessed on 4 November 2022).

life

Article

Assessing the Suitability of a Virtual 'Pink Warrior' for Older Breast Cancer Survivors during COVID-19: A Pilot Study

Maria C. Swartz [1,*], Michael C. Robertson [2], Ursela Christopherson [2], Stephanie J. Wells [1], Zakkoyya H. Lewis [3], Jinbing Bai [4], Michael D. Swartz [5], H. Colleen Silva [6], Eloisa Martinez [7] and Elizabeth J. Lyons [2]

1 Department of Pediatrics Research, Division of Pediatrics, The University of Texas MD Anderson Cancer Center, Houston, TX 77030, USA
2 Department of Nutrition, Metabolism & Rehabilitation Sciences, The University of Texas Medical Branch, Galveston, TX 77550, USA
3 Department of Kinesiology & Health Promotion, California State Polytechnic University, Pomona, CA 91768, USA
4 Nell Hodgson Woodruff School of Nursing, Emory University, Atlanta, GA 30322, USA
5 Department of Biostatistics and Data Science, School of Public Health, The University of Texas Health, Houston, TX 77030, USA
6 Department of Surgery, The University of Texas Medical Branch, Galveston, TX 77550, USA
7 Sealy Center on Aging, The University of Texas Medical Branch, Galveston, TX 77550, USA
* Correspondence: mchang1@mdanderson.org

Abstract: The COVID-19 pandemic impacted the conduct of in-person physical activity (PA) interventions among older survivors of BC, who need such interventions to stay active and prevent functional decline. We tested the feasibility of virtually delivering an exergame-based PA intervention to older BC survivors. We enrolled 20 female BC survivors ≥55 years and randomly assigned them to two groups. The intervention group (Pink Warrior 2) received 12 weekly virtual exergame sessions with behavioral coaching, survivorship navigation support, and a Fitbit for self-monitoring. The control group received 12 weekly phone-based survivorship discussion sessions and wore a Mi Band 3. Feasibility was evaluated by rates of recruitment (≥0.92 participants/center/month), retention (≥80%), and group attendance (≥10 sessions), percentage of completed virtual assessments, and number of technology-related issues and adverse events. Intervention acceptability was measured by participants' ratings on a scale of 1 (strongly disagree) to 5 (strongly agree). The recruitment rate was 1.93. The retention and attendance rates were 90% and 88% (≥10 sessions), respectively. Ninety-six percent completed virtual assessments without an adverse event. Acceptability was high (≥4). The intervention met benchmarks for feasibility. Additional research is needed to further understand the impact of virtually delivered PA interventions on older BC survivors.

Keywords: physical activity; exergaming; breast neoplasms; physical function; telehealth

Citation: Swartz, M.C.; Robertson, M.C.; Christopherson, U.; Wells, S.J.; Lewis, Z.H.; Bai, J.; Swartz, M.D.; Silva, H.C.; Martinez, E.; Lyons, E.J. Assessing the Suitability of a Virtual 'Pink Warrior' for Older Breast Cancer Survivors during COVID-19: A Pilot Study. *Life* **2023**, *13*, 574. https://doi.org/10.3390/life13020574

Academic Editor: Daniele Giansanti

Received: 14 January 2023
Revised: 13 February 2023
Accepted: 15 February 2023
Published: 18 February 2023

1. Introduction

The National Cancer Institute considers individuals as survivors of breast cancer (BC) beginning from the time of diagnosis through the end of life [1]. According to this definition, there were 4.1 million survivors of BC living in the United States in 2022 [2]. Although it is encouraging that the survival rate has increased, substantial evidence shows that cancer and cancer treatment can exacerbate age-related declines in physical function [3,4]. Thus, aging survivors of BC are at elevated risk for poor physical function [4] and, consequently, are at higher risk of premature death [5].

Physical activity (PA) has emerged as a key strategy to prevent functional decline and improve quality of life [6]. PA is defined as "any bodily movement that results in energy expenditure" over a period [7]. However, PA engagement in survivors of BC remains suboptimal despite strong evidence of its beneficial effect on physical function, health-related quality of life, and mortality. The prevalence of insufficient PA in breast cancer survivors ranged widely depending on the location and the study population, age, and treatment status, and how PA was measured. For example, a recent study in 2042 women pre and post diagnosed with breast cancer in Germany found that approximately 50% of the participants were insufficiently active pre and post diagnosed, while only 18.2% were sufficiently active [8]. In the United States, a recent study of 1340 female at high risk of developing breast cancer found that one year after treatment, 31.6% were not engaging in recreational/leisure time PA [9].

Adherence to the recommended PA guideline is needed to fully realize the benefits of PA in decreasing symptom burden and improving physical function in BC survivors. Nevertheless, BC survivors reported various perceived barriers to engage in sufficient PA. These include fatigue, pain, limited mobility, and other cancer treatment-related side effects, lack of motivation, time, social support, and limited access to PA programs [10–13]. Barriers such as lack of motivation, time, social support, and limited access to PA programs are common regardless of disease type [11,14]. However, the added cancer-related symptom burden, such as fatigue, pain, and limited mobility from cancer or cancer treatment, can further impact how BC survivors perceive PA-related barriers and respond to PA interventions [11,12,15,16]. Previous exergame interventions primarily focused on using exergame itself to impact health, but it was not paired with behavioral coaching to enhance self-management skills to overcome PA-related barriers [17,18]. Thus, there is a need to pilot test PA interventions designed to provide both self-management skills to overcome PA-related barriers and enhance survivors' motivation to engage in PA despite the experience of cancer-related symptoms.

We previously tested the feasibility of promoting PA in an in-person group setting using exergames along with PA behavioral coaching and BC support discussions among survivors of BC aged 18 years and older [19]. PA behavioral coaching was designed to provide self-management skills to overcome PA-related barriers [19]. Exergame was chosen as a tool to promote PA to help reframe PA as a pleasurable activity [20,21]. Accumulating evidence indicates that targeting a person's motivation and reframing the internal reaction to PA as a fun activity may lead to a more effective intervention [22–27]. Our pilot study's results indicated high levels of acceptance of using exergames and being active in a group setting [19]. Similar to Wurz and colleagues' findings, feedback from survivors indicated that attending in-person sessions remained a barrier for survivors in all stages of their cancer care continuum (e.g., limited ability to travel because of cancer treatment side effects, traffic) [13]. Thus, there is a need to test the use of videoconference platforms to deliver the intervention to increase participation in group-based exercise, and, in turn, increase physical function capabilities.

The COVID-19 pandemic accelerated advances in videoconferencing technology and made it a more common method of communication via smartphones [28]. Additionally, the COVID-19 pandemic also significantly impacted the conduct and participation of in-person PA interventions targeting medically complex populations, such as older survivors of BC [29]. Thus, we adapted a previously tested exergame- and group-based PA intervention [19] to be delivered via videoconference platform. We also adapted our physical function assessments to be conducted via a videoconference platform [30,31]. The goal was to prevent a decline in physical function among older survivors of BC. Therefore, the overall purpose of this study was to test the feasibility of virtually delivering an exergame- and group-based PA intervention in a sample of older survivors of BC during the COVID-19 pandemic.

2. Materials and Methods

We followed the CONSORT (Consolidated Standards of Reporting Trials) 2010 statement for randomized pilot and feasibility trials to report our pilot study (Figure 1) [32]. We conducted a phase II feasibility pilot study [33]. This was a prospective two-group feasibility pilot study in which we used a 1:1 group allocation. The purpose of conducting a randomized pilot and feasibility trial was to increase the likelihood of a successful larger randomized controlled trial by testing the logistics of planned trial [34].

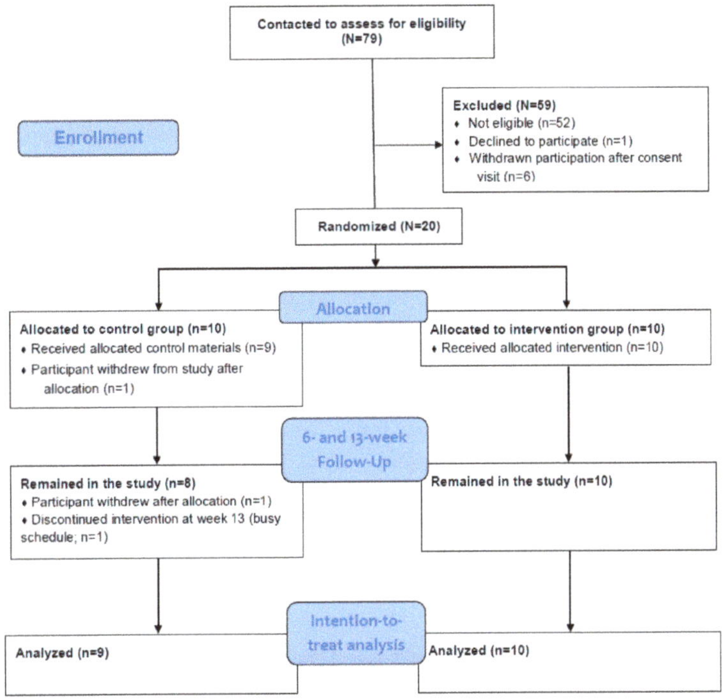

Figure 1. CONSORT pilot and feasibility flow diagram.

The study aimed to assess the feasibility of a remotely delivered exergame- and group-based PA intervention (Pink Warrior 2) for older survivors of BC. Primarily, the study sought to determine the feasibility of recruitment of the target population, retention, safety of remote physical function assessment, and adherence to the PA intervention. A secondary aim of the study was to explore the potential influence of the intervention on physical function outcomes over 12 weeks.

2.1. Participants

Due to COVID-19 pandemic restrictions in clinical settings, participants were recruited from emails and phone calls were drawn from the cancer center's tumor registry and clinical visit lists between August 2020 and October 2021. An introduction email was sent about the study. Potential participants who expressed interest were contacted by phone to screen for eligibility.

The following were the inclusion criteria to enroll in the study: self-reported female gender, age 55 years or older; having had a primary BC diagnosis; being able to speak, read, and write in English; being able to move arms and legs as well as ambulate; currently being insufficiently inactive (<150 min of moderate-intensity activity per week); having a smartphone, tablet, or computer; and having daily access to reliable internet.

2.2. Randomization

Participants were randomly assigned to the intervention group (Pink Warrior 2) or the attention control group. The intervention group received a virtually delivered PA intervention that combined exergame group play, PA behavioral coaching, and BC support (e.g., survivorship guidance). The attention control group participated in weekly telephone and group-based BC support using the BC support discussion materials. The randomization process described by Lyons et al. was used in the current study [19,35]. Briefly, a graduate student who did not assist with the assessments used a web-based app [19,36] to produce a random sequence of treatment assignment to intervention or attention control group using a 1:1 allocation strategy. Each assignment was recorded on a single piece of paper. Then, the graduate student wrapped each treatment assignment paper with a carbon paper and a piece of aluminum foil and then sealed the wrapped assignment in an opaque envelope, which was similar to Swartz et al. [19]. The carbon paper is used to provide an audit trail. The foil was used to ensure that the group assignment was concealed from the assessor before opening the envelope. Lastly, the graduate student shuffled the stack of 20 sealed envelopes and numbered them sequentially by the study ID number and also initialed each envelope to notate the person who had prepared it. The assessor, who was not involved in preparing the envelope, would sign and date the envelope before opening it and save the allocation information in the study file [19].

2.3. Procedures

Detailed procedures have been published elsewhere [19]. Briefly, all participants went through four visits virtually for informed consent and assessment. A summary of the study flow is presented in Figure 2. Similar to our in-person design, the study's duration was 13 weeks, but the Pink Warrior 2 intervention lasted only 12 weeks [19]. Unlike the previous study [19], the current study conducted all study visits virtually using SecureVideo (https://securevideo.com/, accessed on 20 December 2022). The SecureVideo is a HIPAA-compliant telehealth platform. They use the 256-bit AES-encrypted signaling and media stream, and it makes connections to web applications and API through HTTPS only, using TLS 1.3 or 1.2 encryption for in-transit encryption [37]. They also use BitLocker for the 128-bit-AES-encryption for the full database encryption [37]. Additional information on how SecureVideo meets HIPAA standards is provided in the their support center's About SecureVideo Accounts and Services page [37]. Furthermore, SecureVideo was selected because it included advanced scheduling tools that allowed researchers to pre-schedule the sessions and send automatic reminders to individuals 1-day and 2-h before each session. This feature helped to ensure that participants receive adequate reminders for scheduled virtual visits to promote adherence. Moreover, SecureVideo was reviewed and cleared to be used for the current study by the information security team at the University of Texas Medical Branch. One of the security features provided by the SecureVideo platform was that they provided an individualized unique link through email or text message for each participant to log in to ensure a secure connection and helped the team avoid video-teleconferencing hijacking (also known as Zoom bombing) [38]. Each of the individualized unique links can only be used by one participant. However, the team also turned on the waiting room feature, so only participants who the team recognized could enter into the main room. This was done to further minimize video-teleconferencing hijacking risk. Following the informed consent visit (visit 1), participants were mailed a research-grade activity monitor (ActiGraph GT9X Link accelerometer) to wear for a week as well as baseline questionnaires to complete and return by mail before the baseline assessment visit (visit 2). Participants also provided permission for medical record data extraction as well as an SMS text message, email, or phone call to schedule study visits, and for reminders to be sent prior to assessment and study sessions. The medical record data extraction, SMS text message, email, or phone call were used as additional methods to minimize missed appointments for data collection and promote adherence for attending the group sessions. Additionally, checklists were developed for all study visits, and an Excel spreadsheet was created to track study visits and reminders to follow up on questionnaires and Actigraphs. The research coordinators were trained on the assessments and co-developed the tracking excel spreadsheets. These additional steps were carried out with the intent to minimize missing data.

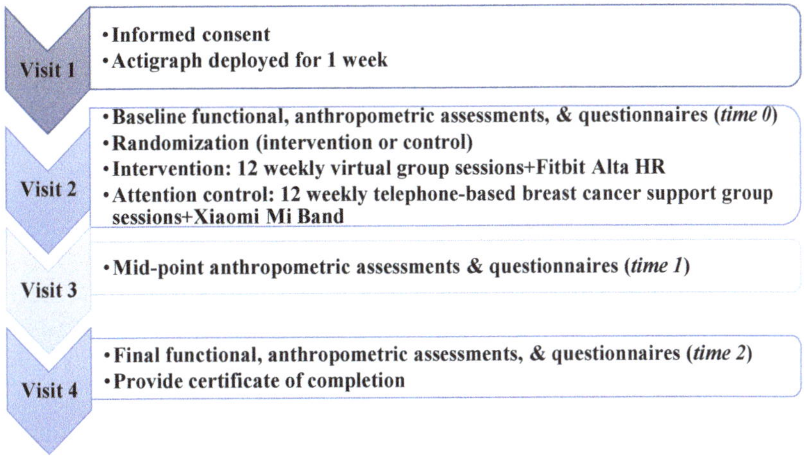

Figure 2. Study flow diagram.

Using our tracking sheet, we noticed unexpected mailing delays during the COVID-19 pandemic, so the team added additional time between visit 1 and visit 2. After approximately 3 to 4 weeks after visit 1, the study team scheduled visit 2 virtually to conduct a full baseline physical function assessment (time 0). The team also switched to a different courier service (FedEx) to further minimize mailing delays. Participants were randomly assigned to either the intervention group or the attention control group at time 0. All materials were mailed to the participants prior to visit 2 (Figure 3). A study orientation regarding the assigned group along with technology setup was completed at the end of the visit 2. Because of limited staffing resources, we could not conduct a blinded assessment. Furthermore, the study's design precluded the blinding of patients to their assigned groups. Visit 3 was the assessment halfway through the study (time 1), and visit 4 (time 2) was the final assessment.

Figure 3. Pink Warrior 2 Assessment and Study Materials.

2.4. Ethics Approval

The Institutional Review Board approved all procedures (protocol 16-0040-02), and our study was registered at ClinicalTrials.gov before study activation (NCT04259905).

2.5. Intervention

Detailed description of the intervention has been published elsewhere [19]. Briefly, participants who were assigned to the intervention group took part in a remotely delivered 12-week multicomponent PA intervention. Similar to the in-person design [19], each virtual group session comprised three components: (1) a PA behavioral coaching segment, (2) an exergame-based activity demonstration and practice segment, and (3) a BC support discussion segment. The duration of each weekly structured virtual group session was scheduled to be 60 min. Approximately 15–20 min are spent on the behavioral coaching segment, the exergame-based segment took approximately 30–35 min, and the BC support discussion segment took approximately 10 min.

The Pink Warrior 2 PA behavioral coaching materials were adapted from materials from the Active Living After Cancer (ALAC) program, the details of which were published previously [19,39,40]. Briefly, the Pink Warrior 2 used the same behavioral coaching materials that were developed for the in-person study [19]. The behavioral coaching materials were developed based on the constructs of Social Cognitive Theory and Self-Determination Theory [15,19,23–26,41]. Under Social Cognitive Theory, we targeted the self-efficacy and self-regulation constructs because both constructs have shown to be associated with the initiation and an increase in PA [22,42]. Under Self-Determination Theory, we specifically targeted the basic psychological needs for autonomy, competence, and relatedness. Self-Determination Theory postulates that by meeting these three basic needs, we would boost the BC survivors' autonomous motivation, which comes from within an individual, to engage in PA [25]. This would then promote PA over time [22,25,26]. Trained facilitators (UC, a graduate student pursuing a PhD and a licensed occupational therapist specialized in hand therapy and/or MCS, the lead investigator) summarized the weekly PA discussion topics that were designed to provide behavior change skills, which were aimed at promoting the adoption of an active lifestyle. Beyond the group discussion, participants were tasked with completing a weekly reflection worksheet corresponding to the weekly PA coaching discussion topics on their own. The goal of the weekly reflection worksheet was to engage participants to practice using the skills discussed in that week's PA coaching session and promote an increase in PA outside of the virtual group sessions. Examples of the weekly reflection worksheet included: setting goals, clarifying values related to PA, and finding support for PA.

The exergame sessions involved a facilitator leading the exercise sessions using console-based exergames (e.g., XBOX 360 Kinect). The game selections for the group sessions were previously published [19]. Briefly, the types of games chosen for our in-person and virtually delivered interventions, in collaboration with an occupational therapist, included mind–body games (e.g., Zen energy and yoga games in Your Shape Fitness Evolved 2012), and fitness-based games (e.g., kickboxing, upper and lower-body training in Your Shape Fitness Evolved 2012 and Zumba) [19]. These exergames were chosen as a way to promote enjoyment, increase self-efficacy, and increase motivation to engage in PA in real life [20,43]. Thus, we have selected a variety of activities for participants to choose from that are similar to what they can find in real life or on the web. Each of the exergame sessions consists of a mix of mind–body games and fitness-based games to keep participants engaged. Each game lasted between 5 and 15 min. The length of time for each game depends on the type of game and the level of difficulty. Prior to playing each game, we would explain what the participants can expect, demonstrated the movements used in the game, and demonstrated the type of modifications they can do while playing the game. For example, we demonstrated how participants can use a chair for support when doing the lunge or squat movements.

Unlike the in-person sessions, participants were not provided a game console to use. The exergames were livestreamed via the SecureVideo platform. A similar technique was used in research conducted by Lin [44] that showed that the livestreaming of exergames did not significantly impact participant's body movement and participation. In our study, the facilitator stood in front of the game console and selected the games based on input from participants. The facilitator's camera was aimed at the TV monitor and zoomed in on the avatar trainer in the game (e.g., yoga, kickboxing, and Zumba) [19]. Participants then followed the avatar trainer to complete the activity. A YouTube video of the games played during the session (e.g., yoga) was provided to the participants to use throughout the week. However, we also encouraged participants to find other activities they would like to explore beyond the exergame-based activities. The intention was to encourage the adoption of an active lifestyle outside of the group sessions. In addition to the weekly exergame group PA session, participants were provided a Fitbit Alta HR activity tracker. A study email and an anonymous Fitbit account were set up for each of the participants to protect their identity. Participants in the intervention group were able to use the device and the associated Fitbit application (app) to track individual and group steps in a private Fitbit group. This component was incorporated to promote social relatedness [45]. Lastly, we also incorporated hand grip-strengthening exercises with and without TheraPutty. Handouts were developed by UC, an occupational therapist whose specialty is hand therapy. Handouts were provided to the participants to promote engagement in grip exercises throughout the week.

In the BC support segment, resources from the National Coalition for Cancer Survivorship and the American Cancer Society were included to elicit survivorship navigation discussions that were used in our in-person intervention [19]. Briefly, we used the Cancer Survival Toolbox materials from the National Coalition for Cancer Survivorship [46]. Additionally, we used the materials provided in the Personal Health Manager kit from the American Cancer Society. The oncology team we worked with provides the Personal Health Manager kit to all patients when they are first diagnosed with breast cancer [19]. The intent of this component was to equip BC survivors with support and credible resources as they navigate through their cancer experience.

2.6. Attention Control Group

Participants assigned to the attention control group (Figure 2) were provided a Xiaomi Mi Band 3 and took part in a weekly telephone-based BC support group for 12 weeks. Attention control was used to ensure the same dose of interaction with a facilitator as intervention participants [47]. A Xiaomi Mi Band 3 was provided as a self-monitoring tool. This activity tracker was selected because the associated app did not include as many behavior change techniques as the Fitbit app [48]. It does provide progressive step goal notifications and a "reach goal" badge when a person reaches the step goal. Similar to the intervention group, a study email and an anonymous Mi Band account were set up for each of the participants to protect their identity. The team also purchased app-based phone numbers for use with the Mi Band app. A master's degree-level research dietitian (SJW) facilitated the BC support group discussions. The attention control group did not receive the behavioral coaching and exergame components of the Pink Warrior 2 intervention. Instead, they were provided same resources from the National Coalition for Cancer Survivorship and the American Cancer Society as the intervention group to elicit survivorship navigation discussions. Each telephone-based BC support group session also lasted about 1 h. This kind of control intervention was selected because of the well-documented effects of attention in studies promoting behavior change [47].

2.7. Primary Outcome Measures

2.7.1. Feasibility

Feasibility was the primary outcome of this pilot trial. Specifically, we evaluated the key elements of the trial, including rates of recruitment, retention, and intervention group attendance and numbers of technology-related issues and participant-reported adverse events. The recruitment rate was defined as the number of participants recruited and randomized per site, per month. The recruitment rate was considered to be feasible if it met the median level of 0.92 participants/center/month [49]. The retention rate was defined as the percentage of participants who completed all endpoint measures. A retention rate was considered to be feasible if 80% or more participants completed the final study assessment [50]. Other aspects of feasibility included the group attendance rate. The benchmark was set at 75% or more participants attending 10 or more sessions for intervention group participants [18,51]. Lastly, we recorded the number of reported technology-related issues and adverse events. The feasibility data were drawn from a database maintenance by the study's research coordinator.

2.7.2. Acceptability

Similar to our in-person study [19], the acceptability of the virtually delivered exergame- and group-based PA intervention was assessed by using items drawn from Vandelanotte et al. [52,53] and Lyons et al. [35]. The acceptability questionnaire was distributed at time 1 and time 2 (Figure 2). The questionnaire consisted of 11 questions to assess participants' agreement on a scale of 1 (strongly disagree) to 5 (strongly agree) regarding the use of exergames, the PA behavioral coaching materials, the BC support discussion topics, and satisfaction with the overall program [19]. The program was deemed acceptable if responses were 4 or higher.

2.8. Secondary Outcome Measures

2.8.1. Objective Physical Function Measures

Details of the remotely assessed physical function measures used in the current study are presented in Table 1. The objective physical function measures we conducted remotely included the Short Physical Performance Battery (SPPB) [54], Timed Up & Go (TUG) [55–57], and the 2-min step test [58,59]. Handouts providing detailed set-up instructions were provided to the participants ahead of time. The selected objective physical function measures have been previously administered remotely in published studies by Blair et al., Guidarelli et al., and Hoenemeyer et al. [30,31,60]. Blair et al. have been using the 30 s chair stand test, which is similar to the five times sit-to-stand test in SPPB and the TUG test via videoconferencing in older survivors of cancer [31]. The trial has been impacted by the COVID-19 pandemic. Guidarelli et al. tested the SPPB and TUG using videoconference in adults with and without cancer. They found good agreement with in-person tests (ICC = 0.88) for TUG and substantial agreement between repeat assessments for total SPPB score (Cohen's kappa of 0.78) [30]. Hoenemeyer et al. also found strong agreement (ICC = 0.74) for TUG and very strong agreement (ICC = 0.87) for the 2-min step test among cancer survivors and their partners [60]. Detailed descriptions of how the assessor set up the assessments are included in the Supplemental Material (Table S1).

Table 1. Details of remotely assessed objective physical function measures.

Name of the Assessment	Description	Detailed Descriptions
The Short Physical Performance Battery (SPPB)	Score range: 0 to 12 Consist of 3 components: 1. Balance (score range from 0 to 4) 2. Timed 10 feet walk (score range from 0 to 4) 3. Timed 5-repeated chair stands (score range from 0 to 4)	1. Balance: side-by-side, semi-tandem, and tandem 2. Timed 10 feet walk: Fastest time of 2 10 feet usual-pace walk 3. Timed 5 repeated chair stands: repeat 5 times of rising from a chair with arms folded across the chest
Timed Up & Go TUG)	Score range: ≤10 s = normal; ≤20 s = good mobility without gait aid; ≤30 s = problems, requires gait aid; ≥14 s is associated with high fall risk	Participants need to stand up from a chair and move as quickly as participants feel safe until the participant passes a tape that is 10 feet from the chair. Then, they turn around and walk back to the chair and sit back down.
Two-minute step test	Norm for 60–64: 75–107 steps march in place for 2 min	Participants need to march in place, but the knee needs to hit the halfway mark between participants' iliac crest and patella height.

2.8.2. Physical Activity Measures

The PA metrics used in this pilot study included mean steps per day and mean minutes of moderate to vigorous PA (MVPA) per day. These measurements were obtained using an ActiGraph GT9X Link accelerometer around the waist for 7 days at each time point. ActiGraph is a validated research-grade 3-axis accelerometer. The Troiano algorithm within the ActiLife software was used to estimate wear time and activity. A cut point of 10 h was used to deemed as a valid wear day. Additionally, we used Keadle et al.'s accelerometer processing data for this phase II pilot study [61]. The Fitbit and Xiamoi Mi Band 3 step counts were used as self-monitoring tools only. They were not used in the PA outcome assessment.

2.8.3. Other Patient Reported Measures

Demographics, such as age, assigned sex at birth, race/ethnicity identity, and cancer diagnosis were self-reported using paper-based questionnaires.

2.9. Statistical Analysis

The goal of this study was to determine the intervention's feasibility and acceptability and to obtain exploratory pilot data to inform the design of a larger intervention study. Thus, this pilot study was not designed to have sufficient power to detect significant differences in physical function outcomes and PA. The study's primary outcomes are feasibility and acceptability. Feasibility consisted of three components. First is the recruitment rate, second is the retention rate, and third is the group attendance rate. As previously indicated, the a priori feasibility benchmark for the recruitment rate is 0.92 participants/center/month [49], the retention rate is set at 80% or more participants completed the final study assessment [18], and group attendance is set at 75% or more attending 10 or more sessions for intervention group participants [51]. As previously indicated, the a priori acceptability benchmark is based on self-reported scores of 4 or higher for all 11 acceptability questions [19].

For assessing and comparing characteristic distributions in our samples, we used Chi-squared and Fisher's exact test as appropriate for categorical data and *t*-tests for continuous variables. Feasibility indicators were assessed with descriptive statistics, namely frequency and percentage. The recruitment rate was calculated based on the number of participants per center per month. The retention rate was calculated as the number of total participants who completed the final assessment divided by total number of participants enrolled and randomized and multiplied by 100. The group attendance rate was calculated as the total number of participants who completed 10 or more sessions divided by the total number of participants for the intervention or control groups and multiplied by 100.

To assess secondary outcomes, we computed the difference between the last measurement and baseline for our continuous data. We report the mean of this difference, the mean baseline, and the mean of the last follow up with their standard deviations. We used these means and standard deviations to compute Cohen's d effect size to facilitate power calculations for future studies. Data were analyzed using SPSS v 24 (IBM Corp., Armonk, NY, USA). Cohen's d (effect size) was calculated using the effect size calculator provided by Lipsey and Wilson [62]. We took an intention-to-treat approach for study analyses and used last-observation-carried forward for missing data.

3. Results

3.1. Participants' Characteristics

Table 2 summarizes the participants' characteristics. Eighty percent of the participants were non-Hispanic white. The mean age was 63.75 (SD 6.35). One participant dropped out immediately after randomization, so we were not able to obtain information from the participant beyond age and race/ethnicity variables. Another participant's baseline questionnaire was lost in the mail after the participant returned the questionnaires to the team via the USPS courier service. Multiple attempts by the team through various routes (e.g., SMS text message, email, or phone call) and at different times of the day were made (approximately 5 times on average per assessment time point) when equipment and/or questionnaires were not returned on time. We also offer to complete the questionnaire on the phone. The intent was to minimize missing data. Despite our best effort, the participant refused to complete the baseline questionnaire again. The BMI on average was 31.89 (6.04), which is considered to be in the obesity range. The majority of participants (89%) had completed active cancer treatment at baseline, and time since diagnosis averaged 96.11 months (i.e., approximately 8 years). The range was from 2 months to 284 months. No patients reported adverse events related to the intervention.

Table 2. Participant characteristics.

Characteristic	**Total (*N* = 20)**	**Intervention (*N* = 10)**	**Control (*N* = 10)**	***p*-Value [a]**
Race/ethnicity (n = 20; n; %)				
Non-Hispanic White	16 (80)	8 (80)	8 (80)	
African American	2 (10)	0 (0)	2 (20)	
Hispanic	1 (5)	1 (10)	0 (0)	0.474
Other	1 (5)	1 (10)	0 (0)	
Stage (n = 18; n; %)				
0	2 (11.1)	1 (10)	1 (12.5)	
I	8 (44.4)	2 (20)	6 (75)	
II	4 (22.2)	4 (40)	0 (0)	0.106
III	4 (22.2)	3 (30)	1 (12.5)	
Treatment type (n = 18; n; %)				
Surgery only	1 (5.6)	1 (10)	0 (0)	
Surgery and chemotherapy	4 (22.2)	2 (20)	2 (25)	
Surgery, chemotherapy, and radiation	8 (44.4)	5 (50)	3 (37.5)	0.904
Surgery and radiation	5 (27.8)	2 (20)	3 (37.5)	
Current treatment status (n = 18; n; %)				
Off treatment	16 (89)	8 (80)	8 (100)	0.477
On treatment	2 (11)	2 (20)	0 (36.67)	
Patient-reported neuropathy (n = 18; n; %)				
Yes	6 (33.3)	4 (40)	2 (25)	0.638
No	12 (66.7)	6 (60)	6 (75)	
Age (n = 20; years, range 55–79; mean; SD)	63.75 (6.35)	64.90 (8.03)	62.60 (4.20)	0.43
Time since diagnosis (n = 18; months; mean; SD)	96.11 (82.61) Range: 2–284 months	113.70 (92.99)	74.13 (66.81)	0.33
BMI (n = 19; kg/m^2; mean; SD)	31.89 (6.04)	33.91 (7.11)	29.66 (3.80)	0.13

[a] *p*-values calculated using Fisher's exact test for categorical variables and the two-sample *t*-test for continuous variables.

3.2. *Feasibility and Acceptability*

3.2.1. Feasibility

Recruitment lasted 14 months at a single site, and the recruitment rate was 1.93 participants/center/month [49]. All 10 participants in the intervention group and eight of the 10 control group participants remained in the study and completed the final assessment (Figure 1). One participant dropped out immediately after randomization into the control group. The other control group participant dropped out right before the final assessment visit (time 2) because of caregiving demands. The mean age of the two participants who dropped out was 64 (SD 9.90), which is similar to the mean age of 18 participants who remained in the study (mean of 63.7 with SD 6.27). Both participants who dropped out were also non-Hispanic white. Among the 18 participants who remained in the study, 77.8% were non-Hispanic white. The overall retention rate was 90% (18/20). As for the group attendance rate, 88% of the intervention group participants attended 10 or more sessions, and 82% of the control group participants participated in 10 or more group calls. On average, the intervention group participants attended 11.4 sessions. As for the virtual assessments, 96% of the 56 virtual assessments were conducted without issues. Two of the 56 virtual assessments experienced internet connectivity issues. To minimize missing data collection, the team would then help to problem solve connectivity issues by turning off the assessor's video or helped participants restart their internet router/connecting phone to a reliable WIFI hotspot.

Eight assessments were delayed (14%). Five were delayed because ActiGraphs were lost in the mail. The team was not able to recover a total of five ActiGraphs. As previously noted, the team switched to a courier service (FedEx) that provided a more reliable tracking service after the team encountered five unexpected mailing delays and loss of equipment using another mailing service. The tracking service also alerted the team to contact the research participants and remind them to complete the questionnaires and wear the ActiGraph for a minimum of 10 h a day over the next 7 days. The tracking system from FedEx also helped the team recover a few ActiGraphs that were accidentally sent to the wrong locations. Beyond loss of equipment, three assessments were delayed because either participants or their family members contracted SARS-CoV-2. We were not able to provide an ActiGraph during an active SARS-CoV-2 infection. These types of delays were not something that the team can control due to infection concerns. To minimize the lag time, the team stayed in communication with the participants on a weekly basis. The ActiGraph and questionnaires were sent out immediately as soon participants or family members recovered and were testing negative for SARS-CoV-2.

A total of two intervention group sessions experienced Zoom-related issues. Under such circumstances, we would resend the invitation links to the participants and troubleshoot on the spot to help the participants connect so they do not miss an intervention session. Overall, participants were able to follow the facilitator via SecureVideo to play the Yoga, Zuma, Kickboxing type of games on Xbox Kinect 360.

3.2.2. Acceptability

Acceptance results were generated based on the 11 questions listed in Figure 4. The results indicated that overall, Pink Warrior 2, the exergame-based physical activity intervention, was acceptable. All 10 BC survivors in the intervention group rated their acceptance as 4 or higher on a 5-point scale for all 11 questions (Figure 4). Specially, they liked the exergame portion (mean of 4.4, SD of 0.84) (Figure 4; Table S2) and found that the activities were appropriate (mean of 4.7, SD of 0.68) (Figure 4; Table S2). Examples of feedback from participants after the intervention are included in the Supplemental Material (Tables S2–S4).

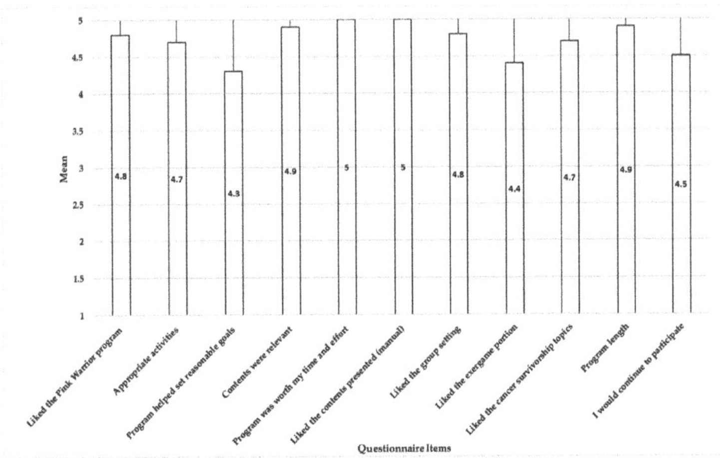

Figure 4. Acceptability of the Pink Warrior 2 intervention (time 2; n = 10).

3.3. Physical Function and PA (Exploratory Results)

Table 3 shows the exploratory results of the physical function and PA assessments for the intervention and control groups. The computed mean at baseline, the mean of the last follow-up, and the mean of the differences between baseline and follow-up with their associated standard deviations by intervention and control groups are presented in Table 3. Intention-to-treat analyses included imputing one measurement using the last measured changes available carried forward for only one participant. The overall analytical sample is n = 19. However, for the TUG and two-minute step test, the overall analytical sample is n = 18. Despite having an assessment checklist and an electronic form, one of the research coordinators did not record the information for TUG and the two-minute step test for one of the participants at baseline.

Table 3. Differences between intervention and control groups (intention-to-treat analysis).

Variables	Intervention			Control			Effect Size
	Baseline Mean (SD)	Follow-Up Mean (SD)	Mean of Difference (SD)	Baseline Mean (SD)	Follow-Up Mean (SD)	Mean of Difference (SD)	Cohen's d
Gait speed (meter/seconds); n = 19	0.76 (0.24)	0.94 (0.17)	0.18 (0.17)	0.89 (0.18)	1.01 (0.15)	0.11 (0.13)	0.46
Total SPPB [a] score; n = 19	8.70 (1.57)	10.30 (1.34)	1.6 (1.17)	9.56 (1.59)	10 (1.12)	0.44 (1.01)	1.06
TUG [b] (seconds); n = 18	10.46 (3.52)	9.78 (3.11)	−0.69 (0.91)	9.12 (1.73)	8.93 (0.85)	−0.01 (2.06)	0.43
Two-minute step test (count); n = 18	62.89 (21.69)	75.0 (24.26)	12.11 (13.83)	75.89 (30.98)	76.11 (28.81)	0.22 (24.11)	0.61
Steps (average steps); n = 19	4652.60 (2659.88)	4423.09 (2016.41)	−229.52 (1905.94)	4268.52 (1721.36)	5838.69 (2767.52)	1570.17 (2355.59)	0.85
MVPA [c] (average minutes); n = 19	9.4689 (9.93)	10.00 (9.13)	0.54 (8.78)	12.07 (13.67)	17.34 (23.09)	5.28 (23.66)	0.27

[a] SPPB: Short Physical Performance Battery; [b] TUG: Timed Up & Go; [c] moderate-vigorous physical activity.

Both groups started the study below the normative values of 0.99 for women 60–69 years old for gait speed [63]. The intervention group showed a greater than 1.0 point increase in SPPB score [64]. The TUG results at baseline indicated that the both groups were at increased risk for the development of disability [65]. In the 2-min step test, the intervention group started below the normative value ranges from 75 to 107 for women 60–69 years old [59]. At follow-up, the intervention group's step counts fell within the normative range. ActiGraph data from both groups showed higher MVPA, but lower step counts were found among the intervention group participants.

4. Discussion

Overall, our results demonstrated that the remotely delivered exergame- and group-based PA intervention was feasible and acceptable in a group of older BC survivors, receiving treatment or not, during the COVID-19 pandemic. The study's feasibility was demonstrated by its recruitment rate (1.93), which was above the benchmark level of 0.92 participants/center/total number of recruitment months; 90% overall retention at the end of the study; 88% adherence rate among individuals in the intervention group; minimal technological difficulties, with 96% of the 56 virtual assessments conducted without problems; and no intervention-related adverse events. The study's acceptability was demonstrated by its mean acceptability scores, which were greater than 4 of 5 for all questions. As for the exploratory aim, both the Pink Warrior 2 intervention and attention control intervention appeared capable of producing increases in PA and function. Surprisingly, the control group participants showed an increase in steps, but the intervention group did not. Both groups had a slight increase in MVPA.

Although our enrollment met the benchmark, it is considered to be on the low end of the enrollment rate range [49]. One of the reasons for the low recruitment rate may be related to a decrease in the willingness to participate in research during the COVID-19 pandemic. Published studies indicated a lower desire to participate in research globally and a lower response rate than the pre-pandemic response rate [66–68]. The retention rate for the current study was at 90%, which is comparable to our previously published study [19] and is within the range (50–100%) of previous exergame-based interventions completed among cancer survivors [18]. Similar to Singh et al., a review that examined PA studies globally (e.g., USA, India, Spain, and Canada), and our previous study [19,50], the present study's control group had a higher dropout rate than the intervention group. In fact, our intervention group's adherence rate (88%) compared favorably to that reported by Singh et al. (81%) [50] and is within the range of retention rates (62–96.6%) of other PA interventions that included cancer survivors who were and were not currently receiving treatment [18,51].

The mean acceptability scores for all questions related to the Pink Warrior 2 intervention were $\geq$4, which was similar to the response to our in-person exergame- and group-based PA intervention [19]. One of the potential reasons why the Pink Warrior 2 intervention was deemed acceptable for all 11 acceptability questions was that we demonstrated modifications participants can do for the movements shown in the exergames. Similar to an exergame study completed in older adults in New Zealand, our BC survivors were able to play either standing or sitting [69]. Thus, we considered the physical abilities of various participants at each session in order to promote participants' competence and desire to engage in doing the exergame activities based on their physical abilities, which can lead to better physical function. An additional potential reason may be related to increased enjoyment in participants when using the exergame as a tool to promote PA. Two reviews by Silva et al. [17] and Tough et al. [18] that examined the exergame interventions for persons with cancer globally (e.g., USA, South Korea, Japan, Germany, and Canada) found exergaming interventions to be more acceptable than standard of care, and they appear to improve balance, physical function, physical performance, PA levels, and reduced pain in persons with cancer. In comparison, our control group response indicated that only the appropriate activities and program length questions reach acceptability scores $\geq$4. The free-text responses suggested that the participants in the Pink Warrior 2 intervention in general felt connected with other participants and liked the intervention. The control group, however, did not like the survivorship navigation materials as much.

Our exploratory analysis of physical function outcomes suggested that the participants benefited from the virtually delivered exergame- and group-based PA intervention. Both intervention and control group participants were below the normative values of 0.99 m/second gait speed for women ages 60–69 years [63]. Both the intervention and control groups showed a clinically important change of $\geq$0.11 m/second [70]. A 0.11 m/second gait speed increase has been shown to be associated with a decreased risk in morbidity and mortality [70]. In addition, a 1.0 point increase in SPPB score was seen in the intervention group (d = 1.06). Brown et al. [71] observed that a 1.0 point increase in SPPB score was associated with a 12% decrease in mortality risk among survivors of cancer [71]. The TUG results at baseline ($\geq$9 s) indicate that both intervention and control group participants were at increased risk for the development of disability [65]. However, both groups improved. The intervention group participants had a 0.69 s decrease in time, while the control group participants had a 0.01 s decrease in time (d = 0.43). In the 2-min step test, the intervention group started below the normative value range (75–107) but reached the normative value at the follow-up assessment [59]. We also want to highlight that the attention control group also showed improvement in physical function. This suggest the possibility of a non-specific effect of the attention control intervention.

Surprisingly, the improvements in these objectively measured outcomes for our intervention group were not matched by the group's average step count and MVPA. This exploratory finding is in contrast with what we found in our in-person intervention, where we found an increase in steps and MVPA among intervention group participants [19]. Potential reasons for current findings may be due to SARS-CoV2–related challenges [72–75] or the use of ActiGraph among participants with slow walking speed [76]. SARS-CoV2 may have been a factor because the literature indicated a worldwide decrease in PA during the initial lockdown [72–74]. Additionally, Bu et al. found that despite the lifting of the initial COVID-19 lockdown, there was a steady increase in the percentage of people who continued to report not engaging in any PA [73]. Furthermore, four participants in the intervention group either had SARS-CoV2 or had family members in the same household who had SARS-CoV2. In contrast, one participant in the control group reported a household member diagnosed with SARS-CoV2. Given that recent reviews indicated that SARS-CoV2 is associated with decreases in PA and mobility [75,77], we hypothesize that this may be the underlying reason for the decrease in PA among our intervention group participants. Another potential reason for the low PA level may be related to the use of ActiGraph. Hergenroeder et al. found that ActiGraph significantly undercounted steps among individuals with slower walking speeds of <1.0 m/second [76]. This finding informs our future selection of activity monitor when designing PA interventions among individuals who may have mobility limitations.

Overall, our pilot trial had several strengths. First, it involved an innovative intervention design of delivering exergame-type activities via livestream. This was accepted by the intervention group participants, which aligns with the finding from Lin [44]. The weekly group PA sessions via the videoconference platform also did not affect the acceptability of the Pink Warrior 2 intervention compared with our in-person group findings [19]. Second, we contributed to the accumulating literature indicating that it is possible to safely conduct objective mobility, aerobic endurance, and functional fitness assessments using a videoconferencing platform [30,31]. Furthermore, supplementing the self-report measures with objective functional measure may enrich study findings, for self-reported and objective functional measures provide related but distinct information regarding an individual's physical function [78]. Additionally, the objective functional measures may be able to capture more physical function limitations than the self-report measures [79].

Our study also had limitations related to the study design. Thus, our results need to be interpreted with caution. First, this pilot study had a small sample size and short duration. Thus, the study lacked the statistical power to detect significant differences in outcome measures and could not perform long-term monitoring of PA maintenance. We are also not able to conduct subgroup analyses to evaluate the potential effect of SARS-CoV2 in our intervention group. Second, we are not able to determine the effect of individual components of the intervention since we aimed to assess the full intervention's feasibility and acceptability. Thus, our focus was on developing the most effective and deliverable multicomponent program remotely rather than specific intervention components. Third, our last observation carried forward has been known to underestimate effect sizes; however, our study only applied the last observation carried forward method for one individual missing one assessment. Therefore, we expect such underestimation, if any, would be minimal for our effect size estimates. Additionally, using an underestimated effect size would still result in adequate power when designing a future study. Last, the pilot trial was conducted in the southeastern Texas area. Therefore, it is limited in generalizability. However, the initial evidence will be used to inform a larger and more generalizable trial. Despite the limitations of our pilot study, we found initial evidence that our remotely delivered exergame- and group-based PA intervention was able to produce moderate to large effect size and clinically important changes on physical function, which provided initial evidence that a larger-sample trial with modifications is warranted (e.g., using a more sensitive activity monitor for populations with slower walking speed).

Our team did face several SARS-CoV2-related challenges. First, the use of a paper-based questionnaire presented a challenge. Although administrating the paper-based questionnaires was successfully implemented previously for our in-person intervention [19], we faced several logistical challenges when using paper-based questionnaires remotely during the COVID-19 pandemic. For example, there were mailing delays and a loss of questionnaire packages. Previously, when we had in-person visits, the team reviewed the questionnaires with the participants in person. Therefore, we were able to obtain missing data information immediately. Despite our efforts to contact the participants via phone, emailing the questionnaires to the participants, or offer to complete the questionnaires at assessments, parts of the four baseline questionnaires (e.g., cancer stage information or treatment information) remained missing. Thus, the team was able to obtain some of the clinical information through the electronic health record if the participant received care within the study team's health system. With the mailing delays or loss of questionnaires, the participants were less willing to complete the full questionnaires either by phone, through emails, or at the assessment times again. Our team has since translated all questionnaires to be administered via REDCap for future studies to minimize missing data issues. Second, we also experienced losses in equipment. We were not able to recover five ActiGraphs. Third, we also experience a delay in assessment due to active SARS-CoV2 infection in the household. Overall, delays in assessments may have impacted the results, such as not having an accurate baseline and not having accurate post-intervention assessments for secondary outcomes. Our team has since changed our courier service to one with better tracking notifications to prevent further mailing delays and equipment loss. Fourth, we also experienced missing objectively collected data for one participant at baseline despite the training of coordinators, a co-development of assessment checklist, and having an electronic form. One of the coordinators was initially hired to conduct the assessments. The team did not find out about the missing data until the coordinator resigned from the position in the midst of the pandemic. Based on this lesson learned, we have since also set up documentations in the REDCap and added skip prevention to minimize missing data issues during data collection for future studies. In spite of challenges, we were able to meet the feasibility and acceptability metrics set a priori.

5. Conclusions

In summary, our findings lend initial evidence that a virtually delivered multicomponent PA intervention that includes exergame group play, PA behavioral coaching, and BC support is feasible and acceptable to older BC survivors regardless of their current treatment status. Additionally, our exploratory findings indicate potential physical function benefits in BC survivors, and consequently, a potential reduction in mortality of the Pink Warrior 2 intervention. However, our initial findings will need to be verified in a larger study. Using such technology can help overcome some of the limitations to PA program access experienced by older adults [37]. Additionally, we contributed to the accumulating evidence indicating that objective physical function measures can be conducted among a population with lower physical function level [30,31]. Future study is warranted to determine the effect of exergame- and group-based PA on physical function in survivors of BC. We also need to explore how to integrate the use of exergame and PA behavior coaching into cancer support groups to extend the reach of evidence-based PA programs to the wider population of cancer survivors. Lastly, our study findings can be used as initial evidence for future studies to explore its application in the clinical setting, and we can take the approach we have used in the current study to develop for use in other populations and diseases.

Supplementary Materials: The following supporting information can be downloaded at: https://www.mdpi.com/article/10.3390/life13020574/s1, Table S1: Detailed descriptions of how the SPPB, TUG, and 2-min step test were conducted virtually; Table S2: Participants' written feedback regarding the Pink Warrior 2 and the support programs; Table S3: Acceptability of the Pink Warrior 2 intervention (time 2; n = 10); Table S4: Acceptability of the UTMB support group program (time 2; n = 7); Table S5: Consort 2010 checklist of information for reporting a pilot or feasibility trial.

Author Contributions: Conceptualization, M.C.S.; methodology, M.C.S., U.C., Z.H.L. and E.J.L.; formal analysis, M.C.S., M.C.R. and M.D.S.; investigation, M.C.S., U.C., S.J.W., E.M., H.C.S. and E.J.L.; resources, M.C.S. and E.J.L.; data curation, M.C.S., S.J.W., E.M. and U.C.; writing—original draft preparation, M.C.S., M.C.R. and M.D.S.; writing—review and editing, U.C., S.J.W., J.B., Z.H.L., E.M. and H.C.S.; visualization, M.C.S. and S.J.W.; supervision, M.C.S. and E.J.L.; project administration, M.C.S., S.J.W. and E.M.; funding acquisition, M.C.S. and E.J.L. All authors have read and agreed to the published version of the manuscript.

Funding: This research was funded by the UTMB Claude D. Pepper Older Americans Independence Center NIH grant (P30 AG024832, PI: E. Volpi). Salary support provided, in part, by a Cancer Center Support Grant (CA16672, PI: P. Pisters, MDACC), from the NCI/NIH.

Institutional Review Board Statement: The study was conducted in accordance with the Declaration of Helsinki and approved by the Institutional Review Board (or Ethics Committee) of University of Texas Medical Branch (protocol code 16-0040-02 and approved on 7 May 2020) for studies involving humans.

Informed Consent Statement: Informed consent was obtained from all subjects involved in the study.

Data Availability Statement: Requests for data may be sent to the corresponding author. Data would be made available following the University of Texas Medical Branch Data Sharing Policy.

Acknowledgments: The authors would like to thank the following individuals for their assistance with participant recruitment, data collection, and data management: Jason Bentley, V Suzanne Klimberg, Sandra Hatch, Bryan Tutt provided editorial support.

Conflicts of Interest: The authors declare no conflict of interest.

References

1. Sulik, G. What cancer survivorship means. *AMA J. Ethics* **2013**, *15*, 697–703.
2. American Cancer Society. Cancer Treatment & Survivorship Facts & Figures 2022–2024. Available online: https://www.cancer.org/content/dam/cancer-org/research/cancer-facts-and-statistics/cancer-treatment-and-survivorship-facts-and-figures/2022-cancer-treatment-and-survivorship-fandf-acs.pdf (accessed on 24 August 2022).
3. Schmitz, K.H.; Cappola, A.R.; Stricker, C.T.; Sweeney, C.; Norman, S.A. The Intersection of Cancer and Aging: Establishing the Need for Breast Cancer Rehabilitation. *Cancer Epidemiol. Biomark. Prev.* **2007**, *16*, 866–872. [CrossRef]
4. Cespedes Feliciano, E.M.; Vasan, S.; Luo, J.; Binder, A.M.; Chlebowski, R.T.; Quesenberry, C.; Banack, H.R.; Caan, B.J.; Paskett, E.D.; Williams, G.R.; et al. Long-term Trajectories of Physical Function Decline in Women with and without Cancer. *JAMA Oncol.* **2023**. [CrossRef]
5. Braithwaite, D.; Satariano, W.A.; Sternfeld, B.; Hiatt, R.A.; Ganz, P.A.; Kerlikowske, K.; Moore, D.H.; Slattery, M.L.; Tammemagi, M.; Castillo, A.; et al. Long-term Prognostic Role of Functional Limitations Among Women With Breast Cancer. *JNCI J. Natl. Cancer Inst.* **2010**, *102*, 1468–1477. [CrossRef]
6. BLAIR, C.K.; MOREY, M.C.; DESMOND, R.A.; COHEN, H.J.; SLOANE, R.; SNYDER, D.C.; DEMARK-WAHNEFRIED, W. Light-Intensity Activity Attenuates Functional Decline in Older Cancer Survivors. *Med. Sci. Sport. Exerc.* **2014**, *46*, 1375–1383. [CrossRef]
7. Caspersen, C.J.; Powell, K.E.; Christenson, G.M. Physical activity, exercise, and physical fitness: Definitions and distinctions for health-related research. *Public Health Rep.* **1985**, *100*, 126–131.
8. Jung, A.Y.; Behrens, S.; Schmidt, M.; Thoene, K.; Obi, N.; Hüsing, A.; Benner, A.; Steindorf, K.; Chang-Claude, J. Pre- to postdiagnosis leisure-time physical activity and prognosis in postmenopausal breast cancer survivors. *Breast Cancer Res.* **2019**, *21*, 117. [CrossRef]
9. Cannioto, R.A.; Hutson, A.; Dighe, S.; McCann, W.; McCann, S.E.; Zirpoli, G.R.; Barlow, W.; Kelly, K.M.; DeNysschen, C.A.; Hershman, D.L.; et al. Physical Activity Before, During, and After Chemotherapy for High-Risk Breast Cancer: Relationships With Survival. *JNCI J. Natl. Cancer Inst.* **2020**, *113*, 54–63. [CrossRef]
10. Hefferon, K.; Murphy, H.; McLeod, J.; Mutrie, N.; Campbell, A. Understanding barriers to exercise implementation 5-year post-breast cancer diagnosis: A large-scale qualitative study. *Health Educ. Res.* **2013**, *28*, 843–856. [CrossRef]
11. Gomes, M.L.B.; Pinto, S.S.; Domingues, M.R. Barriers to physical activity in women with and without breast cancer. *ABCS Health Sci.* **2020**, *45*, e020022. [CrossRef]
12. Rogers, L.Q.; Courneya, K.S.; Shah, P.; Dunnington, G.; Hopkins-Price, P. Exercise stage of change, barriers, expectations, values and preferences among breast cancer patients during treatment: A pilot study. *Eur. J. Cancer Care* **2007**, *16*, 55–66. [CrossRef]
13. Wurz, A.; St-Aubin, A.; Brunet, J. Breast cancer survivors' barriers and motives for participating in a group-based physical activity program offered in the community. *Support. Care Cancer* **2015**, *23*, 2407–2416. [CrossRef]
14. Spiteri, K.; Broom, D.; Hassan Bekhet, A.; Xerri de Caro, J.; Laventure, B.; Grafton, K. Barriers and Motivators of Physical Activity Participation in Middle-Aged and Older Adults—A Systematic Review. *J. Aging Phys. Act.* **2019**, *27*, 929–944. [CrossRef]
15. Bandura, A. Social cognitive theory of self-regulation. *Organ. Behav. Hum. Decis. Process.* **1991**, *50*, 248–287. [CrossRef]
16. Whitehead, S.; Lavelle, K. Older Breast Cancer Survivors' Views and Preferences for Physical Activity. *Qual. Health Res.* **2009**, *19*, 894–906. [CrossRef]
17. Silva, A.P.; Oliveira, E.M.d.; Okubo, R.; Benetti, M. Utilização de exergames e seus efeitos sobre a saúde física de pacientes com diagnóstico de câncer: Uma revisão integrativa. *Fisioterapia e Pesquisa* **2020**, *27*, 443–452. [CrossRef]
18. Tough, D.; Robinson, J.; Gowling, S.; Raby, P.; Dixon, J.; Harrison, S.L. The feasibility, acceptability and outcomes of exergaming among individuals with cancer: A systematic review. *BMC Cancer* **2018**, *18*, 1151. [CrossRef]

19. Swartz, M.C.; Lewis, Z.H.; Deer, R.R.; Stahl, A.L.; Swartz, M.D.; Christopherson, U.; Basen-Engquist, K.; Wells, S.J.; Silva, H.C.; Lyons, E.J. Feasibility and Acceptability of an Active Video Game-Based Physical Activity Support Group (Pink Warrior) for Survivors of Breast Cancer: Randomized Controlled Pilot Trial. *JMIR Cancer* **2022**, *8*, e36889. [CrossRef]
20. Lieberman Debra, A.; Chamberlin, B.; Medina, E.; Franklin Barry, A.; Sanner Brigid, M.; Vafiadis Dorothea, K. The Power of Play: Innovations in Getting Active Summit 2011. *Circulation* **2011**, *123*, 2507–2516. [CrossRef]
21. Costa, M.T.S.; Vieira, L.P.; Barbosa, E.O.; Mendes Oliveira, L.; Maillot, P.; Ottero Vaghetti, C.A.; Giovani Carta, M.; Machado, S.; Gatica-Rojas, V.; Monteiro-Junior, R.S. Virtual Reality-Based Exercise with Exergames as Medicine in Different Contexts: A Short Review. *Clin. Pract. Epidemiol. Ment Health* **2019**, *15*, 15–20. [CrossRef]
22. Fortier, M.S.; Williams, G.C.; Sweet, S.N.; Patrick, H. Self-Determination theory: Process models for health behavior change. In *Emerging Theories in Health Promotion Practice and Research*, 2nd ed.; Jossey-Bass/Wiley: Hoboken, NJ, USA, 2009; pp. 157–183.
23. Milne, H.M.; Wallman, K.E.; Guilfoyle, A.; Gordon, S.; Corneya, K.S. Self-determination theory and physical activity among breast cancer survivors. *J. Sport Exerc. Psychol.* **2008**, *30*, 23–38. [CrossRef]
24. Ryan, R.M.; Deci, E.L. Self-determination theory and the facilitation of intrinsic motivation, social development, and well-being. *Am. Psychol.* **2000**, *55*, 68–78. [CrossRef]
25. Teixeira, P.J.; Carraca, E.V.; Markland, D.; Silva, M.N.; Ryan, R.M. Exercise, physical activity, and self-determination theory: A systematic review. *Int. J. Behav. Nutr. Phys. Act* **2012**, *9*, 78. [CrossRef]
26. Teixeira, P.J.; Marques, M.M.; Silva, M.N.; Brunet, J.; Duda, J.L.; Haerens, L.; La Guardia, J.; Lindwall, M.; Lonsdale, C.; Markland, D.; et al. A classification of motivation and behavior change techniques used in self-determination theory-based interventions in health contexts. *Motiv. Sci.* **2020**, *6*, 438–455. [CrossRef]
27. Stevens, C.J.; Baldwin, A.S.; Bryan, A.D.; Conner, M.; Rhodes, R.E.; Williams, D.M. Affective Determinants of Physical Activity: A Conceptual Framework and Narrative Review. *Front. Psychol.* **2020**, *11*, 568331. [CrossRef]
28. Zimmerling, A.; Chen, X. Innovation and possible long-term impact driven by COVID-19: Manufacturing, personal protective equipment and digital technologies. *Technol. Soc.* **2021**, *65*, 101541. [CrossRef]
29. Ross, K.M.; Carpenter, C.A.; Arroyo, K.M.; Shankar, M.N.; Yi, F.; Qiu, P.; Anthony, L.; Ruiz, J.; Perri, M.G. Impact of transition from face-to-face to telehealth on behavioral obesity treatment during the COVID-19 pandemic. *Obesity* **2022**, *30*, 858–863. [CrossRef]
30. Guidarelli, C.; Lipps, C.; Stoyles, S.; Dieckmann, N.F.; Winters-Stone, K.M. Remote administration of physical performance tests among persons with and without a cancer history: Establishing reliability and agreement with in-person assessment. *J. Geriatr. Oncol.* **2022**, *13*, 691–697. [CrossRef]
31. Blair, C.K.; Harding, E.; Herman, C.; Boyce, T.; Demark-Wahnefried, W.; Davis, S.; Kinney, A.Y.; Pankratz, V.S. Remote Assessment of Functional Mobility and Strength in Older Cancer Survivors: Protocol for a Validity and Reliability Study. *JMIR Res. Protoc.* **2020**, *9*, e20834. [CrossRef]
32. Eldridge, S.M.; Chan, C.L.; Campbell, M.J.; Bond, C.M.; Hopewell, S.; Thabane, L.; Lancaster, G.A. CONSORT 2010 statement: Extension to randomised pilot and feasibility trials. *BMJ* **2016**, *355*, i5239. [CrossRef]
33. Czajkowski, S.M.; Powell, L.H.; Adler, N.; Naar-King, S.; Reynolds, K.D.; Hunter, C.M.; Laraia, B.; Olster, D.H.; Perna, F.M.; Peterson, J.C.; et al. From ideas to efficacy: The ORBIT model for developing behavioral treatments for chronic diseases. *Health Psychol.* **2015**, *34*, 971–982. [CrossRef]
34. Kendall, J.M. Designing a research project: Randomised controlled trials and their principles. *Emerg. Med. J.* **2003**, *20*, 164–168. [CrossRef]
35. Lyons, E.J.; Swartz, M.C.; Lewis, Z.H.; Martinez, E.; Jennings, K. Feasibility and Acceptability of a Wearable Technology Physical Activity Intervention With Telephone Counseling for Mid-Aged and Older Adults: A Randomized Controlled Pilot Trial. *JMIR Mhealth Uhealth* **2017**, *5*, e28. [CrossRef]
36. Haahr, M. Random. org: True Random Number Service. Available online: Random.org (accessed on 1 October 2022).
37. Adams, M.M.; Davis, P.G.; Gill, D.L. A Hybrid Online Intervention for Reducing Sedentary Behavior in Obese Women. *Front. Public Health* **2013**, *1*, 45. [CrossRef]
38. Setera, K. FBI Warns of Teleconferencing and Online Classroom Hijacking During COVID-19 Pandemic. Available online: https://www.fbi.gov/contact-us/field-offices/boston/news/press-releases/fbi-warns-of-teleconferencing-and-online-classroom-hijacking-during-covid-19-pandemic (accessed on 26 December 2022).
39. Basen-Engquist, K.; Taylor, C.L.C.; Rosenblum, C.; Smith, M.A.; Shinn, E.H.; Greisinger, A.; Gregg, X.; Massey, P.; Valero, V.; Rivera, E. Randomized pilot test of a lifestyle physical activity intervention for breast cancer survivors. *Patient Educ. Couns.* **2006**, *64*, 225–234. [CrossRef]
40. Tami-Maury, I.M.; Liao, Y.; Rangel, M.L.; Gatus, L.A.; Shinn, E.H.; Alexander, A.; Basen-Engquist, K. Active Living After Cancer: Adaptation and evaluation of a community-based physical activity program for minority and medically underserved breast cancer survivors. *Cancer* **2022**, *128*, 353–363. [CrossRef]
41. Michie, S.; West, R.; Campbell, R.; Brown, J.; Gainforth, H. *ABC of Behaviour Change Theories*; Silverback Publishing: Great Britain, UK, 2014.
42. Michie, S.; Ashford, S.; Sniehotta, F.F.; Dombrowski, S.U.; Bishop, A.; French, D.P. A refined taxonomy of behaviour change techniques to help people change their physical activity and healthy eating behaviours: The CALO-RE taxonomy. *Psychol. Health* **2011**, *26*, 1479–1498. [CrossRef]

43. Maillot, P.; Perrot, A.; Hartley, A. Effects of interactive physical-activity video-game training on physical and cognitive function in older adults. *Psychol. Aging* **2012**, *27*, 589–600. [CrossRef]
44. Lin, J.H. "Just Dance": The Effects of Exergame Feedback and Controller Use on Physical Activity and Psychological Outcomes. *Games Health J.* **2015**, *4*, 183–189. [CrossRef]
45. Tate, D.F.; Lyons, E.J.; Valle, C.G. High-Tech Tools for Exercise Motivation: Use and Role of Technologies Such as the Internet, Mobile Applications, Social Media, and Video Games. *Diabetes Spectr.* **2015**, *28*, 45–54. [CrossRef]
46. Walsh-Burke, K.; Marcusen, C. Self-Advocacy Training for Cancer Survivors. *Cancer Pract.* **1999**, *7*, 297–301. [CrossRef]
47. LaFave, S.E.; Granbom, M.; Cudjoe, T.K.M.; Gottsch, A.; Shorb, G.; Szanton, S.L. Attention control group activities and perceived benefit in a trial of a behavioral intervention for older adults. *Res. Nurs. Health* **2019**, *42*, 476–482. [CrossRef]
48. Lewis, Z.H.; Cannon, M.; Rubio, G.; Swartz, M.C.; Lyons, E.J. Analysis of the Behavioral Change and Utility Features of Electronic Activity Monitors. *Technologies* **2020**, *8*, 75. [CrossRef]
49. Walters, S.J.; Bonacho Dos Anjos Henriques-Cadby, I.; Bortolami, O.; Flight, L.; Hind, D.; Jacques, R.M.; Knox, C.; Nadin, B.; Rothwell, J.; Surtees, M.; et al. Recruitment and retention of participants in randomised controlled trials: A review of trials funded and published by the United Kingdom Health Technology Assessment Programme. *BMJ Open* **2017**, *7*, e015276. [CrossRef]
50. Singh, B.; Spence, R.R.; Steele, M.L.; Sandler, C.X.; Peake, J.M.; Hayes, S.C. A Systematic Review and Meta-Analysis of the Safety, Feasibility, and Effect of Exercise in Women With Stage II+ Breast Cancer. *Arch. Phys. Med. Rehabil.* **2018**, *99*, 2621–2636. [CrossRef]
51. Ormel, H.L.; van der Schoot, G.G.F.; Sluiter, W.J.; Jalving, M.; Gietema, J.A.; Walenkamp, A.M.E. Predictors of adherence to exercise interventions during and after cancer treatment: A systematic review. *Psycho-Oncology* **2018**, *27*, 713–724. [CrossRef]
52. Vandelanotte, C.; De Bourdeaudhuij, I. Acceptability and feasibility of a computer-tailored physical activity intervention using stages of change: Project FAITH. *Health Educ. Res.* **2003**, *18*, 304–317. [CrossRef]
53. Vandelanotte, C.; De Bourdeaudhuij, I.; Brug, J. Acceptability and feasibility of an interactive computer-tailored fat intake intervention in Belgium. *Health Promot. Int.* **2004**, *19*, 463–470. [CrossRef]
54. Puthoff, M.L. Outcome measures in cardiopulmonary physical therapy: Short physical performance battery. *Cardiopulm. Phys. Ther. J.* **2008**, *19*, 17–22. [CrossRef]
55. Bohannon, R.W. Reference Values for the Timed Up and Go Test: A Descriptive Meta-Analysis. *J. Geriatr. Phys. Ther.* **2006**, *29*, 64–68. [CrossRef]
56. Shumway-Cook, A.; Brauer, S.; Woollacott, M. Predicting the Probability for Falls in Community-Dwelling Older Adults Using the Timed Up & Go Test. *Physical. Ther.* **2000**, *80*, 896–903. [CrossRef]
57. Podsiadlo, D.; Richardson, S. The timed "Up & Go": A test of basic functional mobility for frail elderly persons. *J. Am. Geriatr. Soc.* **1991**, *39*, 142–148. [CrossRef]
58. Rikli, R.E.; Jones, C.J. Development and Validation of a Functional Fitness Test for Community-Residing Older Adults. *J. Aging Phys. Act.* **1999**, *7*, 129–161. [CrossRef]
59. Rikli, R.E.; Jones, C.J. Functional Fitness Normative Scores for Community-Residing Older Adults, Ages 60–94. *J. Aging Phys. Act.* **1999**, *7*, 162–181. [CrossRef]
60. Hoenemeyer, T.W.; Cole, W.W.; Oster, R.A.; Pekmezi, D.W.; Pye, A.; Demark-Wahnefried, W. Test/Retest Reliability and Validity of Remote vs. In-Person Anthropometric and Physical Performance Assessments in Cancer Survivors and Supportive Partners. *Cancers* **2022**, *14*, 1075. [CrossRef]
61. Keadle, S.K.; Shiroma, E.J.; Freedson, P.S.; Lee, I.M. Impact of accelerometer data processing decisions on the sample size, wear time and physical activity level of a large cohort study. *BMC Public Health* **2014**, *14*, 1210. [CrossRef]
62. Lipsey, M.W.; Wilson, D.B. *Practical Meta-Analysis*; Sage Publications, Inc.: Thousand Oaks, CA, USA, 2001; p. ix, 247.
63. BOHANNON, R.W. Comfortable and maximum walking speed of adults aged 20–79 years: Reference values and determinants. *Age Ageing* **1997**, *26*, 15–19. [CrossRef]
64. Perera, S.; Mody, S.H.; Woodman, R.C.; Studenski, S.A. Meaningful change and responsiveness in common physical performance measures in older adults. *J. Am. Geriatr. Soc.* **2006**, *54*, 743–749. [CrossRef]
65. Makizako, H.; Shimada, H.; Doi, T.; Tsutsumimoto, K.; Nakakubo, S.; Hotta, R.; Suzuki, T. Predictive Cutoff Values of the Five-Times Sit-to-Stand Test and the Timed "Up & Go" Test for Disability Incidence in Older People Dwelling in the Community. *Phys. Ther.* **2017**, *97*, 417–424. [CrossRef]
66. Mirza, M.; Siebert, S.; Pratt, A.; Insch, E.; McIntosh, F.; Paton, J.; Wright, C.; Buckley, C.D.; Isaacs, J.; McInnes, I.B.; et al. Impact of the COVID-19 pandemic on recruitment to clinical research studies in rheumatology. *Musculoskelet. Care* **2022**, *20*, 209–213. [CrossRef]
67. de Koning, R.; Egiz, A.; Kotecha, J.; Ciuculete, A.C.; Ooi, S.Z.Y.; Bankole, N.D.A.; Erhabor, J.; Higginbotham, G.; Khan, M.; Dalle, D.U.; et al. Survey Fatigue During the COVID-19 Pandemic: An Analysis of Neurosurgery Survey Response Rates. *Front. Surg.* **2021**, *8*, 690680. [CrossRef]
68. Cardel, M.I.; Manasse, S.; Krukowski, R.A.; Ross, K.; Shakour, R.; Miller, D.R.; Lemas, D.J.; Hong, Y.-R. COVID-19 Impacts Mental Health Outcomes and Ability/Desire to Participate in Research Among Current Research Participants. *Obesity* **2020**, *28*, 2272–2281. [CrossRef]
69. Taylor, L.M.; Maddison, R.; Pfaeffli, L.A.; Rawstorn, J.C.; Gant, N.; Kerse, N.M. Activity and Energy Expenditure in Older People Playing Active Video Games. *Arch. Phys. Med. Rehabil.* **2012**, *93*, 2281–2286. [CrossRef]

70. Bohannon, R.W.; Glenney, S.S. Minimal clinically important difference for change in comfortable gait speed of adults with pathology: A systematic review. *J. Eval. Clin. Pract.* **2014**, *20*, 295–300. [CrossRef]
71. Brown, J.C.; Harhay, M.O.; Harhay, M.N. Physical function as a prognostic biomarker among cancer survivors. *Br. J. Cancer* **2015**, *112*, 194–198. [CrossRef]
72. Brown, M.; O'Connor, D.; Murphy, C.; McClean, M.; McMeekin, A.; Prue, G. Impact of COVID-19 on an established physical activity and behaviour change support programme for cancer survivors: An exploratory survey of the Macmillan Move More service for Northern Ireland. *Support. Care Cancer* **2021**, *29*, 6135–6143. [CrossRef]
73. Bu, F.; Bone, J.K.; Mitchell, J.J.; Steptoe, A.; Fancourt, D. Longitudinal changes in physical activity during and after the first national lockdown due to the COVID-19 pandemic in England. *Sci. Rep.* **2021**, *11*, 17723. [CrossRef]
74. Tison, G.H.; Avram, R.; Kuhar, P.; Abreau, S.; Marcus, G.M.; Pletcher, M.J.; Olgin, J.E. Worldwide Effect of COVID-19 on Physical Activity: A Descriptive Study. *Ann. Intern. Med.* **2020**, *173*, 767–770. [CrossRef]
75. Said, C.M.; Batchelor, F.; Duque, G. The Impact of the COVID-19 Pandemic on Physical Activity, Function, and Quality of Life. *Clin. Geriatr. Med.* **2022**, *38*, 519–531. [CrossRef]
76. Hergenroeder, A.L.; Barone Gibbs, B.; Kotlarczyk, M.P.; Kowalsky, R.J.; Perera, S.; Brach, J.S. Accuracy of Objective Physical Activity Monitors in Measuring Steps in Older Adults. *Gerontol. Geriatr. Med.* **2018**, *4*, 2333721418781126. [CrossRef]
77. Park, A.H.; Zhong, S.; Yang, H.; Jeong, J.; Lee, C. Impact of COVID-19 on physical activity: A rapid review. *J. Glob. Health* **2022**, *12*, 05003. [CrossRef]
78. Wittink, H.; Rogers, W.; Sukiennik, A.; Carr, D.B. Physical Functioning: Self-Report and Performance Measures Are Related but Distinct. *Spine* **2003**, *28*, 2407–2413. [CrossRef]
79. Brach, J.S.; VanSwearingen, J.M.; Newman, A.B.; Kriska, A.M. Identifying Early Decline of Physical Function in Community-Dwelling Older Women: Performance-Based and Self-Report Measures. *Physical. Ther.* **2002**, *82*, 320–328. [CrossRef]

life

MDPI

Article

Remote Psychotherapy during the COVID-19 Pandemic: A Mixed-Methods Study on the Changes Experienced by Austrian Psychotherapists

Michael Stadler [1,†], Andrea Jesser [1,*,†], Elke Humer [1], Barbara Haid [2], Peter Stippl [2], Wolfgang Schimböck [2], Elisabeth Maaß [3], Helmut Schwanzar [3], Daniela Leithner [3], Christoph Pieh [1] and Thomas Probst [1]

1 Department for Psychosomatic Medicine and Psychotherapy, University for Continuing Education Krems, 3500 Krems, Austria
2 Austrian Federal Association for Psychotherapy, 1030 Vienna, Austria
3 Österreichische Gesellschaft Für Wissenschaftliche, Klientenzentrierte Psychotherapie und Personorientierte Gesprächsführung (ÖGWG), 4020 Linz, Austria
* Correspondence: andrea.jesser@donau-uni.ac.at
† These authors contributed equally to this work.

Abstract: The outbreak of the COVID-19 pandemic and associated measures to contain the SARS-CoV-2 coronavirus required a change in treatment format from face-to-face to remote psychotherapy. This study investigated the changes experienced by Austrian therapists when switching to psychotherapy at a distance. A total of 217 therapists participated in an online survey on changes experienced when switching settings. The survey was open from 26 June until 3 September 2020. Several open questions were evaluated using qualitative content analysis. The results show that the setting at a distance was appreciated by the therapists as a possibility to continue therapy even during an exceptional situation. Moreover, remote therapy offered the respondents more flexibility in terms of space and time. Nevertheless, the therapists also reported challenges of remote therapy, such as limited sensory perceptions, technical problems and signs of fatigue. They also described differences in terms of the therapeutic interventions used. There was a great deal of ambivalence in the data regarding the intensity of sessions and the establishment and/or maintenance of a psychotherapeutic relationship. Overall, the study shows that remote psychotherapy seems to have been well accepted by Austrian psychotherapists in many settings and can offer benefits. Clinical studies are also necessary to investigate in which contexts and for which patient groups the remote setting is suitable and where it is potentially contraindicated.

Keywords: remote psychotherapy; psychotherapy via telephone; psychotherapy via videoconferencing; tele-health; e-mental-health; COVID-19; pandemic; psychotherapy; qualitative psychotherapy research; mixed-methods psychotherapy research

Citation: Stadler, M.; Jesser, A.; Humer, E.; Haid, B.; Stippl, P.; Schimböck, W.; Maaß, E.; Schwanzar, H.; Leithner, D.; Pieh, C.; et al. Remote Psychotherapy during the COVID-19 Pandemic: A Mixed-Methods Study on the Changes Experienced by Austrian Psychotherapists. *Life* **2023**, *13*, 360. https://doi.org/10.3390/life13020360

Academic Editor: Daniele Giansanti

Received: 28 December 2022
Revised: 23 January 2023
Accepted: 25 January 2023
Published: 29 January 2023

1. Introduction

The outbreak of the COVID-19 pandemic necessitated a sharp modification in the psychotherapeutic treatment format [1]. To contain the spread of the SARS-CoV-2 coronavirus, restrictive measures such as restrictions on outdoor activities, quarantine and social distancing were implemented. Psychotherapists and patients alike were faced with the challenge of adapting to a new, virtual setting within a very short time [2,3]. The number of patients treated psychotherapeutically via telephone or the internet increased sharply. During the first curfew in 2020, the number of patients treated via telephone on average per week in Austria increased by 979% and via the internet by 1561% (both $p < 0.001$) [4]. Study results confirmed that the infectious disease COVID-19 has a clear impact on the practice of psychotherapy in Austria [3].

In Austria, psychotherapy via digital media was not considered lege artis until that time [5] and was also not covered by health funds. The legal regulations changed with the outbreak of the COVID-19 pandemic and the associated need to restrict socio-physical contact [6]. This is not the least since study results indicate that psychotherapy via the internet can be regarded as equally effective to psychotherapy in face-to-face contact [7,8]. There seems to be no significant difference in effectiveness between the treatment modes of face-to-face setting, real-time video conferencing and telephone [9]. Psychotherapy via videoconferencing has already been described as promising and tends to be equivalent in patients with anxiety disorders [10]. In a study by Bouchard et al. [11] involving 71 patients with panic disorder and agoraphobia, a very strong therapeutic alliance was demonstrated in video conferencing, which did not differ from treatments in face-to-face settings. Similarly, research findings suggest the effectiveness of treatment via telephone for depressive symptoms [12–15]. Moreover, dropout rates might be lower for telephone therapies than for treatments in face-to-face settings [16].

While there was still a general skepticism among therapists towards psychotherapy at a distance before the COVID-19 pandemic, despite the positive results of efficacy research [17,18], this attitude has since changed due to experiences gained during the first months of the pandemic [19,20]. While the digital treatment setting did not play a substantial role in healthcare before COVID-19, the outbreak of the COVID-19 pandemic also pushed former skeptics to work at a distance [21]; it seems that quite often, the experiences made in the process have led to a re-evaluation of remote psychotherapy [20]. At the same time, treatment at a distance is not considered by therapists to be fully comparable to the face-to-face setting [19].

In the remote setting, the therapeutic alliance between patients and therapists was described by therapists as impaired, for example, due to the loss of the physical exchange or the lack of various sensory impressions, and was experienced as more superficial and businesslike. Moreover, therapists were confronted with technical challenges and security-related issues regarding the safeguarding of confidentiality in the online setting [22]. Furthermore, therapists, psychologists, and social workers also expressed concerns about building and maintaining the therapeutic alliance [23]. Therapists seem to perceive greater differences between treatment settings in face-to-face contact and at a distance than patients [24]. Moreover, therapists' satisfaction with psychotherapy via videoconferencing seems to be related to their level of professional maturity and experience, as older therapists with previous experience in the video-based setting have a more positive attitude towards remote psychotherapy [25]. Furthermore, study results indicate that therapeutic interventions also differ between settings, and interventions of different psychotherapeutic orientations are more distinct in face-to-face contact than in remote psychotherapy [26].

2. Materials and Methods

2.1. Study Design and Procedure

The primary objective of the current study was to collect data on the changes experienced by Austrian therapists when switching from face-to-face to remote psychotherapy and/or from remote to face-to-face psychotherapy during the first year of the COVID-19 pandemic. Specifically, we wanted to find whether psychotherapists experienced changes in the therapeutic relationship (research question 1), whether they experienced changes in the content of the sessions (research question 2), whether they noticed changes in the intensity of the sessions (research question 3), whether the structure of the sessions changed in their practice (research question 4), and how they experienced the digital setting and the lack of physical presence (research question 5). A further aim was to investigate potential differences in these experiences with respect to the sociodemographic and professional characteristics of psychotherapists.

Following an exploratory research approach, we conducted a cross-sectional online survey among Austrian psychotherapists authorized to provide psychotherapeutic treatment to patients. This includes psychotherapists in training in recognized training institutions, who are already working under supervision after the fourth year of training and psychotherapists registered at the Austrian Federal Ministry of Social Affairs, Health, Care, and Consumer Protection. Registered psychotherapists have completed their training as psychotherapists and are officially licensed to treat patients. Health insurance companies only refund psychotherapy sessions provided by registered therapists. The survey was set up with REDCap (Research Electronic Data Capture) [27,28] and was open from 26 June 2020 until 3 September 2020. By this time, the COVID-19 measures imposed by the government had been largely relaxed. A previous initial lockdown in Austria from March 16 to 30 April 2020, mandated that Austrians were only allowed to leave their homes for certain activities, such as covering important basic needs, caring for others in need or going to work. During this time, many therapists switched from face-to-face psychotherapy to remote sessions [4].

The survey included a total of 128 questions covering basic sociodemographic variables; the number of patients who were switched from face-to-face to remote psychotherapy or from remote to face-to-face psychotherapy; therapists' experiences with digital media; the type of media used; an assessment of the various therapeutic interventions used in the different settings; as well as several open-ended questions about the therapeutic relationship, content, intensity, and structure of remote as compared to face-to-face sessions and therapists' experience of the lack of physical presence and the spatial distance between themselves and the patient. Open-ended questions were formulated to elicit as wide a range of perceptions as possible. We estimated that it would take psychotherapists 15–20 min to complete the questionnaire in its entirety. In the present study, only the open-ended questions from the survey were analyzed. Quantitative findings are published elsewhere [26].

The study received approval from the ethics committee and the data protection officer of the University for Continuing Education Krems (EK GZ 27/2018-202). We follow the APA Journal Article Reporting Standards for Qualitative Research in Psychology [29] in the presentation of our research.

2.2. Participants

Austria has a long tradition of psychotherapy and a wide range of 23 accredited psychotherapy schools [30]. They can be classified into four orientations. The largest orientation is the humanistic orientation (37.8% of the psychotherapists in Austria), followed by the psychodynamic orientation (25.9% of the psychotherapists in Austria), the systemic orientation (24.3% of the psychotherapists in Austria) and the behavioral orientation (12.0% of the psychotherapists in Austria).

A link to the online survey was sent to all registered psychotherapists by the last author in cooperation with the Austrian Federal Association for Psychotherapy, which supported the study. Continuing education credits points were awarded as an incentive for participation. In addition, psychotherapists in training who were already treating patients under supervision were invited to participate in the survey. Email lists for psychotherapists in training were provided by the Austrian Federal Association for Psychotherapy. In addition, a link was sent to psychotherapy students from the University for Continuing Education Krems, which is one of several institutions offering training as a psychotherapist. In total, $n = 222$ respondents participated in the survey. All participants gave electronic informed consent after reading the data protection declaration. Five therapists did not experience a change of treatment format and were therefore excluded from further analyses, resulting in a final sample of $n = 217$ therapists.

2.3. Measures

The study comprised 10 open-ended questions.

- Q1: In your own words, please describe how the therapeutic relationship with your patients changed as a result of the switch from psychotherapies in personal contact to psychotherapies via digital media.
- Q2: In your own words, please describe how the content of sessions changed as a result of the switch from psychotherapies in personal contact to psychotherapies via digital media.

- Q3: In your own words, please describe how the intensity of sessions changed as a result of the switch from psychotherapies in personal contact to psychotherapies via digital media.
- Q4: In your own words, please describe how the structure of sessions changed as a result of the switch from psychotherapies in personal contact to psychotherapies via digital media.
- Q5: How do you experience the lack of physical presence in remote psychotherapy sessions?
- Q6: How do you experience the spatial distance and remaining in your own space (not in the therapy room) when you conduct psychotherapy via digital media?

Questions 1–4 were also asked regarding the switch back from the remote setting to face-to-face psychotherapy. As many psychotherapists referred to remote psychotherapy in their responses, these responses were also included in the analysis and coded together with the respective question addressing the switch from face-to-face to remote psychotherapy.

2.4. Data Analysis

We used a conventional approach to qualitative content analysis [31]. In conventional content analysis, categories are derived from the data rather than from theory. It is generally applied in study designs that aim to describe a phenomenon about which little theory or literature is yet available.

Out of 217 respondents, 63 answered all open-ended questions, 143 answered at least one, and 11 did not fill in any free text field. Many of the answers were very detailed. We received only a few keyword-like responses, as is usually the case with open-ended survey responses. In sum, we received 1448 free text comments: 308 describing changes in the therapeutic relationship, 275 describing changes regarding the content of sessions, 265 describing changes in intensity, 238 describing changes in the structure of sessions, 192 focusing on experiences regarding the lack of physical presence in remote psychotherapy and 170 addressing the spatial distance. Overall, responses to Q1 were the most comprehensive. Respondents addressed various aspects of their own accord that not only had to do with the therapeutic relationship but also related to the subsequent questions. As a result, some answers were repeated in later questions. They were only coded if new aspects were addressed.

At the beginning of the coding process, two coders read through the whole data set to familiarize themselves with the material. Subsequently, data were imported into Atlas.ti for coding [32], and one coder read through the material again, inductively defining categories in the process. After coding 30% of the material, the second coder coded the same material with the list of categories and category definitions provided by the first coder. To enhance reliability [33], we assessed the agreement of how the two coders coded the data set [34]. Percentage agreement was high at 94.3%, and inter-coder agreement using Krippendorff c-α-binary = 0.985. Any citation on which the coders disagreed was discussed between the two coders, and the category definitions were expanded in this process. In addition, this step of the coding process created larger thematic clusters to which categories were assigned. Afterward, the second coder coded the entire data set, documenting the cases in which assignment to a category was not clear. These cases were coded together.

Chi-squared tests were conducted to analyze potential differences in the frequency of main categories reported by psychotherapists in terms of sociodemographic (years of age: $\leq$40, 41–50, 51–60, >60; gender: female, male) and professional characteristics (years in the profession: $\leq$5, 6–10, >10; psychotherapeutic orientation: psychodynamic, humanistic, systemic, behavioral). Differences in the frequencies of subcategories were only analyzed with respect to gender, as the number of coded text passages in the subgroups of the different orientations, age and experience groups was insufficient to make reliable inferences about the population of psychotherapists. To analyze differences in the length of the free text answers, *t*-tests (gender) and univariate ANOVAs (age group, professional experience group, psychotherapeutic orientation) were applied. Statistical analyses were performed in SPSS version 26 (IBM Corp, Armonk, NY, USA). p-values of $\leq$0.05 were considered statistically significant (2-sided tests).

3. Results

3.1. Sample Description

A total of 77% of respondents were female, and 23.0% were male. They were M = 50.66 (SD = 9.65) years old, and while most of them were certified psychotherapists in Austria (91.2%), 8.8.% worked under supervision in the last part of their training to become psychotherapists. Regarding their professional experience, M was 10.61 years (SD = 9.50) (value was set to 0 for psychotherapists working under supervision). Overall, 46.1% of respondents practiced humanistic psychotherapy, 22.6% practiced psychodynamic psychotherapy, 20.7% belonged to the systemic orientation and 10.6% to the behavioural orientation (10.6%).

Most respondents worked in their private practice (96.8%), and 39.6% had gathered experience with remote psychotherapy before COVID-19. Most psychotherapists used psychotherapy via telephone (88.5%) or videoconferencing (76.5%). Psychotherapy via email was used by 22.6% of respondents, and 9.2% used chats or other digital media (2.8%). M = 11.21 (SD = 10.12) patients were switched from face-to-face psychotherapy to remote psychotherapy, and M = 9.62 (SD = 10.34) patients were switched from remote to face-to-face psychotherapy.

3.2. Results of Qualitative Analysis

The analysis resulted in seven main categories, which each comprise several subcategories. Figure 1 represents the main categories and their frequency in relation to the number of respondents. We chose not to report the number of coded text passages but always reported the number of respondents per category. Some respondents commented more often on a topic. These responses are reported descriptively but are not reflected in the frequencies. The order of the categories is not based on the frequency of the categories but was chosen so that the presentation of the contents of the subcategories builds on each other as coherently as possible.

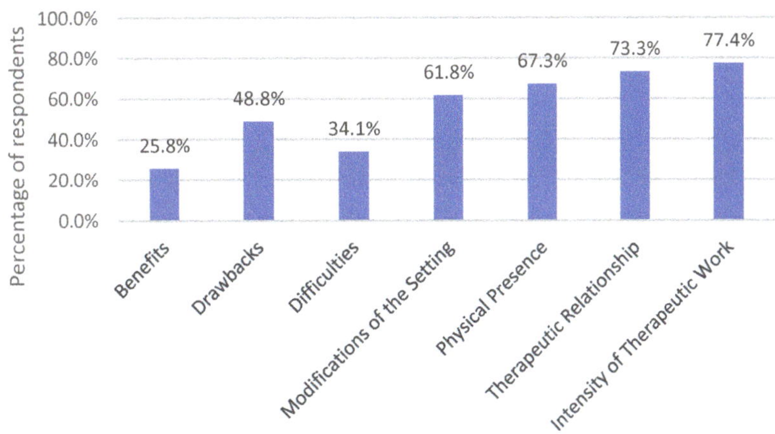

Figure 1. Main categories of the qualitative content analysis and the percentage of respondents reporting one or more experiences in each of the main categories. The percentages of the main categories may differ from the sum of the percentages in the individual subcategories (Figures 2–8) because it may be that a respondent reported experiences in several subcategories (e.g., technical problems and distraction) within one main category (e.g., drawbacks) and thus appears in each of these subcategories but is only counted once per main category.

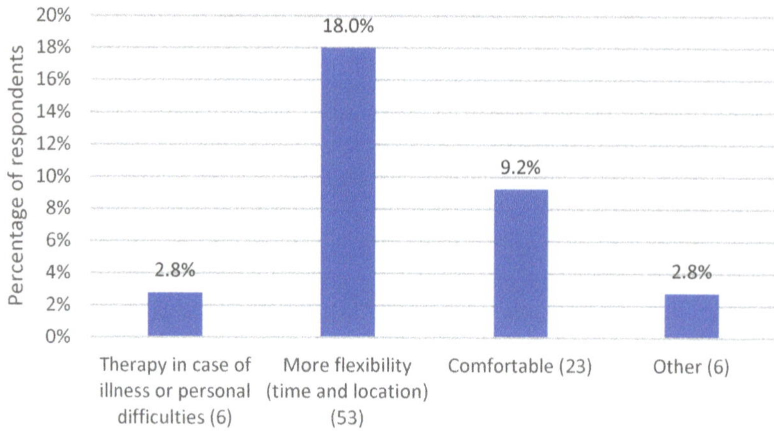

Figure 2. Percentage of respondents who experienced various benefits of remote psychotherapy. The number in parentheses after the subcategory name indicates the number of coded text passages.

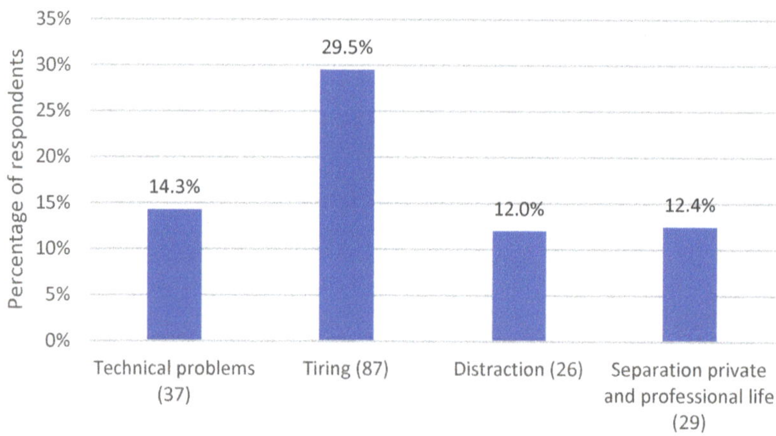

Figure 3. Percentage of respondents who experienced various drawbacks of remote psychotherapy. The number in parentheses after the subcategory name indicates the number of coded text passages.

Differences in the frequencies of two main categories emerged between female and male psychotherapists (Table 1): female psychotherapists reported difficulties more often (38%) compared to their male colleagues (20%; $p = 0.02$). Additionally, more female psychotherapists expressed experiences related to the therapeutic relationship (79% vs. 52%; $p < 0.001$).

Frequencies of all investigated main categories did not differ among age groups ($p \geq 0.08$), groups of professional experience ($p \geq 009$) and among the four therapeutic orientations ($p \geq 0.06$).

Female psychotherapists provided longer comments (mean number of characters (M) = 932.6, SD = 751.0) than male psychotherapists (M = 595.8, SD = 630.1), t (94.5) = 3.116; $p = 0.002$. The length of the text answers was neither associated with the age, the professional experience, nor the therapeutic orientation of the participating psychotherapists ($p \geq 0.11$).

3.2.1. Benefits

A total of $n = 56$ respondents (25.8%) reported the benefits of remote psychotherapy. Figure 2 displays the allocated subcategories.

N = 6 (2.8%) commented positively that therapies could also be provided in the case of illness. For high-risk patients and pregnant patients, the risk of infection through travel and face-to-face contact in the practice was eliminated. An important subcategory, mentioned by *n* = 39 (18%) respondents, addresses the flexibility enabled by remote psychotherapy. Respondents reported that they were able to offer appointments more flexibly than usual, even at short notice. This proved beneficial in crises or for patients who needed a higher frequency of sessions. The possibility of fitting sessions in between appointments proved helpful for psychotherapists and patients in scheduling sessions, as did the fact that there was no need to travel to and from the practice. Mothers of younger children, in particular, benefited from increased flexibility. Spatial independence ensured that business trips, study abroad, and even vacations were no longer an obstacle to offering or attending psychotherapy. Other *n* = 20 (9.2%) respondents observed remote sessions to be more comfortable both for themselves and their patients. Clothing, food, drink, and not having to go out in bad weather were mentioned, as well as being unobserved on the phone. *N* = 6 (2.8%) mentioned other advantages, for example, having more resources available at home, such as books for consultation, or being able to maintain a professional distance more easily.

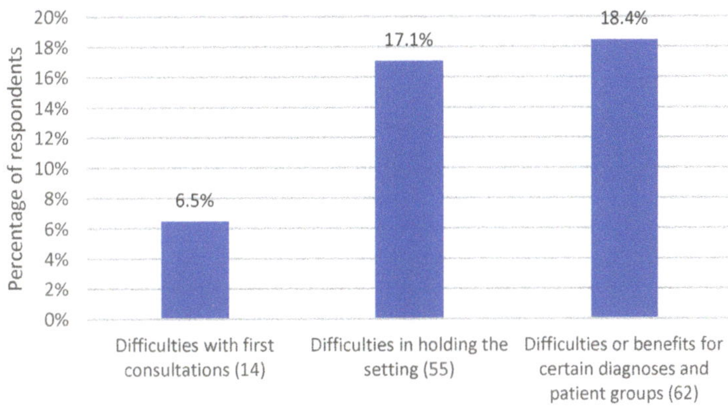

Figure 4. Percentage of respondents who experienced various difficulties of remote psychotherapy. The number in parentheses after the subcategory name indicates the number of coded text passages.

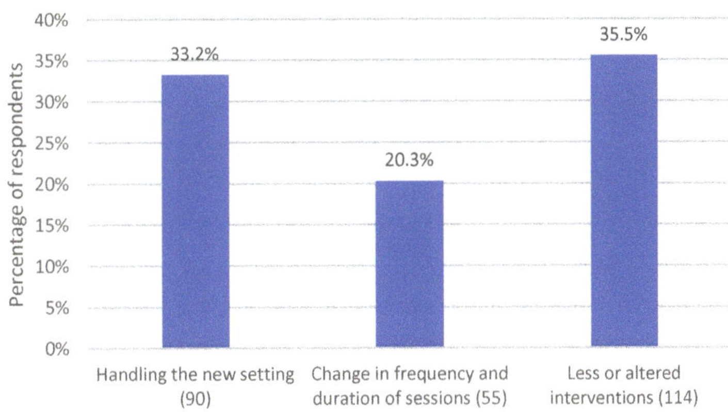

Figure 5. Percentage of respondents who reported modifications of the setting. The number in parentheses after the subcategory name indicates the number of coded text passages.

Frequencies of all investigated subcategories did not differ between female and male psychotherapists ($p \geq 0.112$).

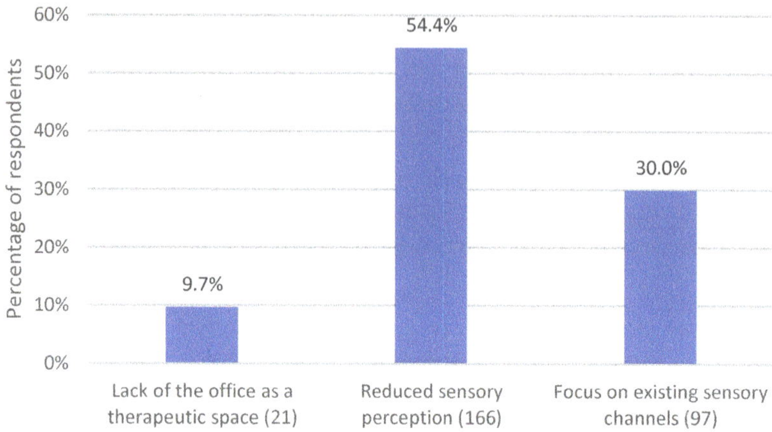

Figure 6. Percentage of respondents who commented on diverse experiences with the lack of physical presence. The number in parentheses after the subcategory name indicates the number of coded text passages.

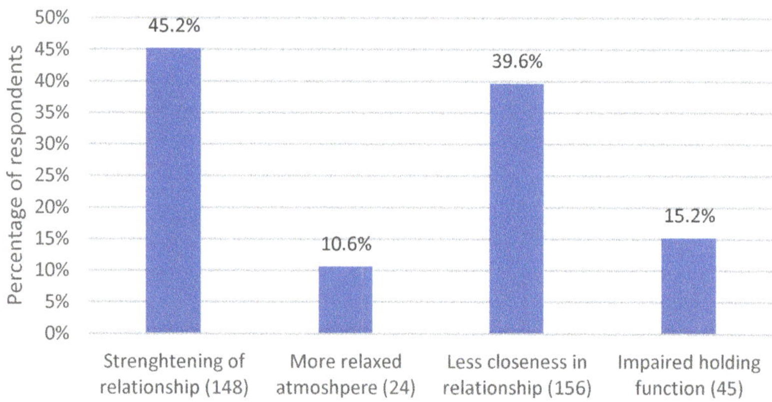

Figure 7. Percentage of respondents who commented on aspects relating to the quality and intensity of the psychotherapeutic relationship. The number in parentheses after the subcategory name indicates the number of coded text passages.

3.2.2. Drawbacks

$N = 106$ respondents (48.8%) reported the drawbacks of remote psychotherapy. Subcategories are displayed in Figure 3.

$N = 31$ (14.3%) respondents reported technical problems: connection problems, delays or interruptions in the transmission of sound and images and poor sound and image quality. These interfered with therapy processes and made it difficult for therapists to tune in to their patients. Therapists were also required to ensure the functionality of the technology and to comply with data protection regulations, which was an additional challenge for some.

Other ($n = 64$ (29.5%)) respondents noted that remote psychotherapy made them feel more exhausted. They reported fatigue from longer screen time and distractions at home. They also described that it required more concentration (1) to compensate for the lack of perceptions and capture patients' emotions (passive) and (2) to convey empathy through verbal communication only, in case of sessions on the phone (active).

Respondent 170 mentioned how *"it was more exhausting to find the "right words" because all other sensory channels were eliminated"*. Additionally, respondent 172 commented that *"Over the phone, it was difficult and required a lot of concentration to capture emotions only through the spoken word"*.

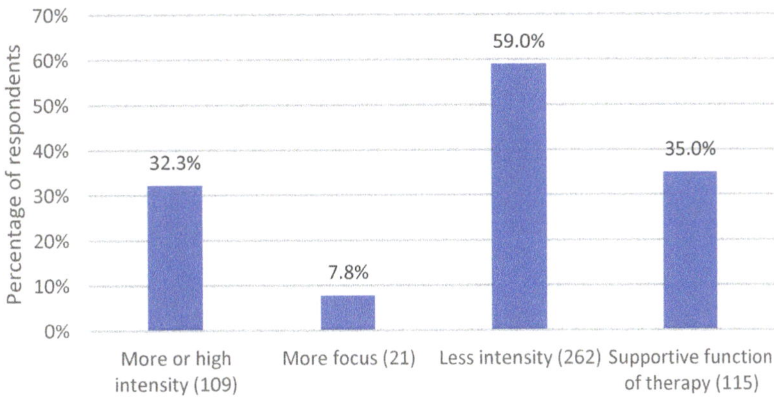

Figure 8. Percentage of respondents who commented on their perceptions regarding the intensity of psychotherapeutic work. The number in parentheses after the subcategory name indicates the number of coded text passages.

Table 1. Main categories of the qualitative content analysis by gender.

Main Category	Female (n = 167)	Male (n = 50)	Statistics
Benefits, % (N)	25.1% (42)	28.0% (14)	χ^2 (1) = 0.163; p = 0.686
Drawbacks, % (N)	48.5% (81)	48.0% (24)	χ^2 (1) = 0.004; p = 0.950
Difficulties, % (N)	37.7% (63)	20.0% (10)	χ^2 (1) = 5.415; p = 0.020
Modifications of the Setting, % (N)	64.7% (108)	50.0% (25)	χ^2 (1) = 3.491; p = 0.062
Physical Presence, % (N)	68.3% (114)	64.0% (32)	χ^2 (1) = 0.318; p = 0.573
Therapeutic Relationship, % (N)	79.0% (132)	52.0% (26)	χ^2 (1) = 14.214; p < 0.001
Intensity of Therapeutic Work, % (N)	79.0% (132)	70.0% (35)	χ^2 (1) = 1.774; p = 0.183

In this context, n = 26 (12%) respondents observed that they or their patients were more distracted in the remote setting. Other people in the household and pets were described as distractions for both therapists and patients. Therapists perceived patients as distracted if they ate or drank during psychotherapy sessions. They experienced themselves as distracted by glimpses into the patients' private spaces, by the environment at home or by seeing their own faces on the screen during videoconferencing sessions. N = 27 (12.4%) respondents, who worked from home, reported that it was challenging to maintain a separation between their personal and professional life. It was more difficult for them to distance themselves from work, pay attention to a healthy work-life balance, and maintain clear work schedules. It was also harder to adopt a therapeutic stance.

No differences between female and male psychotherapists in the frequencies of mentioned drawbacks were observed ($p \geq 0.372$).

3.2.3. Difficulties

N = 74 (34.1%) respondents addressed the difficulties encountered in remote psychotherapies. Figure 4 illustrates the subcategories.

N = 14 (6.5%) mentioned having worked remotely only with patients who had been in therapy for some time. First consultations were described as difficult. In particular, establishing a holding psychotherapeutic relationship was mentioned as a challenge. *Non-committal* (respondent 79), *businesslike* (respondent 151), and *insecure* (respondent 208) were adjectives therapists used to describe their experience.

Other n = 37 (17.1%) respondents described difficulties in upholding the therapeutic setting. The temporal frame was an issue for some. Therapists reported they had to contact patients more often, for example, to remind them of sessions. For other respondents, the focus was on the therapeutic space. They expressed a sense of losing control in the remote setting. As respondent 193 expressed, *"The setting was eroding"*. Although it was possible for therapists to choose the room in which they located themselves, it was not possible to exert any influence on the spatial conditions at the patients' homes. Disturbances by family members were mentioned very frequently as well as *"undignified conditions"* (respondent 178), e.g., in case patients had to attend a session from their bathroom or from behind a paravan. Patients had to take care of adequate conditions "on their side" themselves, which required more self-responsibility. Therapists also described it as difficult not having any influence on the technical connection or on how patients participated in the session, e.g., lying in bed, eating, etc.

Finally, we generated the category "difficulties or benefits for certain diagnoses and patient groups", which was addressed by n = 40 (18.4%) respondents. Not only difficulties but also benefits were subsumed under this category. Several respondents mentioned difficulties with remote psychotherapies for patients with structural deficits who needed a lot of stabilization. There appears to be little consensus and very contradictory statements regarding the different types of mental disorders. For anxiety disorders and schizophrenia spectrum disorders, staying in one's own room was described as building confidence and reducing anxiety by some respondents. Other respondents found it more difficult to treat these patients remotely. Some respondents also saw advantages in treating traumatized patients in a remote setting. They described that patients were able to address traumatic events for the first time in the remote setting, as it helped them to distance themselves from the event and reduced feelings of shame. Not being seen during their account had a disinhibiting effect. Other respondents experienced working with traumatized patients as more difficult in the remote setting because patients could not be supported so well, for example, when dissociating. As respondent 151 put it:

> *"Traumatic experiences were addressed. But for me as a therapist, the patient in tears on the phone is an experience that I do not wish to repeat. I had a strong feeling that I could not fulfill my responsibility as a therapist. Even without touching the patient in such situations, I am convinced that my physical presence alone and my staying present are important for the patient. Also, the thought came to my mind: what do I do when he/she throws down the phone and—yes, what does he/she do? jumps out of the window, runs into the street without looking left or right..."*

Another patient group that was mentioned several times was the group of children and adolescents. Therapists found it more difficult to work remotely with children, as the remote setting made it more difficult to engage in playing and relied a lot on verbal communication. There were more positive observations regarding remote psychotherapy with adolescents. It was noted that remote formats are familiar to young people and are therefore a good way to get in touch.

Among the experienced difficulties of remote psychotherapy, gender differences became visible for the subcategory "difficulties or benefits for certain diagnosis and patient groups", with female psychotherapists reporting more often respective experiences (21%)

than male psychotherapists (8%; χ^2 (1) = 4.383; p = 0.036). For the remaining two categories, no differences were observed with respect to gender ($p \geq 0.144$).

3.2.4. Modifications of the Setting

N = 134 (61.8%) respondents made statements regarding how they dealt with changes in the setting and how they adapted to the new situation. Subcategories are shown in Figure 5.

The subcategory "handling the new setting", mentioned by n = 72 (33.2%) respondents, describes statements from respondents about how the move to remote psychotherapy initially caused a sense of uncertainty among therapists and among patients. For the most part, respondents described continuing to hold sessions from their practice rooms. This was felt to help separate work from private life, maintain a professional attitude, and provide continuity for patients in the form of a familiar space.

> *"The beginning was structured by me, when I started the Zoom call and let the (waiting) person in. It was important to me to maintain continuity and stability in the sense that I told and showed the patients that I was sitting in the usual armchair in the practice room. Their familiar space thus continued to exist, only they were not spatially there, I was connected to them via telephone or Zoom".* (respondent 78)

Alternatively, some therapists switched to their home office. In this case, the background visible on the screen was arranged in such a way as to create a professional context (e.g., removal of personal items and pictures, covering glass doors to other living spaces, etc.). In addition, respondents explained that it was important for them to convey a sense of security to the patients. This also included establishing the new setting, e.g., discussing from where patients participated in the session and whether undisturbed communication was possible in this environment, how to handle the software and deal with technical failures, how data protection regulations were complied with, who could provide help in the event of a crisis and who could also be reached by the therapist, etc. Respondents mentioned how they had to be more demanding that patients ensure adequate setting conditions "on their side" or adhere to setting conditions, such as starting times. As respondent 85 stated: *"Discussing "rules of conduct" in advance is important (e.g., pat. not just hanging up, closing laptop)".*

In a few cases, respondents described that patients went for a walk during the session because they could not establish an undisturbed atmosphere at home. Respondents also discussed the possible advantages and disadvantages of remote psychotherapy with patients and inquired about patient expectations. Some therapists also mentioned new rituals they introduced for beginning and ending a session via videoconferencing or on the phone.

Another subcategory mentioned by n = 44 (20.3%) respondents concerns changes in the frequency and duration of sessions. Some respondents explicitly mentioned not changing anything about the structure of the sessions (day of the week, time, duration). However, another part of the respondents reported more frequent or less frequent, more irregular or more regular and shorter sessions. In addition to videoconferencing and telephone sessions, some therapists also communicated with patients in writing (text messaging, email, chat).

A very large subcategory mentioned by n = 77 (35.5%) respondents concerns the use of therapeutic interventions. The majority of respondents described having to forgo many interventions in the remote setting. In particular, interactive interventions (e.g., role-playing, constellation work, etc.), body-based interventions (demonstrations, movement, EMDR, hypnotic trances, autogenous states of relaxation), art therapy interventions (visualizations, sand play, sculptures, etc.), therapeutic play in work with children, animal-assisted interventions, and work with objects or with guided affective imagery were mentioned. In contrast, other interventions were used more frequently, such as the assignment of homework. The therapeutic conversation also gained importance in remote psychotherapy, as much of the content was addressed verbally. Some respondents described that interventions were more difficult to apply but could be adapted for the remote setting. For example, some body-based interventions could be delivered in an adapted form, such as autogenic states of relaxation or hypnotic trances, as could interactive interventions, such as role-playing and guided affective imagery. Some respondents reported using handouts, exercise sheets, and audio that they gave to patients to take home. The overall impression was that respondents used interventions primarily in a stabilizing or resource-strengthening way and focused on techniques that activated cognition. In contrast, they worked in a less confrontational, less regression-promoting, and less emotion-activating manner. Trauma-specific interventions were also used with caution.

Handling the new setting was mentioned by more female (38%) than male (18%) psychotherapists (χ^2 (1) = 6.752; p = 0.009). For the other two subcategories, no differences were observed between male and female therapists ($p \geq 0.714$).

3.2.5. Lack of Physical Presence

N = 146 (67.3%) respondents named categories related to the absence of physical presence. The subcategories are displayed in Figure 6.

N = 21 (9.7%) respondents noted that the office was lacking as both a physical and an intrapsychic space in remote psychotherapies. They observed that patients missed the time in the office away from their usual contexts. Also missing was the journey to and from the office as a mental space for reflective engagement with what patients wanted to talk about or had worked through in the session. Sessions took place more *"in-between"* (respondent 65). As one respondent put it: *"Patients reported that it is unusual when the journey home can no longer be experienced and one is back in "real" life from one second to the next. Processing what was discussed suffers"*. (respondent 30) Therapists also missed rituals that had shaped the therapeutic encounter in the office, such as inviting patients in, shaking hands, offering a drink or passing a handkerchief.

A majority (n = 118 (54.4%)) of respondents made mention of impaired sensory perception in remote psychotherapies. They referred to the perception of nonverbal communication signals and body language, such as facial expressions, gestures, posture, movements, ideomotor activity, and breathing. They also reported altered acoustics, lack of smell, and eye contact. It was mentioned that it became more difficult to gather diagnostic information, to emotionally tune in to the patient, and to assess the effect of interventions. Respondent 94 commented, *"The distance made it more difficult to perceive, to sense, to observe"*. The difficulty of assessing the atmosphere was mentioned in particular. The assessment of silence was mentioned several times in this context, as here by respondent 59:

> *"A young woman wanted to stay in contact via telephone—in this case it was difficult, especially for me, to assess her reactions without having an image (silence—is she thinking about what has been said or is she crying quietly??? Difficult to assess; asking was disruptive in the process)"*

In addition, body-oriented psychotherapists pointed out the lack of (inter-)bodily perception. Respondent 53 described that *"physical encounters support the process of emotional processing. Traumatic experiences can be better processed through therapeutic physical proximity"*.

N = 65 (30%) respondents reported how they tried to replace missing sensory perceptions by focusing on existing sensory channels. They attached particular importance to attentive listening, the perception of speech melody, tonality and subtleties in speech (formulations, choice of language and words, pauses, speaking pace, volume, etc.). Respondent 179 noted, *"the lack of physical presence focused my attention on listening and the words used and was just as intense"*. Other respondents remembered how an imaginary image of the patient was formed during telephone contact. Therapists also observed that they used their voice and speech more consciously to stay in contact with their patients. In the case of video conferencing, respondents described how they paid close attention to what was visible on the screen. Respondent 73 observed, *"only a section of the patient is visible, but you focus on details that are otherwise not present to this extent"*. Respondent 54 recounted:

> *"The "large format" of the upper half of the body during video chat, with the visibility of subtle changes in facial expressions, had its own "physical" presence for me. When I was on the phone and the patient's voice was close to my head, I also experienced a special*

> *kind of presence. Even when chatting, I had perceptions of the patients' bodily presence caused by what they wrote"*

Also visible were glimpses into the private spaces of patients, which were used by many therapists as additional diagnostic information (as described by respondent 159) or to bring themes into therapy (as described by respondent 110). *"I now know a lot more about the patient's living environment, which was readily opened up to me as well—of high diagnostic relevance!"* (respondent 159). *"The personal environment of the patient was more concrete for me and thus possible to include directly"*. (respondent 110). In addition, respondents described how they obtained missing perceptions by asking for them. For example, respondent 40 noted, *"missing observations were discussed verbally"*. However, this places a high demand on patients to put their perceptions into words, as respondent 195 thematized: *"Patients are extraordinarily challenged in verbalizing their emotions"*.

Male psychotherapists reported more often about the lack of the office as a therapeutic space (18.0%) than female psychotherapists (7.2%; χ^2 (1) = 5.148; $p = 0.023$). For the other two subcategories, no differences were observed between male and female therapists ($p \geq 0.080$).

3.2.6. Psychotherapeutic Relationship

A major category mentioned by $n = 159$ (73.3%) respondents is the "psychotherapeutic relationship". It includes as subcategories various aspects related to the quality and intensity of the therapeutic relationship. The four subcategories are illustrated in Figure 7.

$N = 98$ (45.2%) respondents described that the therapeutic relationship was strengthened or even intensified by the fact that patients experienced that their therapists were there for them even in the crisis and that psychotherapies were continued in the remote setting. Respondents stated that patients were very *"grateful"*, *"happy"*, *"relieved"*, *"unburdened"* or reacted *"positively"*. For example, respondent 28 voiced:

> *"I offered all my clients to use the new forms immediately after the announcement of the ÖBVP (the Austrian Federal Association for Psychotherapy, which informed psychotherapists that sessions were to be held remotely if possible), and this was received with "gratitude" or "relief". Some were afraid/worried about having to "go through the crisis alone". The quick provision of alternatives certainly had a positive influence on the relationship ("She doesn't leave me alone", "She is also there for me in the general crisis")"*

Repeatedly, the shared experience of the crisis was considered as uniting, as was the fact that patients, as well as therapists, were in lockdown, attended the sessions from home and sometimes both were navigating (technical) "uncharted territory". For example, respondent 42 described:

> *"Conversations were more personal because of Corona—in the sense that you share the lockdown situation. We are more or less in the same boat, and have similar difficulties (small apartments, bad WiFi, no childcare—so children who "barge in", etc.)—these are things you simply catch through Corona, because especially in the beginning everything was new, untested, spontaneous, complicated by external circumstances. (...) It was also more personal because of the way of communication: the patient is sitting comfortably at home, with a cup of coffee or tea, in familiar surroundings, without makeup and in her sweatpants, and she is just happy to be able to have contact with someone, due to Corona. This changes the nature of the conversation. I, as a therapist, of course tried to have a professional ambiance, yet I was also at home and in a similar situation"*

In addition, many respondents described how relational closeness and intimacy were generated in remote contact. They used adjectives such as *"open"*, *"confidential"*, *"personal"*, *"holding"*, *"trusting"*, *"strengthening"*, *"reliable"*, *"intense"*, *"consolidated"*, *"stable"*, *"cooperative"*, *"deepened"*, *"connected"*, *"secure"* and *"intimate"* to describe their and their patient's relational experience. Sometimes the closeness in remote contact was described as a "special" or "different" kind of closeness than that in face-to-face contact.

In this context, $n = 23$ (10.6%) respondents observed that in the remote setting, the atmosphere was more relaxed, and there was less negative transference in the therapeutic relationship. They attributed this to the spatial separation and to the fact that patients were at home in their safe environment.

In contrast, $n = 86$ (39.6%) respondents mentioned that they experienced less closeness in the psychotherapeutic relationship during remote sessions. "*Superficial*", "*difficult*", "*distant*", "*impersonal*", "*noncommittal*", "*flattened*", "*fragile*", "*lonely*", "*alienated*", "*cold*", "*businesslike*", "*less palpable*", "*less immediate*", "*foreign*", "*uncertain*" and "*reserved*" were adjectives used to describe relational experiences in remote sessions. In many cases, this was attributed to the fact that the other person is more difficult to "grasp" emotionally in remote contact and that atmospheric information is lost. Respondent 11 described this as a "*lack of relational immediacy*", and respondent 30 stated, "*I felt like I couldn't grasp the patient as well. It was more difficult to assess the client's emotional situation to the same extent as in a face-to-face conversation*". Respondent 138, in turn, commented "*on the relationship level, it was no longer possible to "tune in" as usual*", and respondent 154 elaborated by stating, "*establishing a presence in the relationship, being empathically accurate and empathizing at the moment and being congruent/immediately involved is more difficult, as a result of which the flow of the relationship often falters*".

In this context, $n = 33$ (15.2%) respondents observed that the holding function is impaired in the remote setting, i.e., respondents see their ability to emotionally support patients in crises, to provide support in difficult situations or to work through difficult issues therapeutically as limited.

Female psychotherapists expressed strengthening of the relationship more frequently than their male colleagues (49% vs. 32%; χ^2 (1) = 4.544; $p = 0.033$). For the other three subcategories, no differences were observed between male and female therapists ($p \geq 0.112$).

3.2.7. Intensity of Psychotherapeutic Work

A final major category mentioned by $n = 168$ (77.4%) respondents subsumes statements about the intensity of psychotherapeutic work and comprises four subcategories, which are displayed in Figure 8.

$N = 70$ (32.3%) respondents experienced high or even higher intensity in remote psychotherapy. This was explained by the fact that emotions can be expressed more openly in the remote setting, and difficult or shameful topics can be raised more easily, as respondent 73 described: "*The distance allowed some patients to be more open because there was less closeness and less shame*". Respondent 193 put it this way: "*also an increased possibility to approach previously avoided contents from a distance*". Patients were described as more disinhibited and open when they participated in sessions from the safety of their home environment. Respondent 79 observed, "*Some appreciated their familiar surroundings and were able to talk about more intimate topics*". Themes activated by the pandemic also came into therapy and could be elaborated, which sometimes deepened the process, as respondent 62 reported: "*Patients perceive the switch* (to remote psychotherapy sessions) *as a form of caring (being concerned about them, making an effort, etc.), which sometimes also evokes memories, longings, deprivations,* etc. *regarding childhood*". A greater density and thus intensity of the conversations was also described, here by respondent 196: "*With many patients, an increase in intensity was noticeable, the conversations were denser and more often led to a mutually satisfactory result*".

In this context, $n = 17$ (7.8%) respondents also mentioned that the therapeutic work was more focused on topics or therapy goals. For example, respondent 177 commented, "*Condensed, rapid delving into all relevant topics of concern*". Respondent 135 observed, "*For many patients, the work was even more to-the-point and focused on change*". Respondents explained this as a result of increased concentration in remote contact and of the need for both parties to verbalize emotions more, as well as to focus attention on the available channels of perception and, in particular, on the spoken word.

However, respondents also made contrary observations. $N = 128$ (59%) respondents described that the intensity of therapeutic sessions decreased in remote psychotherapy, for example, because processes were disrupted by technical difficulties or because it was not possible to use the full range of interventions, or because patients did not engage emotionally and presented only everyday topics. More in-depth or biographical work was avoided, which was experienced by respondents as a flattening of the content. Respondent 6 summed this up with her statement, "*In some conversations, a kind of coffeehouse gossip atmosphere arose for a short time since otherwise you only talk on the phone with friends for such a long time*". Respondent 124 also commented pointedly that patients remained in their "*comfort zone*". Conversations were described as more rational and less emotional.

COVID-19 as a topic and the issues the pandemic raised (coping with everyday life, fears, dealing with COVID-19 preventive measures, job loss, etc.) were the focus of remote psychotherapies. Respondent 94 commented, "*Conversations became more superficial. It became almost impossible to explore topics in depth. The topics were limited to current events and Covid measures, and the original goal of the therapy was neglected. The intensity of the conversations decreased a lot*". Respondent 32 also noted that "*ongoing processes and reflections were interrupted*". At the same time, it was emphasized several times that the engagement with daily events was not necessarily due to the switch to remote psychotherapy but was due to the crisis. "*It wasn't the switch that changed the issues, it was the crisis that changed the issues*" (respondent 137).

In this context, $n = 76$ (35%) respondents stated that for them, the supportive function of therapy was the primary focus of remote contacts during the pandemic. This involved crisis intervention and counseling in the "*here & now*" (respondent 52) to relieve stress. Respondents described how they worked in a more supportive, resource-oriented and structuring way and were more directive and "*less exploratory*" (respondent 11).

No gender differences became evident in the frequencies of all reports related to the subcategories relating to the intensity of therapeutic work ($p \geq 0.064$).

4. Discussion

This study aimed to survey the changes experienced by Austrian psychotherapists when switching from face-to-face to remote psychotherapy in the first year of the COVID-19 pandemic.

An important finding of the analysis is that neither therapeutic orientation nor years of professional experience had any influence on perceived changes when switching from face-to-face to remote psychotherapy or vice versa during the pandemic. This raises the assumption that differences between therapeutic orientations are sometimes given too much weight. As has already been shown in research on psychotherapy outcomes, different therapeutic orientations are similarly effective, and differences in effectiveness are due to factors other than therapeutic orientation [35,36].

This study further showed that working from home, especially the elimination of travel time to and from the office, allowed the surveyed therapists more flexibility in time management. This result is reflected in other studies [37–39]. Therapists with younger children, in particular, benefited from the greater flexibility. On the other hand, the elimination of time spent traveling also led to a loss of mental space for patients for reflective discussion before and after the session. This was also observed by Ahlström et al. [22]. In particular, the male therapists in our study reported missing the office as a therapeutic space. This could be because women found working from home more convenient, especially due to childcare responsibilities, and therefore did not miss face-to-face practice as much. In fact, another Austrian study showed that male psychotherapists treated more patients on average in face-to-face contact than female psychotherapists during the COVID-19 pandemic, which suggests that they continued working from their office or returned to their offices more rapidly [40].

Challenges mentioned by many respondents were the occurrence of technical problems and a reduced perception of sensory impressions. Technical problems were also reported in other studies from the same period [41,42]. As also noted by Jesser et al. [43] and Eichenberg et al. [44], therapists tried to compensate for the lack of non-verbal communication by focusing on other channels of perception. The respondents described this as exhausting and tiring, a finding that was also echoed by other authors [21,39,43]. The lack of a non-verbal level affected the therapeutic process. Respondents had difficulties in fine-tuning, found it harder to empathize, changed interventions and/or felt that an element of diagnostics was missing. Bayles et al. [45] argue that the quality of information is diminished in that therapeutic action is based on implicit and procedural non-verbal communication and that non-verbal information transmitted by the body in the setting at a distance is limited. Roesler [46] describes non-verbal information as essential in the process of mutual understanding. The loss or distortion of non-verbal elements has an impact on the patient's emotional security [46]. The accompanying lack of affective nuance can emotionally weaken the therapist's experience of working with the patient [41]. Respondents in our study also described a sense of loss of control in the remote setting. While they were able to choose their own space, they had no control over where their patients were or under what conditions they were attending. For many patients, finding an undisturbed space for confidential communication proved to be a challenge. This was also found by other authors [41,42,47]. The remote setting could also challenge patients to take more responsibility for themselves, which could be beneficial for patients with more moderate disorders. Simpson et al. [48], for example, described the "democratizing effect" of remote therapies, which enable patients to become more active in their own "territory". Furthermore, Jesser et al. [43] worked out that in a setting at a distance, successes can also be experienced more independently of the therapist.

In many cases, psychotherapy via video conferencing also offered our respondents insight into the patient's private environment and thus provided interesting additional information. Similarly, Jesser et al. [43] and Simpson et al. [48] described these insights as a unique opportunity to get a first-hand picture of the patients' life circumstances described in the sessions. Respondents did, however, describe the distraction caused by other people or animals in the household as a challenge, a finding that is also consistent with findings from other studies [42]. Furthermore, there is the challenge of separating private and professional life, as Liberati et al. [49] and Shklarski et al. [50] also noted. A study by Békés et al. [51] concluded that therapists who faced more challenges when switching to the digital setting tended to be younger. This could be related to family responsibilities. Therapists with young children face challenges in creating a space where they have the opportunity to engage with their patients in a focused and empathetic way. However, we found no evidence in our study that therapist age had an impact on perceived challenges of the remote setting, or on other observed changes related to the remote setting.

Respondents in our study, and female therapists in particular, mentioned the difficulties as well as the advantages of remote psychotherapy for certain diagnoses and patient groups. The observation that women seem to be more thoughtful about the difficulties and advantages of remote therapy for different patient groups and how to navigate this new setting could be attributed to women being more reflective and communicative in the study or in general [52]. Indeed, women provided, on average, 57% longer comments on free text questions vs. men. Respondents indicated that remote therapy, especially the setting via telephone, had proved helpful for patients with anxiety disorders. This finding is consistent with that of Jesser et al. [43], where respondents described that patients seemed more confident in remote treatment from their homes. Evidence from the research suggested the effectiveness of remote therapies for depression and/or anxiety disorders [10,12–15,53]. For the first time, according to our respondents, it was also possible for patients to address traumatic events in the setting at a distance. The remote treatment helped the patients to distance themselves from the events; furthermore, the setting was experienced as less fraught with feelings of shame. Previous study results already indicated the effectiveness of treating post-traumatic stress disorder (PTSD) in a distance setting [54,55] and described it as a viable alternative compared to the face-to-face setting [56]. By contrast, other respondents in our study considered the remote treatment of traumatized patients to be more difficult, as it did not enable patients to be supported as well, e.g., in the case of dissociative disorders. Other evidence from research can also be found for this [43,57]. In this context, we might need to consider that the COVID-19 pandemic itself and its associated constraints constituted a traumatic experience for some people. From the literature, we know that people were disposed of different resources protecting them against the traumatic experience of the pandemic. Killgore et al. [58] found that resilience was higher among people who, for example, maintained more social relationships, engaged in outdoor activities, and exercised more. Further research could examine whether patients experienced psychotherapy as helpful in coping with the pandemic and how patients with more or less resilience benefited differently from remote psychotherapies.

It was more difficult for our surveyed therapists to provide psychotherapeutic treatment for children in the setting at a distance. They could only accompany during play without actively participating or intervening. This is consistent with other research findings, which already pointed out that significant elements (e.g., creative opportunities) are lost or cannot be used in the remote treatment of children [50,59]. Instead, therapists focused more on their patients' verbal communication, facial expressions, and tone of voice [60]. The ambiguity of the findings highlights the need for further research. How can the respective therapeutic methods be adapted to the remote treatment format [6], and are there possible contraindications for certain diagnoses or patient groups?

Our results suggest a higher variability in the duration and frequency of sessions in remote therapy. This would suggest that, in addition to psychotherapeutic work, crisis intervention and counseling settings have been given more space in the respondents' range of activities. Further research would be needed to determine to what extent the changed settings could be used for genuine therapeutic work or whether the focus of the work had shifted.

A significant issue indicated by our research is the abandonment or restriction of the use of therapeutic interventions in distance therapy. This was also observed by Cantone et al. [61]. Notermans et al. [62] found that interventions that intend to activate intense or aversive feelings are avoided in the setting of remote therapy. Probst et al. [26] also concluded that therapeutic interventions are considered more typical for face-to-face psychotherapy than for psychotherapy at a distance. This could be explained by the fact that in training, the use of therapeutic interventions has so far been taught exclusively in the context of face-to-face psychotherapy [26]. Further research is needed to determine whether genuine interventions can be adapted for remote therapy. Therapists may have been uncertain about using certain interventions in the remote setting due to a lack of experience. This could be counteracted by offering training on remote treatment that is rooted in education and training contexts.

We noticed a great ambivalence in our respondents' answers regarding the relational experiences and intensity experienced in the remote sessions. The continuation of therapy during the period of restrictions on outdoor activities was described by respondents as having a confidence-building and relationship-strengthening effect. Female therapists, in particular, described a strengthening of relationships. Arguing from a sociological perspective, this finding could be explained by women taking on more nurturing roles in society [63]. Research has shown that women shouldered much of the increased demands of housework and childcare during the pandemic [64]. It could be hypothesized that women are also more likely to take a nurturing role in therapy. Indeed, psychotherapy research has shown that female psychotherapists are more loyal, more optimistic, and less critical than their male colleagues and also more able to put their own person in the background [65]. While male therapists tend to use more confrontational techniques, female therapists intervene more empathically [66]. As a result, women may also be more likely to perceive the gratitude of their patients, which was particularly important during the pandemic. Huscsava et al. [42] and Bouchard et al. [11] also came to the conclusion that therapeutic relationships were strengthened by the continuation of psychotherapy in a remote setting during the pandemic. One narrative review already published in 2014 was able to show that for patients, the therapeutic relationship in psychotherapy via videoconferencing does not differ from the face-to-face setting [67]. Stoll et al. [68] also rated the therapeutic relationship in the online setting as equal or even better compared to the face-to-face setting. Some of the respondents also perceived a high or higher intensity. Emotions were expressed more openly; furthermore, difficult or embarrassing topics could be addressed more easily. A study by Stefan et al. [21] already provided indications that patients can open up more easily about embarrassing topics over the telephone. It seems that this effect is not limited to psychotherapy via telephone, as patients also felt more confident and less intimidated to talk openly about their emotional state and problems in the setting of videoconferencing [67]. In this context, Russell [69] pointed to the disinhibitory effect in online settings, which leads to some patients opening up more emotionally in video conferencing or telephone settings. Furthermore, Roesler [46] described the intensification effect, which often occurs in the context of virtual interaction. In this situation, information that is only transmitted in a restricted way is completed through the use of fantasy in an imaginative process that also includes the processes of projection and transference [46].

On the other hand, some respondents described a decrease in the intensity of the therapeutic sessions in the setting of distance therapy. Other authors also reported that the therapeutic work became more superficial in terms of content [20,22,42] and that the topics were increasingly oriented toward the patients' everyday life [42]. Some of the respondents also perceived less closeness in the psychotherapeutic relationship in remote therapy and/or experienced the establishment of a sustainable psychotherapeutic relationship as challenging. Psychotherapists interviewed in the study by Stefan et al. [21] also described the therapeutic relationship as more superficial. The respondents in our study also reported limited possibilities of being able to emotionally support the patient in the setting at a distance. Therapists, according to Germain et al. [70], may feel that they can only support their patients in a limited way (e.g., because they cannot offer a handkerchief). Huscsava et al. [42] concluded that therapists feel more insecure in the event of a crisis due to limited options for taking action.

The hypothesis put forward by Roesler [46] is that using technological means to interact psychotherapeutically leads to a fundamental change in interpersonal encounters, the intrinsic rules and consequences of which are still not understood sufficiently well. Given the ambivalence and ambiguity of the empirical findings found in various studies [43,51], it is clear that further research and, in particular, observational studies are needed to better understand interaction in the remote setting, especially in the absence of pandemic conditions.

There are several limitations to this study. Firstly, it is a non-randomized study with some confounding factors that might influence the results (e.g., experiences of tele-psychotherapy mainly relate to the time during the COVID-19 restrictions). Secondly, the cross-sectional design did not allow for obtaining therapists' experiences session by session, which in turn could lead to recall bias in the retrospective assessment of the change experienced when making the switch to remote therapy. Thirdly, only psychotherapists who had entered a valid email address in the Austrian list of psychotherapists were reached. Fourth, the survey was conducted online, which could lead to the higher participation of therapists with a higher preference for psychotherapy via videoconferencing. Fifth, it may not be possible to generalize the results to other countries since e-mental health services already have a long tradition in other countries, and therapists' attitudes and experiences may therefore differ. Finally, it would be interesting to investigate possible changes in therapists' attitudes toward the setting at a distance over time.

5. Conclusions

As a result of the COVID-19 pandemic, the forced and abrupt change in psychotherapeutic treatment format from face-to-face settings to remote psychotherapy faced psychotherapists with unique and complex challenges [50]. Our study showed that remote psychotherapy can be an option to ensure continuity in case of a crisis. Furthermore, the setting offers spatial and temporal flexibility, which means that appointments can be offered more quickly in case of the need for a higher frequency of sessions or in case of crises. Our study indicates that for some disorders (e.g., anxiety disorders), treatment at a distance does have benefits. Further clinical studies are needed to identify how these patients benefit from distance treatment.

At the same time, it was found that the change of setting led to feelings of insecurity on the part of the therapists and that the range of therapeutic interventions was not fully utilized. This underscores the relevance of further research on how the therapeutic methodology can be adapted to the remote setting and for which patients there might be a contraindication. Because remote treatment has not been included in the process of professionalization so far, we see a need to expand the training and further education offered to therapists accordingly. At any rate, the pandemic situation has shown that to fulfill the duty of care toward patients, new ways are needed to ensure psychotherapeutic care [48]. Treatment at a distance could constitute an alternative to counteract the already existing underprovision of psychotherapeutic care in Austria.

Author Contributions: Conceptualization, T.P.; methodology, T.P., A.J. and E.H.; formal analysis, A.J., M.S. and E.H.; investigation, T.P.; data curation, T.P.; writing—original draft preparation, A.J. and M.S.; writing—review and editing, E.H., B.H., P.S., W.S., E.M., H.S., D.L., C.P. and T.P.; visualization, A.J.; supervision, A.J. and T.P.; project administration, T.P. All authors have read and agreed to the published version of the manuscript.

Funding: This research received no external funding.

Institutional Review Board Statement: The study was conducted following the Declaration of Helsinki and approved by the Ethics Committee and the data protection officer of the University for Continuing Education Krems (EK GZ 27/2018-202).

Informed Consent Statement: Informed consent was obtained from all subjects involved in the study.

Data Availability Statement: The raw data supporting the conclusions of this article will be made available by the authors upon reasonable request after signing a confidentiality agreement.

Acknowledgments: Open Access Funding by the University for Continuing Education Krems.

Conflicts of Interest: The authors declare no conflict of interest.

References

1. Ronen-Setter, I.H.; Cohen, E. Becoming "Teletherapeutic": Harnessing Accelerated Experiential Dynamic Psychotherapy (AEDP) for Challenges of the Covid-19 Era. *J. Contemp. Psychother.* **2020**, *50*, 265–273. [CrossRef]
2. Beck-Hiestermann, F.M.L.; Kästner, D.; Gumz, A. Onlinepsychotherapie in Zeiten der Corona-Pandemie. *Psychotherapeut* **2021**, *66*, 372–381. [CrossRef] [PubMed]
3. Probst, T.; Haid, B.; Schimböck, W.; Stippl, P.; Humer, E. Psychotherapie auf Distanz in Österreich während COVID-19. Zusammenfassung der bisher publizierten Ergebnisse von drei Onlinebefragungen. *Psychother. Forum.* **2021**, *25*, 30–36. [CrossRef]
4. Probst, T.; Stippl, P.; Pieh, C. Changes in Provision of Psychotherapy in the Early Weeks of the COVID-19 Lockdown in Austria. *Int. J. Environ. Res. Public Health* **2020**, *17*, 3815. [CrossRef] [PubMed]
5. Austrian Federal Ministry of Health. *Internetrichtlinie für Psychotherapeutinnen und Psychotherapeuten*; Austrian Federal Ministry of Health: Vienna, Austria, 2005; p. 21.
6. Eichenberg, C. Online psychotherapy in times of the corona pandemic. *Psychotherapeut* **2021**, *66*, 195–202. [CrossRef] [PubMed]
7. Carlbring, P.; Andersson, G.; Cuijpers, P.; Riper, H.; Hedman-Lagerlöf, E. Internet-Based vs. Face-to-Face Cognitive Behavior Therapy for Psychiatric and Somatic Disorders: An Updated Systematic Review and Meta-Analysis. *Cogn. Behav. Ther.* **2018**, *47*, 1–18. [CrossRef]
8. Andersson, G.; Cuijpers, P.; Carlbring, P.; Riper, H.; Hedman, E. Guided Internet-Based vs. Face-to-Face Cognitive Behavior Therapy for Psychiatric and Somatic Disorders: A Systematic Review and Meta-Analysis. *World Psychiatry* **2014**, *13*, 288–295. [CrossRef]
9. Day, S.X.; Schneider, P.L. Psychotherapy Using Distance Technology: A Comparison of Face-to-Face, Video, and Audio Treatment. *J. Couns. Psychol.* **2002**, *49*, 499–503. [CrossRef]
10. Berryhill, M.B.; Halli-Tierney, A.; Culmer, N.; Williams, N.; Betancourt, A.; King, M.; Ruggles, H. Videoconferencing Psychological Therapy and Anxiety: A Systematic Review. *Fam. Pract.* **2019**, *36*, 53–63. [CrossRef] [PubMed]

11. Bouchard, S.; Allard, M.; Robillard, G.; Dumoulin, S.; Guitard, T.; Loranger, C.; Green-Demers, I.; Marchand, A.; Renaud, P.; Cournoyer, L.-G.; et al. Videoconferencing Psychotherapy for Panic Disorder and Agoraphobia: Outcome and Treatment Processes from a Non-Randomized Non-Inferiority Trial. *Front. Psychol.* **2020**, *11*, 2164. [CrossRef]
12. Andrews, G.; Basu, A.; Cuijpers, P.; Craske, M.G.; McEvoy, P.; English, C.L.; Newby, J.M. Computer Therapy for the Anxiety and Depression Disorders Is Effective, Acceptable and Practical Health Care: An Updated Meta-Analysis. *J. Anxiety Disord.* **2018**, *55*, 70–78. [CrossRef] [PubMed]
13. Castro, A.; Gili, M.; Ricci-Cabello, I.; Roca, M.; Gilbody, S.; Perez-Ara, M.Á.; Seguí, A.; McMillan, D. Effectiveness and Adherence of Telephone-Administered Psychotherapy for Depression: A Systematic Review and Meta-Analysis. *J. Affect. Disord.* **2020**, *260*, 514–526. [CrossRef] [PubMed]
14. Johansson, R.; Ekbladh, S.; Hebert, A.; Lindström, M.; Möller, S.; Petitt, E.; Poysti, S.; Larsson, M.H.; Rousseau, A.; Carlbring, P.; et al. Psychodynamic Guided Self-Help for Adult Depression through the Internet: A Randomised Controlled Trial. *PLoS ONE* **2012**, *7*, e38021. [CrossRef] [PubMed]
15. Heckman, T.G.; Heckman, B.D.; Anderson, T.; Lovejoy, T.I.; Markowitz, J.C.; Shen, Y.; Sutton, M. Tele-Interpersonal Psychotherapy Acutely Reduces Depressive Symptoms in Depressed HIV-Infected Rural Persons: A Randomized Clinical Trial. *Behav. Med.* **2017**, *43*, 285–295. [CrossRef]
16. Mohr, D.C.; Vella, L.; Hart, S.; Heckman, T.; Simon, G. The Effect of Telephone-Administered Psychotherapy on Symptoms of Depression and Attrition: A Meta-Analysis. *Clin. Psychol. Sci. Pract.* **2008**, *15*, 243–253. [CrossRef] [PubMed]
17. Pierce, B.S.; Perrin, P.B.; McDonald, S.D. Path Analytic Modeling of Psychologists' Openness to Performing Clinical Work with Telepsychology: A National Study. *J. Clin. Psychol.* **2020**, *76*, 1135–1150. [CrossRef]
18. Langarizadeh, M.; Tabatabaei, M.S.; Tavakol, K.; Naghipour, M.; Rostami, A.; Moghbeli, F. Telemental Health Care, an Effective Alternative to Conventional Mental Care: A Systematic Review. *Acta Inform. Med.* **2017**, *25*, 240–246. [CrossRef]
19. Humer, E.; Stippl, P.; Pieh, C.; Pryss, R.; Probst, T. Experiences of Psychotherapists with Remote Psychotherapy During the COVID-19 Pandemic: Cross-Sectional Web-Based Survey Study. *J. Med. Internet Res.* **2020**, *22*, e20246. [CrossRef]
20. Jesser, A.; Muckenhuber, J.; Lunglmayr, B.; Dale, R.; Humer, E. Provision of Psychodynamic Psychotherapy in Austria during the COVID-19 Pandemic: A Cross-Sectional Study. *Int. J. Environ. Res. Public Health* **2021**, *18*, 9046. [CrossRef]
21. Stefan, R.; Mantl, G.; Höfner, C.; Stammer, J.; Hochgerner, M.; Petersdorfer, K. Remote Psychotherapy during the COVID-19 Pandemic. Experiences with the Transition and the Therapeutic Relationship. A Longitudinal Mixed-Methods Study. *Front. Psychol.* **2021**, *12*, 5500. [CrossRef]
22. Ahlström, K.; von Below, C.; Forsström, D.; Werbart, A. Therapeutic Encounters at the Onset of the COVID-19 Pandemic: Psychodynamic Therapists' Experiences of Transition to Remote Psychotherapy. *Psychoanal. Psychother.* **2022**, *36*, 256–274. [CrossRef]
23. McClellan, M.J.; Florell, D.; Palmer, J.; Kidder, C. Clinician Telehealth Attitudes in a Rural Community Mental Health Center Setting. *J. Rural Ment. Health* **2020**, *44*, 62–73. [CrossRef]
24. Ertelt, T.W.; Crosby, R.D.; Marino, J.M.; Mitchell, J.E.; Lancaster, K.; Crow, S.J. Therapeutic Factors Affecting the Cognitive Behavioral Treatment of Bulimia Nervosa via Telemedicine versus Face-to-Face Delivery. *Int. J. Eat. Disord.* **2011**, *44*, 687–691. [CrossRef] [PubMed]
25. Cioffi, V.; Cantone, D.; Guerriera, C.; Architravo, M.; Mosca, L.L.; Sperandeo, R.; Moretto, E.; Longobardi, T.; Alfano, Y.M.; Continisio, G.I.; et al. Satisfaction Degree in the Using of VideoConferencing Psychotherapy in a Sample of Italian Psychotherapists during COVID-19 Emergency. In Proceedings of the 2020 11th IEEE International Conference on Cognitive Infocommunications (CogInfoCom), Mariehamn, Finland, 23–25 September 2020; pp. 000125–000132. [CrossRef]
26. Probst, T.; Haid, B.; Schimböck, W.; Reisinger, A.; Gasser, M.; Eichberger-Heckmann, H.; Stippl, P.; Jesser, A.; Humer, E.; Korecka, N.; et al. Therapeutic Interventions in In-person and Remote Psychotherapy: Survey with Psychotherapists and Patients Experiencing In-person and Remote Psychotherapy during COVID-19. *Clin. Psychol. Psychother.* **2021**, *28*, 988–1000. [CrossRef]
27. Harris, P.A.; Taylor, R.; Thielke, R.; Payne, J.; Gonzalez, N.; Conde, J.G. Research Electronic Data Capture (REDCap)—A Metadata-Driven Methodology and Workflow Process for Providing Translational Research Informatics Support. *J. Biomed. Inform.* **2009**, *42*, 377–381. [CrossRef]
28. Harris, P.A.; Taylor, R.; Minor, B.L.; Elliott, V.; Fernandez, M.; O'Neal, L.; McLeod, L.; Delacqua, G.; Delacqua, F.; Kirby, J.; et al. The REDCap Consortium: Building an International Community of Software Platform Partners. *J. Biomed. Inform.* **2019**, *95*, 103208. [CrossRef]
29. Levitt, H.M.; Bamberg, M.; Creswell, J.W.; Frost, D.M.; Josselson, R.; Suárez-Orozco, C. Journal Article Reporting Standards for Qualitative Primary, Qualitative Meta-Analytic, and Mixed Methods Research in Psychology: The APA Publications and Communications Board Task Force Report. *Am. Psychol.* **2018**, *73*, 26–46. [CrossRef]
30. Heidegger, K.-E. Psychotherapy in Austria. Available online: https://www.europsyche.org/situation-of-psychotherapy-in-various-countries/austria/ (accessed on 25 May 2020).
31. Hsieh, H.-F.; Shannon, S.E. Three Approaches to Qualitative Content Analysis. *Qual. Health Res.* **2005**, *15*, 1277–1288. [CrossRef]
32. Atlas.ti. Atlas.ti: The Qualitative Data Analysis & Research Software 2018. Available online: https://atlasti.com/de/ (accessed on 27 January 2023).

33. Woods, M.; Paulus, T.; Atkins, D.P.; Macklin, R. Advancing Qualitative Research Using Qualitative Data Analysis Software (QDAS)? Reviewing Potential Versus Practice in Published Studies Using ATLAS.Ti and NVivo, 1994–2013. *Soc. Sci. Comput. Rev.* **2016**, *34*, 597–617. [CrossRef]
34. Krippendorff, K. *Content Analysis: An Introduction to Its Methodology*, 4th ed.; Sage: Los Angeles, CA, USA, 2019; ISBN 978-1-5063-9566-1.
35. Luborsky, L.; Singer, B.; Luborsky, L. Comparative Studies of Psychotherapies. Is It True That "Everywon Has One and All Must Have Prizes"? *Arch. Gen. Psychiatry* **1975**, *32*, 995–1008. [CrossRef]
36. Smith, M.L.; Glass, G.V. Meta-Analysis of Psychotherapy Outcome Studies. *Am. Psychol.* **1977**, *32*, 752–760. [CrossRef]
37. Leukhardt, A.; Heider, M.; Reboly, K.; Franzen, G.; Eichenberg, C. Videobasierte Behandlungen in der psychodynamischen Psychotherapie in Zeiten der COVID-19-Pandemie. *Psychotherapeut* **2021**, *66*, 398–405. [CrossRef] [PubMed]
38. Höfner, C.; Mantl, G.; Korunka, C.; Hochgerner, M.; Straßer, M. Psychotherapie in Zeiten Der COVID-19-Pandemie: Veränderung Der Arbeitsbedingungen in Der Versorgungspraxis. *Feedback* **2021**, *1–2*, 23–37.
39. McBeath, A.; du Plock, S.; Bager-Charleson, S. The Challenges and Experiences of Psychotherapists Working Remotely during the Coronavirus Pandemic. *Couns. Psychother. Res.* **2020**, *20*, 394–405. [CrossRef] [PubMed]
40. Humer, E.; Pieh, C.; Kuska, M.; Barke, A.; Doering, B.K.; Gossmann, K.; Trnka, R.; Meier, Z.; Kascakova, N.; Tavel, P.; et al. Provision of Psychotherapy during the COVID-19 Pandemic among Czech, German and Slovak Psychotherapists. *Int. J. Environ. Res. Public Health* **2020**, *17*, 4811. [CrossRef]
41. Markowitz, J.C.; Milrod, B.; Heckman, T.G.; Bergman, M.; Amsalem, D.; Zalman, H.; Ballas, T.; Neria, Y. Psychotherapy at a Distance. *Am. J. Psychiatry* **2021**, *178*, 240–246. [CrossRef]
42. Huscsava, M.; Plener, P.; Kothgassner, O.D. Teletherapy for Adolescent Psychiatric Outpatients: The Soaring Flight of so Far Idle Technologies during the COVID-19 Pandemic. *Digit. Psychol.* **2020**, *1*, 32–35. [CrossRef]
43. Jesser, A.; Muckenhuber, J.; Lunglmayr, B. Psychodynamic Therapist's Subjective Experiences with Remote Psychotherapy During the COVID-19-Pandemic—A Qualitative Study with Therapists Practicing Guided Affective Imagery, Hypnosis and Autogenous Relaxation. *Front. Psychol.* **2022**, *12*, 6196. [CrossRef]
44. Eichenberg, C.; Raile, P.; Becher, S.; Dapeci, C.; Pacher, J.; Rach, P.J.M.; Schiller, B.; Wimmer, E.; Winter, L. Online- und Telepsychotherapie: Über den Wechsel des Settings während der COVID-19-Pandemie. *Psychother. Wiss.* **2021**, *11*, 71–79. [CrossRef]
45. Bayles, M. Is Physical Proximity Essential to the Psychoanalytic Process? An Exploration through the Lens of Skype? *Psychoanal. Dial.* **2012**, *22*, 569–585. [CrossRef]
46. Roesler, C. Tele-Analysis: The Use of Media Technology in Psychotherapy and Its Impact on the Therapeutic Relationship. *J. Anal. Psychol.* **2017**, *62*, 372–394. [CrossRef]
47. Haas, L.J.; Benedict, J.G.; Kobos, J.C. Psychotherapy by Telephone: Risks and Benefits for Psychologists and Consumers. *Prof. Psychol. Res. Pract.* **1996**, *27*, 154–160. [CrossRef]
48. Simpson, S.; Richardson, L.; Pietrabissa, G.; Castelnuovo, G.; Reid, C. Videotherapy and Therapeutic Alliance in the Age of COVID-19. *Clin. Psychol. Psychother.* **2021**, *28*, 409–421. [CrossRef] [PubMed]
49. Liberati, E.; Richards, N.; Parker, J.; Willars, J.; Scott, D.; Boydell, N.; Pinfold, V.; Martin, G.; Dixon-Woods, M.; Jones, P. Remote Care for Mental Health: Qualitative Study with Service Users, Carers and Staff during the COVID-19 Pandemic. *BMJ Open* **2021**, *11*, e049210. [CrossRef] [PubMed]
50. Shklarski, L.; Abrams, A.; Bakst, E. Navigating Changes in the Physical and Psychological Spaces of Psychotherapists during COVID-19: When Home Becomes the Office. *Pract. Innov.* **2021**, *6*, 55–66. [CrossRef]
51. Békés, V.; Aafjes–van Doorn, K.; Prout, T.A.; Hoffman, L. Stretching the Analytic Frame: Analytic Therapists' Experiences with Remote Therapy during COVID-19. *J. Am. Psychoanal. Assoc.* **2020**, *68*, 437–446. [CrossRef]
52. Merchant, K. How Men and Women Differ: Gender Differences in Communication Styles, Influence Tactics, and Leadership Styles. CMC Sr. Theses 2012. Available online: https://scholarship.claremont.edu/cmc_theses/513/ (accessed on 27 January 2023).
53. Stubbings, D.R.; Rees, C.S.; Roberts, L.D.; Kane, R.T. Comparing In-Person to Videoconference-Based Cognitive Behavioral Therapy for Mood and Anxiety Disorders: Randomized Controlled Trial. *J. Med. Internet Res.* **2013**, *15*, e258. [CrossRef]
54. Kuester, A.; Niemeyer, H.; Knaevelsrud, C. Internet-Based Interventions for Posttraumatic Stress: A Meta-Analysis of Randomized Controlled Trials. *Clin. Psychol. Rev.* **2016**, *43*, 1–16. [CrossRef] [PubMed]
55. Morland, L.A.; Greene, C.J.; Rosen, C.S.; Foy, D.; Reilly, P.; Shore, J.; He, Q.; Frueh, B.C. Telemedicine for Anger Management Therapy in a Rural Population of Combat Veterans with Posttraumatic Stress Disorder: A Randomized Noninferiority Trial. *J. Clin. Psychiatry* **2010**, *71*, 855–863. [CrossRef]
56. Turgoose, D.; Ashwick, R.; Murphy, D. Systematic Review of Lessons Learned from Delivering Tele-Therapy to Veterans with Post-Traumatic Stress Disorder. *J. Telemed. Telecare* **2018**, *24*, 575–585. [CrossRef]
57. Wagner, B. Online-Therapie–eine neue Perspektive in der Psychotherapie für Flüchtlinge und Asylbewerber? *Psychother. Forum* **2016**, *21*, 124–131. [CrossRef]
58. Killgore, W.D.S.; Taylor, E.C.; Cloonan, S.A.; Dailey, N.S. Psychological Resilience during the COVID-19 Lockdown. *Psychiatry Res.* **2020**, *291*, 113216. [CrossRef] [PubMed]

59. Haslinger, M.; Weindl, D.; Peper-Bösenkopf, J.; Haiderer, M.; Singer, V.; Zajec, K. Psychosoziale Versorgung von Kindern Und Jugendlichen Im Ersten Corona-Lock-Down Unter Zuhilfenahme von Telefon Und Online-Tools. Möglichkeiten Und Grenzen. *Psychother. Forum* **2021**, *25*, 124–133. [CrossRef]
60. Erlandsson, A.; Forsström, D.; Rozental, A.; Werbart, A. Accessibility at What Price? Therapists' Experiences of Remote Psychotherapy with Children and Adolescents During the COVID-19 Pandemic. *J. Infant Child Adolesc. Psychother.* **2022**, *21*, 293–308. [CrossRef]
61. Cantone, D.; Guerriera, C.; Architravo, M.; Alfano, Y.M.; Cioffi, V.; Moretto, E.; Mosca, L.L.; Longobardi, T.; Muzii, B.; Maldonato, N.M.; et al. A sample of Italian psychotherapists express their perception and opinions of online psychotherapy during the covid-19 pandemic. *Riv. Psichiatr.* **2021**, *56*, 198–204. [PubMed]
62. Notermans, J.; Philippot, P. Psychotherapy under Lockdown: The Use and Experience of Teleconsultation by Psychotherapists during the First Wave of the COVID-19 Pandemic. *Clin. Psychol. Eur.* **2022**, *4*, 1–19. [CrossRef] [PubMed]
63. Blackstone, A.M. Gender Roles and Society. In *Human Ecology: An Encyclopedia of Children, Families, Communities, and Environments;* Miller, J.R., Lerner, R.M., Schiamberg, L.B., Eds.; ABC-CLIO: Santa Barbara, CA, USA, 2003; pp. 335–338.
64. Chung, H.; Birkett, H.; Forbes, S.; Seo, H. *Working from Home and the Division of Housework and Childcare among Dual Earner Couples during the Pandemic in the UK*; University of Kent: Canterbury, UK, 2020.
65. Peter, B.; Böbel, E.; Hagl, M.; Richter, M.; Kazén, M. Personality Styles of German-Speaking Psychotherapists Differ from a Norm, and Male Psychotherapists Differ from Their Female Colleagues. *Front. Psychol.* **2017**, *8*, 840. [CrossRef] [PubMed]
66. Staczan, P.; Schmuecker, R.; Koehler, M.; Berglar, J.; Crameri, A.; von Wyl, A.; Koemeda-Lutz, M.; Schulthess, P.; Tschuschke, V. Effects of Sex and Gender in Ten Types of Psychotherapy. *Psychother. Res. J. Soc. Psychother. Res.* **2017**, *27*, 74–88. [CrossRef]
67. Simpson, S.G.; Reid, C.L. Therapeutic Alliance in Videoconferencing Psychotherapy: A Review. *Aust. J. Rural Health* **2014**, *22*, 280–299. [CrossRef]
68. Stoll, J.; Müller, J.A.; Trachsel, M. Ethical Issues in Online Psychotherapy: A Narrative Review. *Front. Psychiatry* **2020**, *10*, 993. [CrossRef]
69. Isaacs Russell, G. Remote Working during the Pandemic: A Second Q&A with Gillian Isaacs Russell. *Br. J. Psychother.* **2021**, *37*, 362–379. [CrossRef] [PubMed]
70. Germain, V.; Marchand, A.; Bouchard, S.; Guay, S.; Drouin, M.-S. Assessment of the Therapeutic Alliance in Face-to-Face or Videoconference Treatment for Posttraumatic Stress Disorder. *Cyberpsychol. Behav. Soc. Netw.* **2010**, *13*, 29–35. [CrossRef] [PubMed]

life

Article

Computer Aided COVID-19 Diagnosis in Pandemic Era Using CNN in Chest X-ray Images

Ali Alqahtani [1], Mirza Mumtaz Zahoor [2,3], Rimsha Nasrullah [2], Aqil Fareed [2], Ahmad Afzaal Cheema [2], Abdullah Shahrose [2], Muhammad Irfan [4], Abdulmajeed Alqhatani [5], Abdulaziz A. Alsulami [6], Maryam Zaffar [2,*] and Saifur Rahman [4]

1 Department of Networks and Communications Engineering, College of Computer Science and Information Systems, Najran University, Najran 61441, Saudi Arabia
2 Faculty of Computer Sciences, Ibadat International University, Islamabad 44000, Pakistan
3 Department of Computer and Information Sciences, Pakistan Institute of Engineering and Applied Sciences (PIEAS), Islamabad 44000, Pakistan
4 Electrical Engineering Department, College of Engineering, Najran University, Najran 61441, Saudi Arabia
5 Department of Information Systems, College of Computer Science and Information Systems, Najran University, Najran 61441, Saudi Arabia
6 Department of Information Systems, Faculty of Computing and Information Technology, King Abdulaziz University, Jeddah 21589, Saudi Arabia
* Correspondence: maryam.zaffar82@gmail.com

Citation: Alqahtani, A.; Zahoor, M.M.; Nasrullah, R.; Fareed, A.; Cheema, A.A.; Shahrose, A.; Irfan, M.; Alqhatani, A.; Alsulami, A.A.; Zaffar, M.; et al. Computer Aided COVID-19 Diagnosis in Pandemic Era Using CNN in Chest X-ray Images. *Life* **2022**, *12*, 1709. https://doi.org/10.3390/life12111709

Academic Editors: Daniele Giansanti and Hyunjin Park

Received: 15 September 2022
Accepted: 12 October 2022
Published: 26 October 2022

Abstract: Early detection of abnormalities in chest X-rays is essential for COVID-19 diagnosis and analysis. It can be effective for controlling pandemic spread by contact tracing, as well as for effective treatment of COVID-19 infection. In the proposed work, we presented a deep hybrid learning-based framework for the detection of COVID-19 using chest X-ray images. We developed a novel computationally light and optimized deep Convolutional Neural Networks (CNNs) based framework for chest X-ray analysis. We proposed a new COV-Net to learn COVID-specific patterns from chest X-rays and employed several machine learning classifiers to enhance the discrimination power of the presented framework. Systematic exploitation of max-pooling operations facilitates the proposed COV-Net in learning the boundaries of infected patterns in chest X-rays and helps for multi-class classification of two diverse infection types along with normal images. The proposed framework has been evaluated on a publicly available benchmark dataset containing X-ray images of coronavirus-infected, pneumonia-infected, and normal patients. The empirical performance of the proposed method with developed COV-Net and support vector machine is compared with the state-of-the-art deep models which show that the proposed deep hybrid learning-based method achieves 96.69% recall, 96.72% precision, 96.73% accuracy, and 96.71% F-score. For multi-class classification and binary classification of COVID-19 and pneumonia, the proposed model achieved 99.21% recall, 99.22% precision, 99.21% F-score, and 99.23% accuracy.

Keywords: COVID-19 pandemic; contact tracing; CNN; chest X-ray images; hybrid learning; machine learning; computer-aided diagnosis

1. Introduction

The first case of viral disease COVID-19 [1] was registered in December 2019 in China's city Wuhan, which was subsequently proclaimed in March 2020 as a pandemic by WHO (World Health Organization). Coronavirus is also recognized as SARS-CoV- [2]. It is from the same group as MERS-CoV and SARS-CoV, which were discovered in 2003 & 2015, respectively [3]. As stated by the European Centre for Disease Prevention and Control [4], on 29 July 2021 about 34,435,890 cases were reported as positive for COVID-19, out of which 743,712 deaths were reported. It affects almost every aspect of life including health, education, the economy, etc. A wide part of employees lost their livelihoods due to this outbreak. There are no proper medicines or vaccines discovered yet. However, it can be controlled to some extent through early detection. One of the most common and widely used techniques is RT-PCR [5], which is a real-time detection method. RT-PCR samples are collected through a swab that is inserted into the nose and mouth of the patient to collect the samples. These samples are then sent to labs for testing, but it is a complex, time-consuming manual practice. Automatic detection is an alternative method recommended for the early detection of coronavirus.

Early studies show that chest radiograph images of patients who are infected with coronavirus illustrate some irregularities. The easy availability and accessibility of these radiograph images in areas with limited resources make them better options than PCR [6,7]. However, to examine these radiograph images to detect the infected areas of the lungs, experienced and skilled radiologists are required, but the computer-aided diagnosis system CADx also solves this problem for radiologists. It detects the presence of the virus that causes COVID-19 quickly and with high precision. CADx, using chest X-rays, should have been designed to fight against this virus [8]. Machine learning (ML) and deep learning (DL) play a vital role in the medical field for the detection and treatment of many of the deadliest diseases like brain tumors, chest cancer, etc. During the last couple of decades, deep learning showed vast progress in terms of efficient and accurate predictions. Due to this great ability of generalization, it is able to solve many complex problems of computer vision like image classification, organ detection, disease identification, etc. [9,10].

Deep learning is an advanced field and is a further subclass of machine learning. Its CNN algorithm gives far better results than any other traditional algorithm. One of the best features of CNN is that it automatically extracts features from the images without using handy craft filters. Sometimes, parameter learning through a limited dataset can cause overfitting problems. This problem can be tackled by using pre-trained architectures of CNN like GoogleNet, DenseNet, and VGG-16, which are trained on the ImageNet datasets. By using the transfer learning technique, we can apply the pre-trained architectures to our limited dataset. For this purpose, we have to remove the last few layers of the pre-trained model and then test it on our specific dataset. Proper hyper-parameters and efficient fine-tuning make it a more effective approach [11,12].

In this study, we proposed a COV-Net architecture-based computer-aided diagnosis system for COVID-19 analysis. In the presented work, we used the benchmark dataset which contains images of chest X-rays of viral pneumonia-infected patients, COVID-19 patients, and normal images to train our proposed COV-Net model. In D-HL, boundary homogeneity-related deep features are extracted from the fully connected layer FC-1 of the proposed COV-Net architecture. For structural risk minimization and to enhance the generalization ability of the proposed framework, we used SVM as a classifier for the final prediction. The proposed COV-Net contains four convolutional blocks, and optimized arrangements of layers facilitate better and more efficient learning with fewer parameters as compared to state-of-the-art deep CNNs. The following are the contributions of this study:

1. A new, well fine-tuned CNN architecture named COV-Net with fewer parameters is proposed to diagnose COVID-19 efficiently.
2. Using edges exploitation operation in an optimized structure with the convolutional operator facilitates learning edges-related features of infection patterns in chest X-ray images. It leads to improved detection of COVID-19 in a timely manner.
3. A D-HL-based framework for COVID-19 and pneumonia identification in chest X-ray images was proposed by using new deep CNN and SVM.
4. We exploit the structural and empirical risk error minimization using the proposed COV-Net and ML classifier in hybrid learning (HL) for COVID-19 analysis. In the proposed deep hybrid learning scheme, the learning capability of the proposed CNN is explored and ML classifiers are used to enhance the discrimination proficiency of the proposed framework for chest X-ray analysis.

2. Related Work

Many experiments have been conducted in the last two years to recognize the viral disease COVID-19 by using deep learning methods and traditional machine learning techniques. L. Lin et al. [13] applied a framework that proposed CNN with ResNet50 as its backbone, which extracts 2D as well as 3D features from CT images and then combines them through max-pooling followed by the softmax function, which gives an AUC equal to 0.96. X. Xu et al. [14] devoted classic ResNet architecture to differentiate coronavirus from I-AVP. The model along with the location-attention mechanism provided 86.7% accuracy.

G. Biraja et al. [15] conducted a study to determine uncertainty using drop-weights based on BCNNs. He used a pre-trained model Resnet50-V2 with fully connected layers then applied drop-weights followed by a softmax layer. It gained an accuracy of 89.92%. W. Shuai et al. [16] exploited both static and dynamic data for the detection of patients with the potential to move from a malignant to a critical stage. Static data included a personal and clinical record of the person while a series of CT images served as dynamic data. They combined static data with dynamic and fed them to MLP, which then served as input for the long short-term memory (LSTM). J. Cheng et al. [17] designed a model which classifies four categories including COVID-19, influenza A/B, CAP, and non-pneumonia patients by using the U-Net-34 2D segmentation network. Resent-152 is the backbone of the 2D classification deep learning network and achieved 94.98% accuracy with an area under an AUC of 97.71. J. Shuo et al. [18] created and installed an AI system within four weeks for the detection of COVID-19 to reduce the burden on radiologists and clinicians.

They used UNet++ for lung segmentation of CT images along with ResNet50 and got a specificity of 0.922 and sensitivity of 0.974. N. Ali et al. [19] performed three different types of two-class classifications with four different classes (viral pneumonia, COVID-19, bacterial pneumonia, and normal). Due to the limited availability of the dataset, the transfer learning technique was used, which uses five pre-trained DL architectures: ResNet101, ResNet52, ResNet50, Inception- ResNet-V2, and Inception. Resnet50 showed the best results for all three binary classifications (classification 1: 96.1%, classification 2: 99.3%, classification 3: 99.7%). They also applied approaches with and without pre-training of COVID-CAP, and achieved 95.7% and 98.3% accuracy, respectively. L. Wang et al. [20] created a dataset called COVIDx containing 13,975 chest X-beam images and used the model COVID-Net for recognition of COVID-19, obtaining an accuracy of 92.4%. M. Abed Mohammed et al. [21] associated deep learning models (like DarkNet, GoogleNet, ResNet50, MobileNets V2, and Xception) and traditional ML models (like KNN, decision tree, ANN, SVM with linear kernel, and RBF), and results demonstrated that DL frameworks outperformed ML frameworks and achieved 98.8% accuracy with ResNet50 architecture, while ML model SVM achieved its best accuracy of 95% and 94% with RBF kernel. D.Hemdan et al. [22] used the COVIDX-Net framework which includes seven different CNN models (Visual Geometry Group Network (VGG19), Inception-ResNet-V2, DenseNet121, InceptionV3, ResNetV2, Xception, and MobileNetV2) to categorize COVID-19 negative or positive cases. Architectures VGG19 and DenseNet121 gave almost similar results for the detection of normal and COVID-19 and gave F1-scores of 0.89 and 0.91, respectively.

A. Khandakar et al. [23] developed a vigorous method for the automatic recognition of coronavirus and pneumonia from chest X-ray scans and used pre-trained DL models to maximize the accuracy of detection. H.S. Maghdid et al. [24] purposed a simple CNN model to detect COVID-19 for early diagnosis. V. Chauhan et al. [25] applied transfer learning methodology and used pre-trained models to extract features. The results of pre-trained models were combined with a prediction vector and majority voting was used for the final prediction. T. Rahman et al. [26] have proposed three different schemes of classifications: normal/pneumonia classification, bacterial/viral pneumonia classification, and normal/bacterial/viral pneumonia classification. M. Loey et al. [27] exploited the transfer learning technique with GAN to detect coronavirus using chest X-rays. GAN helped in decreasing the overfitting issue produced by the small dataset and increased the dataset to 30 times more than the original dataset. A. Degerli et al. [28] proposed a novel strategy that not only detects coronavirus but also quantifies the severity by creating infection maps. They tried two configurations: first, they froze the encoder layers, then they permitted them to fluctuate. A. O. Ibrahim et al. [29] proposed an automatic DL structure for coronavirus-infected areas. They trained and tested the proposed model to check the effectiveness and generalization by using slices of 2D CT.

Generally reported work in literature lakes advocated the following points:

1. Most of the work presented in the past has been assessed using only accuracy, but recall, precision, and F-score are better performance measures to evaluate the generalization of the model for the complex dataset.
2. In most of the previous works, only COVID-19 detection is performed. However, simply detecting COVID-19 is insufficient to diagnose other severe abnormalities, e.g., pneumonia.
3. In COVID-19 analysis, the detection rate of infected X-ray images from normal individuals is still challenging because of fewer inter-class variations.

To overcome these limitations, we proposed a multi-class chest X-ray classification method using standardized performance evaluation matrices like recall, precision, F-score, and accuracy for improved diagnosis.

3. Methods and Materials

In the proposed work, COVID-19 detection was performed by using the proposed COV-Net CNN and the ML classifier and included some phases. First, X-ray images went through the preprocessing pipeline, which included data augmentation. At that point, a preprocessed dataset was split into training and testing datasets. We trained our proposed COV-Net-based model by using a training dataset. Training accuracy and loss were computed after every epoch. Testing data were used to evaluate the performance of the proposed method by following the appraisal metrics of accuracy, precision, recall, and F-score. A detailed overview of the proposed methodology is demonstrated in Figure 1.

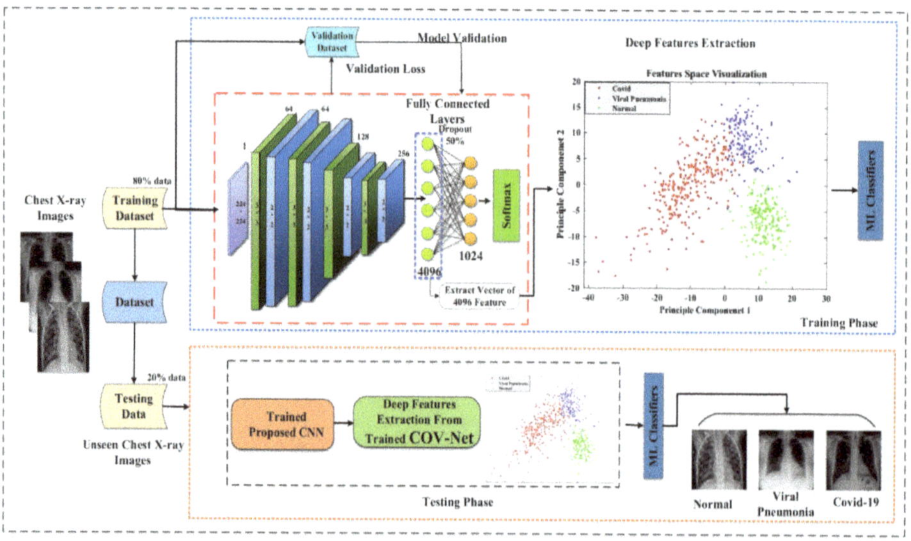

Figure 1. Block-based figure of proposed COVID-19 analysis model.

3.1. Dataset

In this work, we used the chest X-rays dataset. From the dataset, 300 normal chest X-ray pictures, 300 images of viral pneumonia, and only 300 images of coronavirus-infected patients were selected. All images were collected from the publicly available Kaggle repository [30]. The exhibition of the framework greatly depended upon the accuracy of the dataset. For this reason, we first sampled the data before using them. In data sampling, we only used those images that were useful and eliminated falsified images. The dataset contained chest X-rays of three classes (COVID-19/pneumonia/normal). All images are in JPEJ format as shown in Figure 2; the first one shows a normal chest X-ray, the second one shows a COVID-19 X-ray, & the third one shows a pneumonia X-ray.

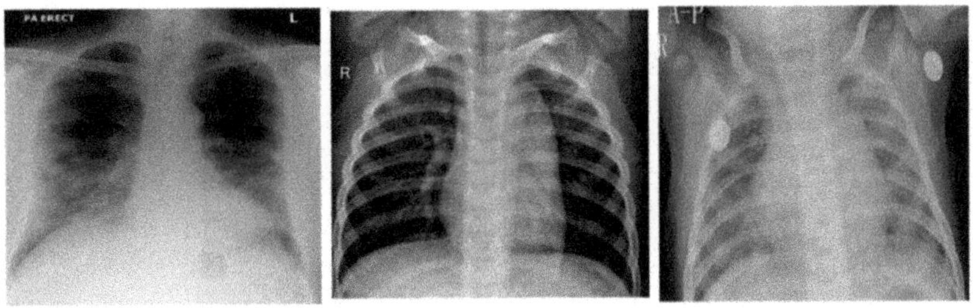

Figure 2. Sample images from dataset of three classes (normal, COVID-19, pneumonia).

3.2. *Data Augmentation*

Data augmentation is a method to increase the data samples during the training of the model. After employing data balancing and sampling, we contained 300 images of each class for better and generalized model training. We applied the data augmentation method to enhance and increase the dataset instances for better training of the model and to avoid overfitting [31]. Different data augmentation methods were applied in random rotation and random horizontal translation, as described in Table 1, which yielded an augmented dataset batch during training of the proposed model.

Table 1. Augmentation parameter details.

Parameters	Values
Random Rotation	[−5, 5]
Random Horizontal Translation	[−0.5, 1]
Random Vertical Translation	[−0.5, 1]

3.3. *Proposed CNN Architecture*

The proposed CNN architecture COV-Net used in this study included four convolutional blocks. Each block was constituted of a convolutional layer, batch normalization, and activation function, namely ReLU, followed by max-pooling as shown in Figure 3. In convolutional layers, filters convolved over the input image. The convolutional function performed the dot product of filter and valued and extracted features from the input images. CNN used a backpropagation algorithm for dynamic feature extraction. One of the advantages of CNN over ANN is that it automatically extracts domain-specific features from the images. By further using an edge operator (max-pooling), it learned profoundly discriminative features to train the model. In the pooling, layer down-sampling was also performed, which enhanced the performance of the model by making a small variation in the input image and by decreasing the non-linear dimensions of the resulting feature maps.

To highlight the features for classification, resulting feature maps were extracted from a fully connected layer. A dropout layer was added at the end to avoid overfitting. Detailedd layer wise description of propsed imodel is illusterated in Figure 4. The cross-entropy function was used as a cost function along with the softmax function. To categorize COVID-19, healthy people, and viral pneumonia, we used traditional ML classifiers, namely, random forest, Naïve Bayes, support vector machine (SVM), k-Nearest Neighbor (k-NN), and ensemble model.

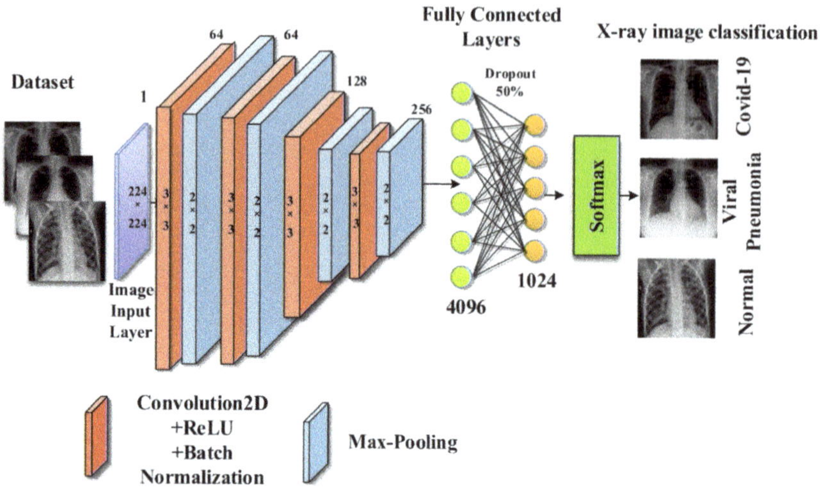

Figure 3. Detailed overview of proposed COV-Net.

#	Name	Type	Activations	Learnables
1	ImageInputLayer 256×256×1 images with 'zerocenter' normalization	Image Input	256×256×1	-
2	Conv-1 64 3×3 convolutions with stride [1 1] and padding 'same'	Convolution	256×256×64	Weights 3×3×1×64 Bias 1×1×64
3	bn_1 Batch normalization	Batch Normalization	256×256×64	Offset 1×1×64 Scale 1×1×64
4	ReLU-1 ReLU	ReLU	256×256×64	-
5	MaxPool_1 2×2 max pooling with stride [2 2] and padding 'same'	Max Pooling	128×128×64	-
6	Conv-2 64 3×3 convolutions with stride [1 1] and padding 'same'	Convolution	128×128×64	Weights 3×3×64×64 Bias 1×1×64
7	bn_2 Batch normalization	Batch Normalization	128×128×64	Offset 1×1×64 Scale 1×1×64
8	ReLU-2 ReLU	ReLU	128×128×64	-
9	MaxPool_2 2×2 max pooling with stride [2 2] and padding 'same'	Max Pooling	64×64×64	-
10	Conv-3 128 3×3 convolutions with stride [1 1] and padding 'same'	Convolution	64×64×128	Weights 3×3×64×128 Bias 1×1×128
11	bn_3 Batch normalization	Batch Normalization	64×64×128	Offset 1×1×128 Scale 1×1×128
12	ReLU-3 ReLU	ReLU	64×64×128	-
13	MaxPool_3 2×2 max pooling with stride [2 2] and padding 'same'	Max Pooling	32×32×128	-
14	Conv-4 256 3×3 convolutions with stride [1 1] and padding 'same'	Convolution	32×32×256	Weights 3×3×128×256 Bias 1×1×256
15	bn_4 Batch normalization	Batch Normalization	32×32×256	Offset 1×1×256 Scale 1×1×256
16	ReLU-4 ReLU	ReLU	32×32×256	-
17	MaxPool_4 2×2 max pooling with stride [2 2] and padding 'same'	Max Pooling	16×16×256	-
18	drop1 50% dropout	Dropout	16×16×256	-
19	fc1 4096 fully connected layer	Fully Connected	1×1×4096	Weights 4096×65536 Bias 4096×1
20	drop2 25% dropout	Dropout	1×1×4096	-
21	fc2 1024 fully connected layer	Fully Connected	1×1×1024	Weights 1024×4096 Bias 1024×1
22	fc3 3 fully connected layer	Fully Connected	1×1×3	Weights 3×1024 Bias 3×1
23	softmax softmax	Softmax	1×1×3	-
24	classoutput crossentropyex	Classification Output	1×1×3	-

Figure 4. Architectural detail of proposed COV-Net.

In this study, we used MATLAB to run the code. In the training phase of our proposed COV-Net model, we used the "rmsprop" function as an optimizer. It is a gradient-based method. It normalized the gradient by balancing the momentum, diminishing the progression for a large gradient to obtain from exploding, and expanding the progression for a small gradient to obtain from vanishing [32]. After an experimental analysis, an optimal learning rate of "0.0001" was selected. To reduce computational complexity, the batch size was set to 16 per epoch, which is a small size. To improve generalization, L1 regularization was used. As a cost function, the cross-entropy function was used along with the softmax function.

3.4. Implementation Details

The "RMSPROP" function was used as an optimizer. In the beginning, "0.0001" learning rate was selected randomly and 50 epochs were used, meaning each photo of training data was examined 50 times. As only 50 epochs were selected due to the limited dataset, we chose a large number of epoch models to move towards overfitting, which means instead of training, the model started removing the available small dataset. For this reason, we chose to lose many epochs to avoid overfitting.

3.5. Initial Training

We split the data into two parts: training and testing; 80% of them were used for training while 20% were set aside for testing according to Pareto's Principle [33]. We saved 10% of the 80% of the training dataset for validation, and the remaining 70% was utilized to train the model. Initial training helped us to check what our model can yield as a baseline model. Before starting training of the model, many different preprocessing techniques were used to boost the performance of the model. As we proposed, a CNN model was used so the training starts from scratch. In our proposed COV-Net model, we used the softmax function along with the cross-entropy cost function for classification.

3.6. Feature Extraction Using Proposed CNN Architecture

In the proposed work, we proposed new CNN architecture to obtain deep features from chest X-rays. The proposed CNN architecture extracts the most discriminative and deep features. The first fully connected layer (FC-1) extracted 4096 features from the images, which we used as a feature vector. Figure 5 shows the resulting feature maps from various layers of the CNN model of sample images of chest X-rays.

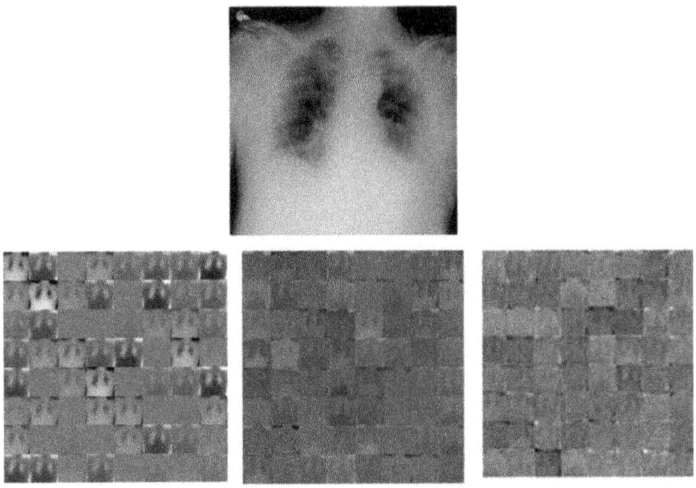

(**a**) Layer 1 (**b**) Layer 3 (**c**) Layer 5

Figure 5. Features maps representation from three different layers of proposed COVID-Net architecture, (**a**) layer1, (**b**) layer3, and (**c**) layer5.

At the primary level, almost complete data that are present in the input image are saved by activations.

(a) As we go to the higher layer, activation started to keep fewer data.
(b) At a deep level, the information became more detailed.

The uprising of the data into a more detailed and higher level was associated with each layer of the proposed CNN COV-Net (the deeper the network, the more composite the data and information). The proposed architecture COV-Net extracted features from input images. We extracted 4096 highlights from the FC-1 layer, and these highlight vectors were fed into different conventional machine learning classifiers as input to discover if the inspected patient was positive for COVID-19, viral pneumonia-infected, or just a normal patient. The dynamic features we used in our proposed model were driven by the FC-1 layer as shown in Table 2.

Table 2. Extracted features detail of proposed architecture.

Features Layer	Feature Dimension
FC-1	1 × 1 × 4096

3.7. *Classification Using Conventional ML Classifiers*

The proposed CNN COV-Net architecture was used to extract features from the augmented dataset. We extracted features from the FC-1 layer, and details are shown in Table 2. After extracting the features, these features were passed as input to conventional ML classifiers to train them. Different ML classifiers like Naïve Bayes, decision trees, KNN, and SVM determine the robustness of the classification. The performance of these models was measured by classifying COVID-19, pneumonia-infected, and healthy patients. The accuracy of classification attained by using conventional ML classifiers performed better than the softmax function. This is because it extracted the most highlighted features from chest X-rays of different patients by using the most abstract feature extraction techniques.

3.7.1. SVM

SVM is a linear model. It can tackle linear and non-linear issues. Its basic idea is that it makes a line to separate two classes. New data components are assigned to one class based on predictive analysis. As a rule, a parallel classifier expects that the data being referred to contain two potential objective variables. It utilizes a procedure called kernel trick to change the data and then find boundaries between them. It groups data and trains models inside really limited levels of extremity, making a three-dimensional order model that simply follows the X/Y prescient axis [34].

$$L(\gamma, \alpha, \beta) = \frac{1}{2} \parallel \gamma \parallel^2 - \sum \sum_{i=1}^{m} \beta i[yi(\gamma \cdot x + \alpha)1] \tag{1}$$

3.7.2. k_NN

This is used for regression as well as classification. Its calculation utilizes highlight closeness to anticipate the upsides of any new information focuses, implying that the new point is allocated a worth dependent on how intently it resembles training dataset points [32].

$$\sqrt{\sum_{i=1}^{n} (qi - pi)^2} \tag{2}$$

3.7.3. Naïve Bayes

It is a group of algorithms that are based on the "Bayes Theorem". They work on the principle that every pair of classifying features is independent [35].

$$P(U|V) = \frac{P(V|U)P(U)}{P(V)} \tag{3}$$

3.7.4. Random Forest

It is fundamentally a supervised method. It is an ensemble model which contains multiple decision trees. It collects results from all decision trees and then, based on the highest voting, makes a decision [36].

$$RFfi_i = \frac{\sum j \in all\ trees\ normfi_{ij}}{T} \tag{4}$$

3.8. Performance Metrics

Classification performance of the model is calculated through different performance metrics, for example, accuracy [37], recall [38], precision [38], and F-score [39], etc. When classifying medical images, we use different terms like false negative, false positive, etc.

3.8.1. Precision

It is the proportion of correct positive predictions to the total positive prediction. It indicates the rate of correct positive predictions. It is calculated as:

$$Precision = \frac{TP}{(TP + FP)} \times 100 \tag{5}$$

3.8.2. Recall

In this, we calculate true positive predictions from total positive predictions that might have been made. It shows a number of missing positive predictions. It is calculated as:

$$Recall = \frac{TP}{(TP + FN)} \times 100 \tag{6}$$

3.8.3. Accuracy

It is the most regular performance measure. It gives correct predictions to the total predictions.

$$Accuracy = \frac{(TN + TP)}{(FP + FN + TP + TN)} \tag{7}$$

3.8.4. F-Score

It shows steadiness between recall and precision.

$$F - score = \frac{2 \times (Precision + Recall)}{Precision + Recall} \tag{8}$$

4. Results

In our research, we presented a CNN model which extracted features from the augmented dataset. We had a small dataset, so we applied the data augmentation method to enhance the dataset. The augmented dataset also played an important part in accuracy improvement because of its high generalization ability. The proposed model was trained with 50 epochs under a batch size of 8. Deep and discriminative features were extracted from the proposed CNN architectures. The extracted features were passed as input to some conventional ML classifiers, e.g., Naïve Bayes, KNN, random forest, and support vector machine. In the event of binary classification of COVID-19 and pneumonia, KNN and SVM achieved 100% accuracy, recall, precision, and F1-score, shown in Tables 3 and 4.

Table 3. Performance comparison using ML classifiers for two classes (Pne = pneumonia, Cov = COVID-19). The Bold shows results of proposed method.

Classifiers	Parameters	Type	TP	FP	FN	Recall (%)	Precision (%)	F1-Score (%)	Accuracy (%)
KNN	K = 2	Cov	60	0	0	100	100	100	100
		Pne	60	0	0				
	K = 3	Cov	60	0	0	100	100	100	100
		Pne	60	0	0				
	K = 4	Cov	60	0	0	100	100	100	100
		Pne	60	0	0				
	K = 5	Cov	59	1	0	99.2	99.2	99.2	99.2
		Pne	60	0	1				
SVM	**Linear**	**Cov**	**60**	**0**	**1**	**99.21**	**99.22**	**99.21**	**99.23**
		Pne	**59**	**1**	**0**				
	RBF	Cov	60	0	1	99.2	96.2	97.7	99.2
		Pne	59	1	0				
	Gaussian	Cov	60	0	1	99.2	99.2	99.2	99.2
		Pne	50	1	0				
	PolyOrder-2	Cov	60	0	0	100	100	100	100
		Pne	60	0	0				
	PolyOrder-3	Cov	60	0	1	99.2	99.2	99.2	99.2
		Pne	59	1	0				
	PolyOrder-4	Cov	60	0	0	100	100	100	100
		Pne	60	0	0				
	PolyOrder-5	Cov	60	0	0	100	100	100	100
		Pne	60	0	0				
Decision tree		Cov	55	5	1	95	95.2	95.1	95.0
		Pne	59	1	5				
Naïve Bayes		Cov	59	1	1	98.3	98.3	98.3	98.3
		Pne	59	1	1				
RF	max no. of splits 5	Cov	59	1	0	99.15	99.2	99.2	99.2
		Pne	60	0	1				

Table 4. Performance comparison of proposed framework using ML classifiers for three classes (Nor = normal, Cov = COVID-19, Pne = pneumonia). The Bold shows results of proposed method.

Classifiers	Parameters	Type	TP	FP	FN	Recall (%)	Precision (%)	F1-Score (%)	Accuracy (%)
K-Nearest Neighbors	K = 2	Cov	56	4	5	92.2	92.2	92.2	92.2
		Pne	56	4	4				
		Nor	54	6	5				
	K = 3	Cov	56	3	3	93.3	93.3	93.3	93.3
		Pne	58	2	5				
		Nor	54	6	3				
	K = 4	Cov	57	3	5	93.3	93.3	93.3	93.3
		Pne	58	2	4				
		Nor	53	7	3				
	K = 5	Cov	53	7	3	92.2	92.4	92.3	92.2
		Pne	58	2	8				
		Nor	55	5	3				
Decision Tree		Cov	50	10	11	81.1	81.1	81.1	81.1
		Pne	48	12	11				
		Nor	48	12	12				
Naïve Bayes		Cov	57	3	6	92.2	92.3	92.2	92.2
		Pne	56	4	3				
		Nor	53	7	5				
Random Forest	max no. of splits 5	Cov	56	4	2	95	95.1	95.1	95.0
		Pne	60	0	5				
		Nor	55	5	2				
SVM	**Linear**	**Cov**	**56**	**4**	**2**	**96.6**	**96.7**	**96.7**	**96.7**
		Pne	**60**	**0**	**2**				
		Nor	**58**	**2**	**2**				
	Gaussian	Cov	57	3	5	94.5	94.5	94.5	94.4
		Pne	58	2	3				
		Nor	55	5	2				
	RBF	Cov	56	4	5	93.9	93.9	93.9	93.9
		Pne	58	2	3				
		Nor	55	5	3				
	Poly- Order3	Cov	57	3	3	95.6	95.6	95.6	95.6
		Pne	60	0	3				
		Nor	55	5	2				
	Poly- Order4	Cov	57	3	3	95	95.1	95.03	95.0
		Pne	60	0	4				
		Nor	54	6	2				
	Poly- Order5	Cov	56	4	3	94.43	94.5	94.5	94.4
		Pne	60	0	5				
		Nor	54	6	2				

We also evaluated our proposed D-HL method with the baseline proposed COV-Net to emphasize the performance improvement of our proposed method. Table 5 proved that our proposed technique enhanced the discrimination strength of our proposed model in accuracy (1.73%) and F-score (1.68%).

Table 5. Proposed hybrid learning method comparison with proposed COV-Net.

Model	Recall	Precision	Accuracy	F-Score
Proposed COV-Net	95.0%	95.07%	95.0%	95.03%
Proposed D-HL-based Framework	96.69%	96.72%	96.73%	96.71%

5. Discussion

In the presented D-HL architecture, the softmax layer was replaced with a machine learning classifier. The CNN learning algorithm utilized empirical risk minimization as a method to reduce false positives and false negatives during training. When the back-propagation algorithm reaches the first hyperplane that separates, the training phase ends, and progress generally stops as a result. Another limitation of CNN is that it frequently assigns one output neuron a high value (around +1) while assigning low values to the other neurons (close to 1).

This makes it very difficult to reject implementation errors. Softmax classifiers provide us with likelihoods for each class label. On the other hand, conventional ML techniques help us develop a robust rejection strategy. The generalization ability of CNN is weaker compared to that of SVM DL approaches, in contrast to conventional ML methods, are the least understandable from an AI aspect and are assumed to as a black box.

We performed classification with three classes as well as with two classes. In three classes, pneumonia, normal, and COVID-19 were included and in binary classifications, we used COVID-19 and pneumonia. SVM gave the highest accuracy with three classes. We achieved outstanding accuracy in binary classification with SVM and KNN. In the case of three classes, SVM gave an accuracy of 96.7%, recall of 96.6%, precision equal to 96.7%, and F1-score equal to 96.7%. Table 3 shows the detailed overview of the proposed CNN architecture with all four conventional ML classifiers. Confusion matrixes based on the performance analysis of classifiers are demonstrated in Figure 5. We compared results obtained by the proposed method with other existing works based on different performance metrics. Apostolopoulos et al. [40] used five different CNN pre-trained architectures to classify between three classes(COVID-19, normal, and pneumonia) and gave a sensitivity of 98.66%, accuracy of 94.72%, and specificity of 96.46%. H.S. Maghdid et al. [24] used the transfer learning technique with the AlexNet model and got 94.1% accuracy, 72% sensitivity, and 100% specificity. S.S Khan et al. [41] applied a convolutional auto-encoder to achieve 0.7652 area under a curve. A. Narin et al. [19] also used five CNN models (ResNet50, ResNet101, ResNet152, inception-ResNetV2, and InceptionV3) to perform binary classification of four classes and achieve an accuracy of 96.1%, recall of 91.8%, specificity of 96.6%, F1-score of 83.5%, and precision of 76.5% with COVID-19 & normal binary classification. R. Kumar et al. [42] performed an experiment with DenseNet & GoogleNet and attained an F-score equal to 0.91, AUC: 0.97. Similarly, Makris A. et al. [43,44] used five different pre-trained CNN models and achieved 95% accuracy. Arora, R. et al [45] proposed stochastic deep learning model using ensemble of slandered convolutional models and evaluate developed model on standard dataset contain three classes: COVID-19, normal and pneumonia and attain an accuracy and AUC of 0.91 and 0.97, respectively. A detailed comparison is illustrated in Figure 6 and Table 6.

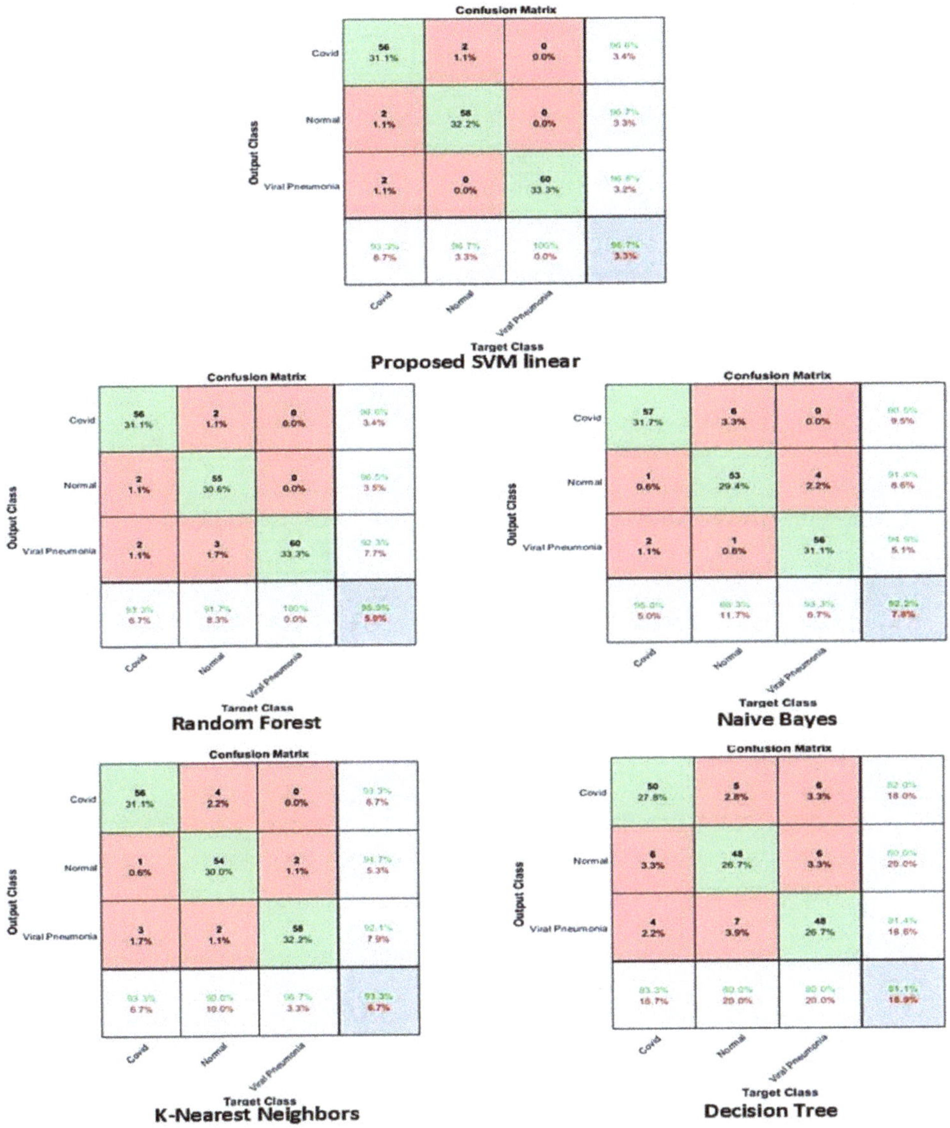

Figure 6. Confusion matrix-based performance analysis of competitive ML classifiers.

Experimental results show that our proposed models outperform all these experiments and achieved 96.69% recall, 96.72% precision, 96.73% accuracy, and 96.71% F-score, as shown in Table 5 and Figure 7.

Certain limitations still apply to our research investigation. The training period for feature extraction was lengthy due to the tiny batch sizes employed to extract the runtime features, which would have typically required a large amount of GPU RAM. Second, the proposed framework must go through a thorough clinical trial before radiologists' professional judgment may be utilized to resolve the patient data.

Table 6. Proposed hybrid learning method comparison with existing techniques on publicly available dataset. The Bold shows results of proposed method.

Author	Methodology	Recall	Precision	Accuracy	F-score
Apostolopoulos et al. (2020) [40]	VGG19, MobileNet, Inception, Xception, Inception ResNet v2.	98.6%	-	94.72%	-
H.S Maghdid et al. (2021) [24]	Transfer learning with AlexNet model	72%	-	94.1%	-
A. Narin et al. (2020) [19]	Pre-trained CNN architectures: ResNet50, ResNet101, ResNet152, inception-ResNetV2 and InceptionV3	91.8%	76.5%	96%	83.5%
Arora, R. et al. (2021) [45]	CNN architecture DenseNet & GoogleNet	91%	-	-	91%
Makris A. et al. (2020) [43]	5 pre-trained CNNs	-	-	95%	-
Proposed DH-L Framework	**Proposed COV-Net with conventional ML classifier**	**96.69%**	**96.72%**	**96.73%**	**96.71%**

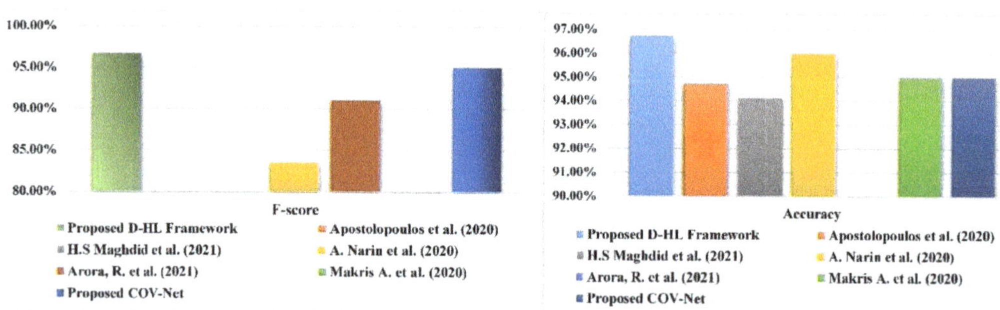

Figure 7. Comparative analysis of proposed COV-Net and D-HL with existing literature using accuracy and F-score [19,24,40,43,45].

6. Conclusions

Well-timed identification of COVID-19 infection is vital to preserve the patient's life and control the further spread of this life-threatening disease. In this study, a new CNN-based scheme for the detection of COVID-19 is proposed. COVID-19 analysis is performed using chest X-ray images containing three categories (pneumonia, COVID-19, and normal). Experimental results proved that the hybrid learning-based framework has shown improved performance compared to other methods. When the proposed framework's performance is compared with the state-of-the-art deep models', it shows that the proposed deep hybrid learning-based method achieved 96.69% recall, 96.72% precision, 96.73% accuracy, and 96.71% F-score for multi-class classification, and for COVID-19 and pneumonia we achieved 99.21% recall, 99.22% precision, 99.21% F-score, and 99.23% accuracy. The proposed COV-Net is less complex than pre-trained and custom-designed networks, and it is feasible to run it on ordinary current PCs. This is conceivable because the algorithm requires fewer resources for both training and execution. Performance analysis is carried out to attain the generalized model and it is likely to assist radiologists in making decisions in their clinical practice.

Author Contributions: Conceptualization, supervision, M.M.Z. and R.N.; methodology, M.M.Z.; software, validation, A.F. and A.S.; formal analysis, M.I. and A.A. (Ali Alqahtani); investigation, A.A. (Abdulmajeed Alqhatani); resources, M.Z. and A.A. (Ali Alqahtani); writing—original draft preparation, R.N., A.A.C. and A.F.; writing—review and editing, S.R. and A.A.A.; project administration,

M.Z. and M.M.Z.; funding acquisition, A.A. (Abdulmajeed Alqhatani), A.A. (Ali Alqahtani); and M.I. All authors have read and agreed to the published version of the manuscript.

Funding: The authors acknowledge the support from the Deanship of Scientific Research, Najran University, Kingdom of Saudi Arabia, for funding this work under the research group funding program grant code number (NU/RG/SERC/11/3).

Institutional Review Board Statement: Not applicable.

Informed Consent Statement: Not applicable.

Data Availability Statement: Data are available in publicly accessible repositories which are described in Section 3.1.

Acknowledgments: The authors acknowledge the support from the Deanship of Scientific Research, Najran University, Kingdom of Saudi Arabia, for funding this work under the research group funding program grant code number (NU/RG/SERC/11/3).

Conflicts of Interest: The authors declare that they have no conflicts of interest to report regarding the present study.

References

1. Roosa, K.; Lee, Y.; Luo, R.; Kirpich, A.; Rothenberg, R.; Hyman, J.; Yan, P.; Chowell, G. Real-time forecasts of the COVID-19 epidemic in China from February 5th to February 24th, 2020. *Infect. Dis. Model.* **2020**, *5*, 256–263. [CrossRef] [PubMed]
2. Coronaviridae Study Group of the International Committee on Taxonomy of Viruses. The species Severe acute respiratory syndrome-related coronavirus: Classifying 2019-nCoV and naming it SARS-CoV-2. *Nat. Microbiol.* **2020**, *5*, 536–544. [CrossRef] [PubMed]
3. Rabaan, A.A.; Al-Ahmed, S.H.; Haque, S.; Sah, R.; Tiwari, R.; Malik, Y.S.; Dhama, K.; Yatoo, M.I.; Bonilla-Aldana, D.K.; Rodriguez-Morales, A.J. SARS-CoV-2, SARS-CoV, and MERS-COV: A comparative overview. *Infez. Med.* **2020**, *28*, 174–184. [PubMed]
4. Evans, R. European Centre for Disease Prevention and Control. *Nurs. Stand.* **2014**, *29*, 30. [CrossRef] [PubMed]
5. Ren, L.-L.; Wang, Y.-M.; Wu, Z.-Q.; Xiang, Z.-C.; Guo, L.; Xu, T.; Jiang, Y.-Z.; Xiong, Y.; Li, Y.-J.; Li, X.-W.; et al. Identification of a novel coronavirus causing severe pneumonia in human: A descriptive study. *Chin. Med. J.* **2020**, *133*, 1015–1024. [CrossRef] [PubMed]
6. Wan, Z.; Zhang, Y.; He, Z.; Liu, J.; Lan, K.; Hu, Y.; Zhang, C. A Melting Curve-Based Multiplex RT-qPCR Assay for Simultaneous Detection of Four Human Coronaviruses. *Int. J. Mol. Sci.* **2016**, *17*, 1880. [CrossRef]
7. Khan, A.; Khan, S.H.; Saif, M.; Batool, A.; Sohail, A.; Khan, M.W. A Survey of Deep Learning Techniques for the Analysis of COVID-19 and their usability for Detecting Omicron. *arXiv* **2022**, arXiv:2202.06372.
8. He, J.-L.; Luo, L.; Luo, Z.-D.; Lyu, J.-X.; Ng, M.-Y.; Shen, X.-P.; Wen, Z. Diagnostic performance between CT and initial real-time RT-PCR for clinically suspected 2019 coronavirus disease (COVID-19) patients outside Wuhan, China. *Respir. Med.* **2020**, *168*, 105980. [CrossRef] [PubMed]
9. Palagi, L.; Pesyridis, A.; Sciubba, E.; Tocci, L. Machine Learning for the prediction of the dynamic behavior of a small scale ORC system. *Energy* **2019**, *166*, 72–82. [CrossRef]
10. Zahoor, M.M.; Qureshi, S.A.; Bibi, S.; Khan, S.H.; Khan, A.; Ghafoor, U.; Bhutta, M.R. A New Deep Hybrid Boosted and Ensemble Learning-Based Brain Tumor Analysis Using MRI. *Sensors* **2022**, *22*, 2726. [CrossRef] [PubMed]
11. Guo, Y.; Liu, Y.; Oerlemans, A.; Lao, S.; Wu, S.; Lew, M.S. Deep learning for visual understanding: A review. *Neurocomputing* **2016**, *187*, 27–48. [CrossRef]
12. Zahoor, M.M.; Qureshi, S.A.; Khan, A.; Rehman, A.U.; Rafique, M. A novel dual-channel brain tumor detection system for MR images using dynamic and static features with conventional machine learning techniques. *Waves Random Complex Media* **2022**, 1–20. [CrossRef]
13. Li, L.; Qin, L.; Xu, Z.; Yin, Y.; Wang, X.; Kong, B.; Bai, J.; Lu, Y.; Fang, Z.; Song, Q.; et al. Using artificial intelligence to detect COVID-19 and community-acquired pneumonia based on pulmonary CT: Evaluation of the diagnostic accuracy. *Radiology* **2020**, *296*, E65–E71. [CrossRef] [PubMed]
14. Xu, X.; Jiang, X.; Ma, C.; Du, P.; Li, X.; Lv, S.; Yu, L.; Chen, Y.; Su, J.; Lang, G. Deep Learning System to Screen novel Coronavirus Disease 2019 Pneumonia. *Engineering* **2020**, *6*, 1122–1129. [CrossRef] [PubMed]
15. Ghoshal, B.; Tucker, A. Estimating Uncertainty and Interpretability in Deep Learning for Coronavirus (COVID-19) Detection. *arXiv* **2020**, arXiv:2003.10769.
16. Wang, S.; Kang, B.; Ma, J.; Zeng, X.; Xiao, M.; Guo, J.; Cai, M.; Yang, J.; Li, Y.; Meng, X.; et al. A deep learning algorithm using CT images to screen for Corona Virus Disease (COVID-19). *Eur. Radiol.* **2021**, *31*, 6096–6104. [CrossRef]
17. Jin, C.; Chen, W.; Cao, Y.; Xu, Z.; Tan, Z.; Zhang, X.; Deng, L.; Zheng, C.; Zhou, J.; Shi, H.; et al. Development and evaluation of an artificial intelligence system for COVID-19 diagnosis. *Nat. Commun.* **2020**, *11*, 5088. [CrossRef]
18. Jin, S.; Wang, B.; Xu, H.; Luo, C.; Wei, L.; Zhao, W.; Hou, X.; Ma, W.; Xu, Z.; Zheng, Z.; et al. AI-assisted CT imaging analysis for COVID-19 screening: Building and deploying a medical AI system in four weeks. *medRxiv* **2020**. [CrossRef]

19. Narin, A.; Kaya, C.; Pamuk, Z. Automatic detection of coronavirus disease (COVID-19) using X-ray images and deep convolutional neural networks. *Pattern Anal. Appl.* **2021**, *24*, 1207–1220. [CrossRef]
20. Wang, L.; Lin, Z.Q.; Wong, A. COVID-Net: A tailored deep convolutional neural network design for detection of COVID-19 cases from chest X-ray images. *Sci. Rep.* **2020**, *10*, 19549. [CrossRef]
21. Mohammed, M.A.; Abdulkareem, K.; Garcia-Zapirain, B.; Mostafa, S.A.; Maashi, M.S.; Al-Waisy, A.S.; Subhi, M.A.; Mutlag, A.A.; Le, D.-N. A Comprehensive Investigation of Machine Learning Feature Extraction and Classification Methods for Automated Diagnosis of COVID-19 Based on X-Ray Images. *Comput. Mater. Contin.* **2021**, *66*, 3289–3310. [CrossRef]
22. El-Din Hemdan, E.; Shouman, M.A.; Karar, M.E. COVIDX-Net: A Framework of Deep Learning Classifiers to Diagnose COVID-19 in X-Ray Images. *arXiv* **2020**, arXiv:2003.11055.
23. Chowdhury, M.E.H.; Rahman, T.; Khandakar, A.; Mazhar, R.; Kadir, M.A.; Bin Mahbub, Z.; Islam, K.R.; Khan, M.S.; Iqbal, A.; Al Emadi, N.; et al. Can AI Help in Screening Viral and COVID-19 Pneumonia? *IEEE Access* **2020**, *8*, 132665–132676. [CrossRef]
24. Maghdid, H.S.; Asaad, A.T.; Ghafoor, K.Z.G.; Sadiq, A.S.; Mirjalili, S.; Khan, M.K.K. Diagnosing COVID-19 pneumonia from x-ray and CT images using deep learning and transfer learning algorithms. In Proceedings of the Multimodal Image Exploitation and Learning 2021, Online, 12–17 April 2021; p. 26. [CrossRef]
25. Nour, M.; Cömert, Z.; Polat, K. A Novel Medical Diagnosis model for COVID-19 infection detection based on Deep Features and Bayesian Optimization. *Appl. Soft Comput.* **2020**, *97*, 106580. [CrossRef]
26. Rahman, T.; Chowdhury, M.E.H.; Khandakar, A.; Islam, K.R.; Mahbub, Z.B.; Kadir, M.A.; Kashem, S. Transfer Learning with Deep Convolutional Neural Network (CNN) for Pneumonia Detection Using Chest X-ray. *Appl. Sci.* **2020**, *10*, 3233. [CrossRef]
27. Loey, M.; Smarandache, F.; Khalifa, N.E.M. Within the Lack of Chest COVID-19 X-ray Dataset: A Novel Detection Model Based on GAN and Deep Transfer Learning. *Symmetry* **2020**, *12*, 651. [CrossRef]
28. Degerli, A.; Ahishali, M.; Yamac, M.; Kiranyaz, S.; Chowdhury, M.E.H.; Hameed, K.; Hamid, T.; Mazhar, R.; Gabbouj, M. COVID-19 infection map generation and detection from chest X-ray images. *Health Inf. Sci. Syst.* **2021**, *9*, 15. [CrossRef]
29. Alirr, O.I. Automatic deep learning system for COVID-19 infection quantification in chest CT. *arXiv* **2020**, arXiv:2010.01982. [CrossRef]
30. Rahman, T.; Chowdhury, M.; Khandakar, A. COVID-19 Radiography Database. Kaggle. Available online: https://www.kaggle.com/datasets/tawsifurrahman/covid19-radiography-database (accessed on 11 October 2022).
31. Perez, L.; Wang, J. The Effectiveness of Data Augmentation in Image Classification using Deep Learning. *arXiv* **2017**, arXiv:171204621.
32. Reddi, S.J.; Kale, S.; Kumar, S. On the Convergence of Adam and Beyond. *arXiv* **2019**, arXiv:1904.09237.
33. Manthiramoorthi, M.; Mani, M.; Murthy, A.G. Application of Pareto's Principle on Deep Learning Research Output: A Scientometric Analysis. In Proceedings of the International Conference on Machine Learning and Smart Technology–ICMLST, Chennai, India, 2021. Available online: https://www.researchgate.net/profile/Ganesa-Murthy-Arasakumar/publication/355753261_Application_of_Pareto's_Principle_on_Deep_Learning_Research_Output_A_Scientometric_Analysis/links/617c1b213c987366c300002b/Application-of-Paretos-Principle-on-Deep-Learning-Research-Output-A-Scientometric-Analysis.pdf (accessed on 11 October 2022).
34. Alpaydın, E. *Machine Learning Textbook: Introduction to Machine Learning*; MIT Press: Cambridge, MA, USA, 2020.
35. Guo, G.; Wang, H.; Bell, D.; Bi, Y.; Greer, K. KNN Model-Based Approach in Classification. In Proceedings of the OTM 2003: On the Move to Meaningful Internet Systems 2003: CoopIS DOA, and ODBASE, Catania, Italy, 3–7 November 2003; Lecture Notes in Computer Science. Springer: Berlin/Heidelberg, Germany, 2003; Volume 2888, pp. 986–996. [CrossRef]
36. Breiman, L. Random forests. *Mach. Lang.* **2001**, *45*, 5–32.
37. Diebold, F.X.; Mariano, R.S. Comparing Predictive Accuracy. *J. Bus. Econ. Stat.* **2002**, *20*, 134–144. [CrossRef]
38. Buckland, M.; Gey, F. The relationship between recall and precision. *J. Am. Soc. Inf. Sci.* **1994**, *45*, 12–19. [CrossRef]
39. Sokolova, M.; Japkowicz, N.; Szpakowicz, S. Beyond Accuracy, F-Score and ROC: A Family of Discriminant Measures for Performance Evaluation. In Proceedings of the AI 2006: Advances in Artificial Intelligence, Hobart, Australia, 4–8 December 2006; pp. 1015–1021.
40. Apostolopoulos, I.D.; Mpesiana, T.A. Covid-19: Automatic detection from X-ray images utilizing transfer learning with convolutional neural networks. *Phys. Eng. Sci. Med.* **2020**, *43*, 635–640. [CrossRef]
41. Khan, S.S.; Khoshbakhtian, F.; Ashraf, A.B. Anomaly Detection Approach to Identify Early Cases in a Pandemic using Chest X-rays. *arXiv* **2021**, arXiv:2010.02814. [CrossRef]
42. Kumar, R.; Arora, R.; Bansal, V.; Sahayasheela, V.J.; Buckchash, H.; Imran, J.; Narayanan, N.; Pandian, G.N.; Raman, B. Classification of COVID-19 from chest x-ray images using deep features and correlation coefficient. *Multimed. Tools Appl.* **2022**, *81*, 27631–27655. [CrossRef] [PubMed]
43. Makris, A.; Kontopoulos, I.; Tserpes, K. COVID-19 detection from chest X-Ray images using Deep Learning and Convolutional Neural Networks. *medRxiv* **2020**. [CrossRef]
44. Irfan, M.; Iftikhar, M.; Yasin, S.; Draz, U.; Ali, T.; Hussain, S.; Bukhari, S.; Alwadie, A.; Rahman, S.; Glowacz, A.; et al. Role of Hybrid Deep Neural Networks (HDNNs), Computed Tomography, and Chest X-rays for the Detection of COVID-19. *Int. J. Environ. Res. Public Health* **2021**, *18*, 3056. [CrossRef]
45. Arora, R.; Bansal, V.; Buckchash, H.; Kumar, R.; Sahayasheela, V.J.; Narayanan, N.; Pandian, G.N.; Raman, B. AI-based diagnosis of COVID-19 patients using X-ray scans with stochastic ensemble of CNNs. *Phys. Eng. Sci. Med.* **2021**, *44*, 1257–1271. [CrossRef]

life

Article

The Impact of Digital Contact Tracing Apps Overuse on Prevention of COVID-19: A Normative Activation Model Perspective

Junwei Cao [1], Dong Liu [2,*], Guihua Zhang [3] and Meng Shang [4,*]

1 School of Business, Yangzhou University, Yangzhou 225127, China
2 Department of Global Business, Yeungnam University, Gyeongsan 38541, Korea
3 Department of Business, Yeungnam University, Gyeongsan 38541, Korea
4 School of Flight, Anyang Institute of Technology, Anyang 455008, China
* Correspondence: bruceliu@yu.ac.kr (D.L.); shangmengdr@163.com (M.S.)

Abstract: During the COVID-19 pandemic, many countries have used digital contact tracing apps (DCTAs) to implement contact tracing. Although the use of DCTAs has contributed to the prevention and control of COVID-19, there are doubts in academia about their actual effectiveness. In this study, the role of DCTAs in the prevention of COVID-19 was analyzed in terms of both the responsibility and inconvenience to life in a large-scale DCTA overuse environment, based on the normative activation model. The findings suggest that the overuse of a DCTA activates people's personal norms by triggering awareness of the consequences and ascription of responsibility, leading people to consistently cooperate with the government to prevent COVID-19. However, the inconvenience of living with DCTA overuse weakens the effect of the awareness of consequences and ascription of responsibility and the role of the ascription of responsibility in influencing personal norms. These effects may bear on people's willingness to consistently cooperate with the government to prevent COVID-19. The results of this study confirm the effectiveness of DCTA in counteracting pandemics from a social responsibility perspective in a large-scale environment where DCTA is used, enriching the literature on DCTA research in the COVID-19 pandemic. The results of this study can also help governments develop and improve policies to prevent COVID-19, as well as improve the DCTAs' operating patterns.

Keywords: digital contact tracing; normative activation model; COVID-19 prevention; prevention intention

Citation: Cao, J.; Liu, D.; Zhang, G.; Shang, M. The Impact of Digital Contact Tracing Apps Overuse on Prevention of COVID-19: A Normative Activation Model Perspective. *Life* **2022**, *12*, 1371. https://doi.org/10.3390/life12091371

Academic Editors: Daniele Giansanti and Denis Harkin

Received: 10 August 2022
Accepted: 31 August 2022
Published: 2 September 2022

1. Introduction

Contact tracing of COVID-19 patients is a very important part of the global fight against COVID-19. Contact tracing, timely detection, and adequate isolation will play a significant role in slowing the spread of COVID-19 [1,2]. To implement contact tracing strategies, many digital contact tracing apps (DCTAs) were developed and widely used globally during the COVID-19 pandemic [3,4]. For example, Korea developed the "Self-isolation Safety Protection App", Singapore the "TraceTogether", Japan the "COCOA", France the "StopCovi", Germany the "Corona Warn", and China the "Health Code" [5–9]. A DCTA will automatically record an individual's travel history, and both the user and public health departments will be notified if the user enters a high-risk area or comes into contact with a suspected COVID-19 patient. The appearance of such apps is considered a public health intervention that could slow the spread of COVID-19 and save lives, as well as protect local health services [10,11].

Although DCTAs have been accorded high expectations, many studies have generally raised doubts about their effectiveness in the prevention of COVID-19 practices [10,12–14]. Many scholars believe that the premise for such apps to have an effect in helping prevent

COVID-19 is the need for mass adoption and continuous use [15,16]; however, the DCTA adoption rate is too low to realize its full potential in most countries [17]. First, several studies have suggested that the inconvenience brought about by a DCTA to people's lives affects its large-scale promotion, especially the privacy issue. A DCTA has issues with extensive personal information collection, multiple processing purposes, uncertain storage times, and vague privacy policies [9]. One survey claimed that many people in the UK refuse to use such apps because of privacy concerns [18]. A survey in Ireland noted that many people refused to use the app because they feared that tech companies or the government would use it to monitor users even after the COVID-19 pandemic was over [19]. Chinese users are concerned about the lack of transparency in the operation of DCTAs, the unclear scope of data storage, and the dependence on private companies to operate them [8]. South African users have also shown doubts about the app's ability to protect privacy [20]. In addition to serious privacy issues, DCTAs have caused many other inconveniences, such as incorrect tracking and problems affecting the normal use of mobile phones [8,21]. Studies have shown that many of the close contacts located by a DCTA did not have any contact with COVID-19 patients, but they were also wrongly traced and isolated [8,22]. The operation of a DCTA has also been shown to affect the users' normal use of their cell phones, such as reducing the phone's running speed and affecting its battery life [21]. Second, several studies have suggested that the potential digital divide issue may also affect the large-scale promotion of a DCTA. A study shows that the digital divide during the COVID-19 pandemic often influences some people to use new technologies to prevent COVID-19 [23]. Age, education, income, health status, and regional differences can lead to a digital divide that directly affects people's widespread use of the app [6,24]. A survey in Germany claimed that females and low-income households have lower rates of downloading the DCTA [6]. In a UK survey, it was found that the use of apps among those over 65 years old was low [25]. In addition, a study of the working population in Japan showed that the usage rate among small-company employees and vendors was low, while that among large-company employees and civil servants was high [7]. The reason this literature doubts the efficacy of DCTA is that most of the current literature consists of studies conducted in a free market environment where people who feel inconvenienced by the use of DCTA would simply stop using it without paying any price. However, research conducted in a government-led environment with large-scale mandatory DCTA use is missing.

In addition, it can be argued that the many inconveniences that a DCTA brings to people's lives are the main issues that lead to doubts about its effectiveness. However, one study confirmed that the benefits of a DCTA could offset its negative effects [26]. The use of a DCTA has been shown to be closely related to individual as well as social interests [4]. This may be reflected in an individual's sense of responsibility to family, friends, and the community (e.g., preventing transmission of the virus to others) [16,27]. Therefore, it is meaningful to explain the role of a DCTA in preventing COVID-19 in practice from the perspective of responsibility [11]. As more countries relax their control policies to prevent COVID-19, the trade-off between the benefits and negative effects of a DCTA will likely influence subsequent COVID-19 prevention behavior. The analysis of the role of a DCTA during the COVID-19 pandemic in terms of both responsibility and inconvenience helps resolve the doubts. Regrettably, such studies are currently lacking.

To address these issues, an environment needs to be found in which the social epidemic is relatively stable, while the use of a DCTA is still mandatory on a large scale. China happens to provide a very good environment for investigation. In terms of the people who use it, the DCTA in China is mandatory, and no one can refuse to use it [8]. In terms of the extent of use, the use of DCTAs in China is also widespread, with various DCTAs developed by the central government and local governments. Many Chinese people are already excessively using DCTAs, as they are required to register and show their tracking information when entering or leaving any place, taking public transportation, and traveling across cities. The overuse of technology often has a negative impact [28]. The overuse of

DCTA technology in China does cause inconvenience to people's lives, such as privacy concerns, data security, error tracking, etc. [8]. China is undoubtedly one of the most successful countries in the world in terms of its performance in preventing COVID-19 [29]. Therefore, this study proposes the following research questions:

RQ1: Does the overuse of a DCTA still inspire a sense of responsibility for COVID-19 prevention?

RQ2: How does the responsibility and inconvenience of the overuse of a DCTA affect people's continued cooperation with the government to prevent COVID-19?

Therefore, this study builds its model based on the normative activation model (NAM) according to the actual research needs. People's psychological states can be better measured by structured scales, and a number of NAM-based studies on COVID-19 have also used the questionnaire-based approach [30,31]. The model is then validated by surveying Chinese residents to ultimately address the proposed questions. The results of this study will not only help policy makers improve the operation of DCTA applications in the post-COVID-19 era of prevention and control but will also contribute to the improvement of national policies related to epidemic prevention and control.

2. Theoretical Background

The normative activation model (NAM) proposed by Schwartz [32] was used to explain altruistic behavior and was extended to explain various pro-social and pro-environmental behaviors, such as energy-saving behavior [33], green consumption behavior [34–36], environmental behavior [36,37], etc. The NAM has been widely used in environmental, psychological, and behavioral research and is among the most important theories for studying the individuals' socially or environmentally responsible behavior [38]. During the COVID-19 pandemic, the NAM has been widely used in studies to analyze people's infection prevention behavior. A study analyzed people's willingness to get vaccinated before traveling from the perspective of the NAM, suggesting that mass media messages activated personal norms by positively influencing people's awareness of the consequences and ascription of responsibility, prompting people to get vaccinated before traveling [30]. Another study analyzed Chinese people's intentions to save masks in the post-COVID-19 era from the perspective of the NAM and confirmed that personal norms had a significantly positive impact on mask-saving behavior and that awareness of the consequences and ascription of responsibility indirectly influenced the intention to save masks through personal norms [31]. Therefore, it is appropriate to analyze the contribution of a DCTA to people's willingness to consistently cooperate with the government in preventing COVID-19 from the perspective of the NAM.

The theory uses awareness of the consequences, attribution of responsibility, and personal norms to explain people's pro-social behavior [32] (Figure 1). The NAM suggests that awareness of the consequences and ascription of responsibility can activate personal norms and thus trigger pro-social behavior. Pro-social behavior is an umbrella term covering a range of behaviors that have a positive impact on society, such as giving help, cooperating, and comforting [39]. Awareness of the consequences means that individuals are aware of the negative consequences of their actions [40]. Ascription of responsibility refers to the reflection of individuals who are responsible for the adverse consequences of their non-participation in pro-social activities [40]. Personal norms are defined as the moral obligations that a person needs to fulfill for a particular behavior. According to the theory, a person's pro-social behavior or intentions are influenced by personal norms, and awareness of the consequences and ascription of responsibility can activate such norms [41]. People are more willing to engage in pro-social behavior when they perceive it as a moral obligation to perform or avoid a particular behavior [41,42].

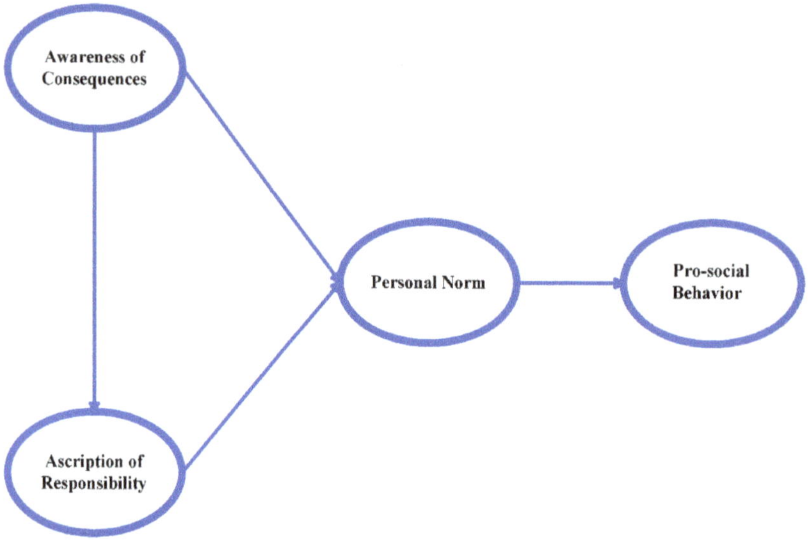

Figure 1. The original norm activation model.

3. Research Model and Hypothesis Development

Based on the NAM, this study develops a research model from the perspective of responsibility and inconvenience to people's lives. It hypothesizes that the overuse of a DCTA can activate people's personal norms by promoting awareness of the consequences and ascription of responsibility, and ultimately, a willingness to consistently cooperate with the government to prevent COVID-19. Meanwhile, the inconvenience caused by the overuse of a DCTA may affect people's normative activation process. The resulting model is shown in Figure 2.

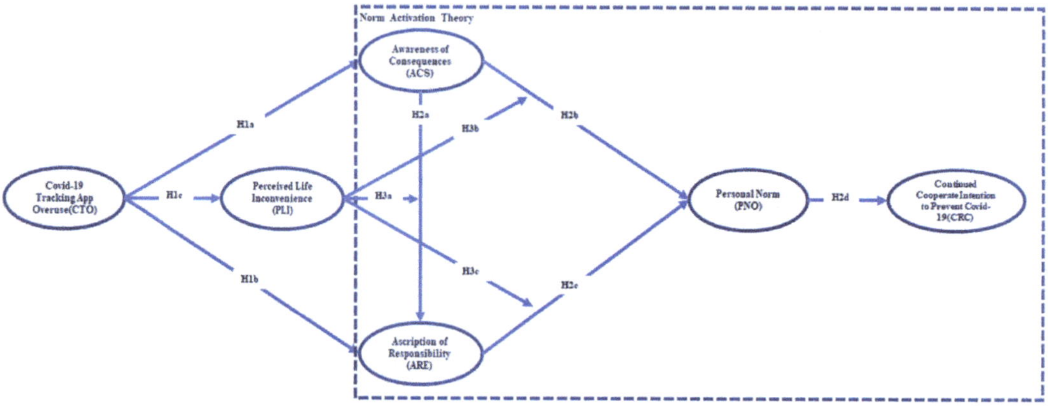

Figure 2. Research model.

The monitoring of individuals will improve self-consequence management [43]. Therefore, the higher the perception of monitoring, the easier it is to perceive the consequences [44]. A DCTA's tracking will make the users always feel monitored, while they will have to remind themselves to deal with possible consequences. For example, if individuals do not take precautions, they may be subject to prolonged isolation, more detailed epidemiological investigation, trajectory disclosure, and other mandatory measures if they become infected or come into close contact with a suspected COVID-19 patient [8]. A study in the UK confirmed that users were extremely concerned about privacy disclosure and stigmatization when using DCTAs during the COVID-19 pandemic [18]. Therefore, the overuse of a DCTA will prompt individuals to carefully consider the consequences and increase their awareness of them. Therefore, the following hypothesis is proposed:

H1a. *The overuse of DCTAs positively affects people's awareness of the consequences during the COVID-19 pandemic.*

Monitoring and responsibility are closely related. A study suggests that parental monitoring activities can trigger students' responsibility for learning [45]. It has also been noted that employee monitoring creates a sense of responsibility in managers [46]. In the prevention and control of the COVID-19 epidemic, the application of DCTAs can monitor the trajectory of each person's action, while the government can be precisely responsible in the event of any trouble. A DCTA also serves an advocacy function that can remind people to take responsibility for their families and communities [16]. Thus, the overuse of a DCTA can leave users in constant fear of being held responsible if their actions have caused the spread of a virus. Therefore, the following hypothesis is proposed:

H1b. *The overuse of DCTAs positively affects people's ascription of responsibility during the COVID-19 pandemic.*

The overuse of technology often has a negative impact [28]. A DCTA, a new technology arising from the COVID-19 pandemic, has been shown to create concerns about privacy, stigmatization, being mis-targeted, and data misuse [8,16,18]. Studies have pointed to concerns about the lack of transparency in the operations of DCTAs, the scope of data storage, the inability to change incorrect "red" codes (representing health risks), the over-reliance on the internet, and the reliance on private companies, such as Alipay and WeChat, to monitor their travel routes [8]. In addition, the overuse of a DCTA can lead to a need for people to present and register their travel tracks at any place, while individuals are often inconvenienced by the rapid changes in the epidemic and the inaccuracy of clients' location displays for daily life. Therefore, the following hypothesis is proposed:

H1c. *The overuse of DCTAs positively affects the inconvenience to daily life during the COVID-19 pandemic.*

The NAM proposes that consequence awareness has a significantly positive effect on the ascription of responsibility and that the two activate personal norms together [39]. Such a relationship has been prevalent during the COVID-19 pandemic. A study suggests that, in people's willingness to be vaccinated, the awareness of consequences creates the ascription of responsibility, while the latter activates personal norms for vaccination [41]. Similarly, a study suggests that people's awareness of the possible consequences allows them to actively take precautions while traveling in recognition of their potential responsibility in preventing COVID-19 [47]. One study analyzed people's behavior during the waste sorting of masks during the COVID-19 pandemic and verified the roles of the awareness of consequences and ascription of responsibility in the activation of personal norms [48]. First, due to the overuse of DCTAs, people will feel that their travels are constantly being monitored and will have a strong awareness of the consequences. Second, the overuse of DCTAs allows for more precise accountability, while a sense of responsibility is attributed when people understand that they will be held accountable for the consequences they cause. Third, people's concerns about the consequences and the ascription of responsibility together contribute to the creation of personal norms, which make people believe that cooperating with COVID-19 prevention is a moral imperative and pro-social behavior. Finally, due to the large-scale and continuous use of DCTAs, the mechanism of influence between awareness of the consequences, ascription of responsibility, and personal norms will persist and may facilitate people's continuous intention to cooperate with the government in preventing COVID-19. Therefore, the following hypotheses are proposed:

H2a. *The awareness of consequences that is caused by the overuse of DCTAs positively influences the ascription of responsibility.*

H2b. *The awareness of consequences that is caused by the overuse of DCTAs positively influences the activation of personal norms.*

H2c. *The ascription of responsibility that is caused by the overuse of DCTAs positively influences the activation of personal norms.*

H2d. *Personal norms have a positive impact on people's intention to consistently cooperate with the government to prevent COVID-19.*

The prospect theory proposes that individuals' preferences and behavior under risk and uncertainty tend to follow an evaluation of their potential gains and losses and that people may be willing to take risks in exchange for benefits in the face of large, perceived benefits [49]. One study confirmed that during the COVID-19 pandemic, the perceived health and privacy risks jointly influenced the perceived benefits, while people would be willing to forego some life conveniences in exchange for health benefits [26,50]. Living with an inconvenience can have a negative impact on people's behavioral intentions [51]. The overuse of DCTAs has caused many inconveniences in people's lives, such as privacy issues and incorrect diagnoses [8]. However, such inconveniences can only weaken people's willingness to continue to cooperate in the prevention of the disease and cannot be a direct deterrent. First, in an environment of government-led mass compulsory use, it is clear that people are more willing to endure the inconvenience of living with COVID-19 than to bear the consequences and responsibility of not cooperating in the prevention of COVID-19, although they are dissatisfied. Second, people use the new COVID-19 prevention technology because of their personal and community interests [4]. When people consider that the consequences of the spread of the epidemic may harm their personal or collective interests, they develop a sense of responsibility attribution, which makes them feel morally obliged to cooperate in the prevention of the epidemic, even if they are slightly dissatisfied. Overall, in the context of government-led mass-mandated use, the inconvenience of living with DCTAs is unlikely to directly affect people's sense of responsibility, sense of consequence, and personal norms, but it can create negative emotions that may weaken the strength of the causality of the responsibility, sense of consequence, and personal norms variables, ultimately affecting people's awareness of COVID-19 prevention. Therefore, the following hypotheses are proposed:

H3a. *The inconvenience to life that is caused by the overuse of DCTAs weakens the contribution of consequence awareness to the ascription of responsibility.*

H3b. *The inconvenience to life that is caused by the overuse of DCTAs weakens the role of consequence awareness in promoting personal norms.*

H3c. *The inconvenience to life that is caused by the overuse of DCTAs weakens the ascription of responsibility in promoting personal norms.*

4. Method

4.1. Questionnaire Design and Survey

The scales for all the variables in the study were designed based on those that have been validated by existing studies. The scales used to measure the variables in the NAM were adapted from a related study conducted on the basis of the NAM (Sang, Yao, Zhang, Wang, Wang, and Liu [36], Kim, Woo, and Nam [38], and He and Zhan [34]). The scale for measuring overuse was adapted from Lee, Kim, Fava, Mischoulon, Park, Shim, Lee, Lee, and Jeon [28]; the scale for measuring the inconvenience to life was adapted from Lee, Kim, Fava, Mischoulon, Park, Shim, Lee, Lee, and Jeon [28]. After the initial questionnaire design was completed, we asked experts in the field to review and revise it and conducted a small-scale pre-test to improve it. Please refer to Appendix A for specific measurement items.

Some of the other design parameters of the scale are as follows. (1) The scale uses a 5-point Likert scale. (2) The questionnaire questions are in English, while the survey was conducted in China; thus, we invited two linguists who were proficient in English to translate the questionnaire from English into Chinese to ensure that the Chinese presentation was error free and easy to understand. (3) We designed reverse questions in the questionnaire to detect invalid questionnaires. (4) Our questionnaire was designed as an anonymous survey, where participants were informed of the purpose of the study, only the necessary data were collected and kept strictly confidential, and respondents were given gifts to participate. (5) In accordance with the regulations of the Research Ethics Committee of Yeungnam University (https://irb.yu.ac.kr/02_gid/gid01.html, accessed on 20 June 2022), no specific ethical review was required for the questionnaire survey of this study.

We selected people living in Shanghai as the population for this study. First, Shanghai is a mega-city in China with a large population, a developed economy, and a rapid diffusion of new technologies and policies; the use of a DCTA to enhance health verification and entrance registration is an important initiative to strengthen COVID-19 prevention in Shanghai. Second, Shanghai had a massive COVID-19 outbreak in March, and after the pandemic was brought under control, the full deployment of "place code" and "health verification machine" devices was quickly made mandatory for citizens to use, while citizens had to scan the QR codes on these devices through their cell phones to complete health verification and tracking registration before entering places (see Figure 3) [52].

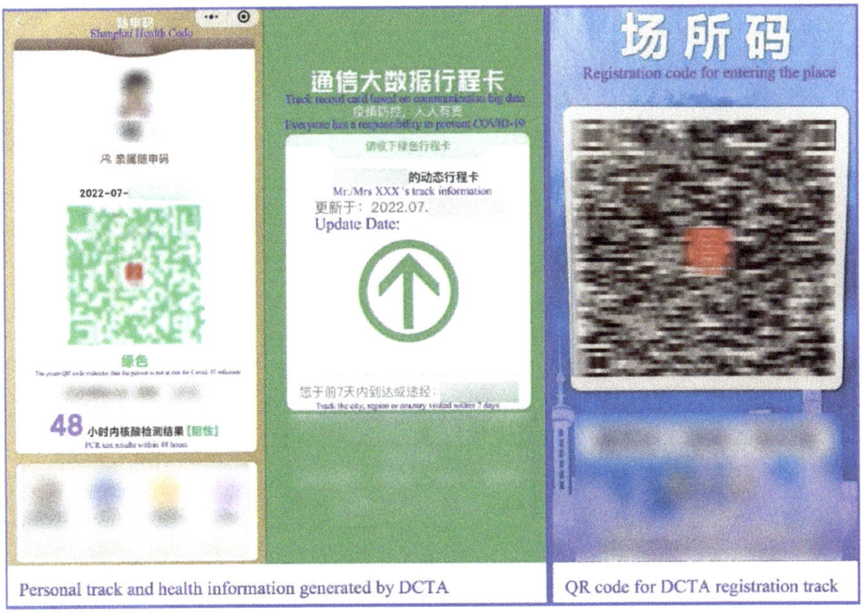

Figure 3. Tracking display/registration QR code generated by the DCTA in Shanghai.

In this study, we randomly joined some instant messaging software chat groups in the Shanghai area and randomly conducted questionnaires among members of them. Participants who completed the survey would receive a CNY 10 shopping coupon. In total, 400 respondents living in Shanghai were randomly surveyed through various SNS platforms from 1 July 2022 to 10 July 2022. Finally, we received a total of 379 questionnaires and obtained 313 valid questionnaires (82.5%) by removing duplicate responses, biased reverse questions, and those with less than 2 min of answer time.

4.2. Structural Equation Model

We first used a descriptive analysis of the demographic characteristics of the sample. Second, we evaluated the indicators related to model quality. Finally, the proposed hypotheses were tested.

The covariance-based structural equation model (CB-SEM) and variance-based partial least squares structural equation modeling (VB-SEM) can both be used to analyze structural equation models. However, the following may be noted. (1) Partial least squares structural equation modeling (PLS-SEM) is more suitable than CB-SEM for measuring structural equation models with more than six latent variables [53]. (2) PLS-SEM is suitable for a wider range of data characteristics than CB-SEM, especially for handling non-normally distributed data [53]. (3) PLS-SEM is more suitable for small-sample measurements and exploratory studies [53].

This is an exploratory study with six latent variables in the research model and a small, effective sample size. Additionally, a multivariate normality analysis was performed on the data collected in this study using a web calculator to measure the distribution of the data (https://webpower.psychstat.org/, accessed on 13 July 2022). The results show Mardia's multivariate skewness (β = 40.707, $p < 0.001$) and multivariate kurtosis (β = 473.530, $p < 0.01$) that suggest multivariate non-normality. In summary, PLS-SEM is more suitable for data analysis in this study [54,55].

5. Results

5.1. Demographics and Bias Test Results

Among the 313 valid questionnaires collected from participants in this study, 125 (39.9%) were male and 188 (60.1%) female; the largest number of people were aged between 30 and 39 years (N = 165, 52.7%), followed by those aged between 40 and 49 years (N = 51, 16.3%). Of the participants, 172 (55%) had a bachelor's degree, and 75 (24%) had master's or doctoral degrees. The vast majority had a monthly income in the range of CNY 10,000–14,999 (N = 146, 46.6%), while 17.9% had a monthly income in the range of CNY 5000–9999 (N = 56). Referring to the data of people's concern about DCTA in China from the Baidu Index (https://index.baidu.com, accessed on 19 August 2022) (Figure 4), this survey result has a certain degree of representativeness.

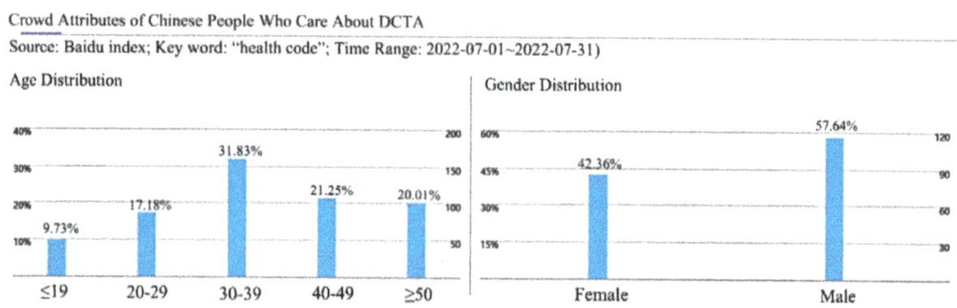

Figure 4. Crowd attributes of Chinese people who care about DCTA.

To detect the non-response bias, a paired *t*-test was performed on the demographic data of the first and last 25 participants in the survey [56]. The results of the test showed no significant differences. Therefore, non-response bias was not a serious problem.

Common method bias (CMB) is also a common problem in surveys, which we measured using two methods. First, this study measured the rate of extraction of single factors according to the method proposed by Podsakoff et al. [56], which was 24.154%, below the threshold of 40%. Second, this study was performed using the full-VIF method of measurement in PLS-SEM to detect CMB [57]. All VIF values were below the threshold of 3.3 [54]. The results of these two tests indicate that CMB was not a serious problem in this study.

5.2. Measurement Model Results

We measured the quality of the model by assessing the composite reliability (CR), average variance extracted (AVE), discriminant validity, and outer loading. As shown in Table 1, the CR and Cronbach's alpha for all the variables exceeded 0.7, indicating that the internal consistency of the data in this study was satisfactory. The AVEs for the variables were all greater than 0.5, while the outer loading exceeded 0.7, indicating that the convergent validity of the data in this study was satisfactory [53].

Table 1. Reliability and validity of constructs.

Latent Variable	Item	Loading	Mean (SD)	Cronbach's α	CR	AVE	R^2
CTO	CTO1 CTO2 CTO3 CTO4	0.841 0.821 0.754 0.731	3.143 (0.804)	0.819	0.866	0.618	-
ACS	ACS1 ACS2 ACS3	0.851 0.915 0.843	3.028 (1.074)	0.839	0.903	0.757	0.111
ARE	ARE1 ARE2 ARE3	0.864 0.750 0.780	3.439 (0.653)	0.715	0.841	0.639	0.063
PLI	PLI1 PLI2 PLI3 PLI4	0.923 0.758 0.927 0.889	3.149 (1.131)	0.898	0.93	0.769	0.020
PNO	PNO1 PNO2 PNO3	0.883 0.894 0.775	3.544 (0.860)	0.818	0.888	0.727	0.125
CRC	CRC1 CRC2 CRC3 CRC4	0.783 0.840 0.781 0.910	3.132 (0.714)	0.855	0.898	0.689	0.024

Abbreviations: CTO—COVID-19 Tracking App Overuse; ACS—Awareness of Consequences; ARE—Ascription of Responsibility; PLI—Perceived Life Inconvenience; PNO—Personal Norm; CRC—Continue to Cooperate Intention to Prevent COVID-19.

We determined the discriminant validity using both Fornell and Larcker's test and the heterotrait–monotrait ratio (HTMT) test. As shown in Table 2, the square root of each variable's AVE is greater than the correlation with other variables [53]. The HTMT values were also all below 0.85. Therefore, the discriminant validity of this study is in accordance with the requirements [53].

Table 2. Discriminant validity.

Fornell–Larcker Criterion						
	ARE	ACS	PLI	PNO	CRC	CTO
ARE	0.799					
ACS	0.186	0.870				
PLI	0.114	−0.024	0.877			
PNO	0.247	0.294	0.253	0.852		
CRC	0.537	0.058	0.099	0.154	0.830	
CTO	0.222	0.333	0.447	0.266	0.174	0.786

Table 2. *Cont.*

Heterotrait–Monotrait Ratio						
	ARE	ACS	PLI	PNO	CRC	CTO
ARE						
ACS	0.236					
PLI	0.141	0.055				
PNO	0.304	0.332	0.294			
CRC	0.636	0.103	0.114	0.173		
CTO	0.224	0.338	0.443	0.301	0.151	

Abbreviations: CTO—COVID-19 Tracking App Overuse; ACS—Awareness of Consequences; ARE—Ascription of Responsibility; PLI—Perceived Life Inconvenience; PNO—Personal Norm; CRC—Continue to Cooperate Intention to Prevent COVID-19.

5.3. Structural Model Results

Before measuring the structural model, we measured co-linearity (ensuring sufficient independence between variables), and the VIF for all the variables was below 3; thus, co-linearity was not a major issue in this study. After ensuring the reliability and validity of the model, we tested the hypotheses using the structural model. The path coefficients and significance test results from the structural model are shown in Table 3. Overuse of DCTAs had a positive and significant effect on awareness of the consequences, ascription of responsibility, and perceived life inconvenience, with H1a, H1b, and H1c being supported. Awareness of the consequences had a significantly positive effect on the ascription of responsibility and personal norms, thus supporting H2a and H2b. Ascription of responsibility had a significantly positive effect on personal norms, supporting H2c. Personal norms had a positive impact on the willingness to consistently cooperate with the government in COVID-19 prevention, supporting H2d. In addition, none of the control variables had a significant effect on the users' intention to consistently cooperate with the government in COVID-19 prevention.

Table 3. Assessment of the structural model.

Hypothesis	β	STDEV	T-Statistic	*p*-Value	Result
H1a: CTO -> ACS	0.333	0.334	6.755	0.000	Support
H1b: CTO -> ARE	0.18	0.181	3.190	0.001	Support
H1c: CTO -> PLI	0.447	0.45	11.93	0.000	Support
H2a: ACS -> ARE	0.126	0.13	1.974	0.048	Support
H2b: ACS -> PNO	0.257	0.259	4.901	0.000	Support
H2c: ARE -> PNO	0.199	0.204	3.711	0.000	Support
H2d: PNO -> CRC	0.157	0.161	2.651	0.008	Support
Edu -> CRC	−0.016	−0.013	0.257	0.797	
Gender -> CRC	0.039	0.046	0.301	0.763	
Income -> CRC	0.035	0.032	0.505	0.613	
Age -> CRC	−0.039	−0.04	0.709	0.478	

Abbreviations: CTO—COVID-19 Tracking App Overuse; ACS—Awareness of Consequences; ARE—Ascription of Responsibility; PLI—Perceived Life Inconvenience; PNO—Personal Norm; CRC—Continue to Cooperate Intention to Prevent COVID-19.

Finally, we evaluated the goodness of fit (GOF) of the model using the standardized root mean square residuals (SRMR). The SRMR value for the model is 0.068, which is less than the threshold value of 0.08. Thus, the fit of the model is satisfactory [58].

5.4. Moderating Effect Results

The perceived life inconvenience was used as a moderating variable; its moderating effect was measured through two steps in this study. First, we measured the significance of the moderating effect; second, we measured the strength of the moderating effect by calculating F^2 as follows: (R^2 interaction model $-$ R^2 main effects model)/(1 $-$ R^2 main effects model). If F^2 is between 0.02 and 0.15, it indicates a small moderating effect; if it is between 0.15 and 0.35, it indicates a moderate moderating effect; and if it exceeds 0.35, it indicates a high moderating effect [59,60].

The moderating effects are shown in Table 4. The perceived life inconvenience significantly reduced the effect of awareness of the consequences on the ascription of responsibility ($\beta = -0.158$, $p < 0.01$), thus supporting H3a. Perceived life inconvenience also significantly reduced the effect of ascription of responsibility on personal norms ($\beta = -0.158$, $p < 0.01$), thus supporting H3c. However, perceived life inconvenience had no significant moderating effect on awareness of the consequences and personal norms, and H3b was rejected ($\beta = -0.078$, n.s.).

Table 4. Moderation effects test.

Hypothesis	R^2 Main Effects Model	R^2 Interaction Model	F^2	β	T-Statistic	*p*-Value	Result
H3a: PLI*ACS -> ARE	0.063	0.091	0.023	−0.158	2.796	0.005	Support
H3b: PLI*ACS -> PNO	0.125	0.215	-	−0.078	1.467	0.142	Reject
H3c: PLI*ARE -> PNO	0.125	0.215	0.072	−0.158	2.645	0.008	Support

Abbreviations: ACS—Awareness of Consequences; ARE—Ascription of Responsibility; PLI—Perceived Life Inconvenience; PNO—Personal Norm.

Slope plots are provided as part of the moderating effect analysis to provide a more visual response to the enhancing/weakening effect of the moderating variable on a specific relationship. We performed slope analysis on the significant moderating relationships. The results are shown in Figures 5 and 6. Perceived life inconvenience significantly reduced the predicted effect of awareness of the consequences on the ascription of responsibility, with a "medium" effect size ($\beta = -0.158$, $p < 0.01$, $0.02 < F^2 = 0.023 < 0.15$). Perceived life inconvenience significantly reduced the impact of ascription of responsibility on personal norms, with a "high" effect size ($\beta = -0.158$, $p < 0.01$, $0.35 < F^2 = 0.072$).

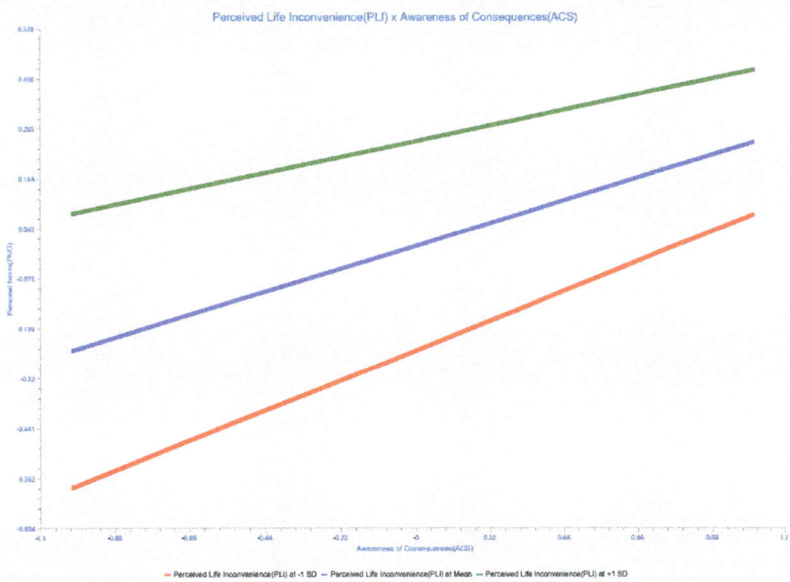

Figure 5. Simple slope analysis (PLI*ACS -ARE).

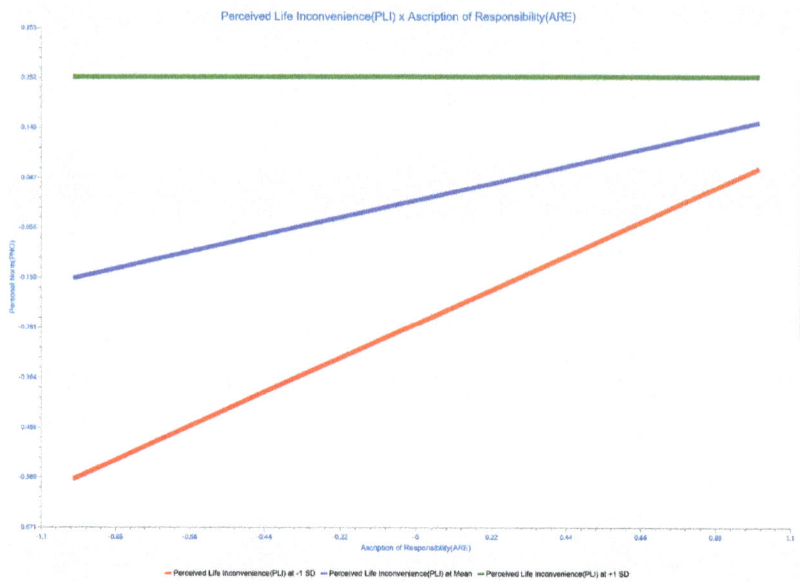

Figure 6. Simple slope analysis (PLI*ARE -NO).

6. Discussion and Conclusions

6.1. *Key Findings*

The results of the study suggest that the overuse of DCTAs triggers awareness of the consequences and ascription of responsibility, that awareness of the consequences is an important antecedent of ascription of responsibility, and that the triggered awareness of the consequences and ascription of responsibility activate people's personal norms. Guided by personal norms, people will continue to cooperate with the government to prevent COVID-19. Such results again validate the NAM theory in the context of the COVID-19 pandemic and are also in general agreement with the results of some studies on DCTA [8,10,11,18,21,30,31]. The tracking ability of a DCTA plays a role in monitoring people for the prevention of COVID-19. First, when a DCTA accurately tracks everyone's travel trajectory and health status, people are worried about being held precisely accountable for the consequences of their bad behavior. Second, the lack of privacy protection in a DCTA may also cause people to worry that they may be stigmatized in the event that they are infected [18]. Finally, the deficiencies of a DCTA's positioning accuracy can raise concerns [8,21]; people are urged to exercise caution to prevent being wrongly identified as a close contact and being investigated. The awareness of consequences and ascription of responsibility that people develop under DCTA monitoring will compel people to consider it a pro-social moral obligation to cooperate with the government to prevent COVID-19, and ultimately, to consistently cooperate with the government to prevent COVID-19.

The overuse of DCTAs has indeed also inconvenienced people in their lives, thus weakening their willingness to actively cooperate with the government in preventing COVID-19 (moderating effects of perceived life inconvenience). People use DCTAs in the spirit of personal and social interests [4]. However, it is human nature to "tend to benefit and avoid harm" [49]. The inconvenience of living with a DCTA can cause people to weigh the pros and cons of preventing COVID-19. Studies have demonstrated that people tolerate privacy risks in DCTA use when the privacy risks they pose are lower than the health risks [26,50]. It is reasonable to infer that when the inconvenience caused by a DCTA exceeds the level of responsibility required, people will choose to take responsibility rather than endure the inconvenience to life. In addition, when the inconvenience caused by a

DCTA results in considerable losses, people may have the feeling that "DCTA has already caused me losses, so what is my obligation and responsibility to cooperate?" However, the reality is that due to the government's strict precautions, the fear of being forcefully held accountable far outweighs the perceived inconvenience of living with a DCTA. Therefore, from the perspective of personal interest, even if people are dissatisfied, they are forced to develop a sense of responsibility and personal norms to cooperate with the government in preventing COVID-19 due to the awareness of the dire consequences. This explains the mechanism by which the inconvenience caused by the overuse of DCTAs plays a moderate-to-high intensity-weakening role in the relationships between awareness of the consequences and ascription of responsibility and between ascription of responsibility and personal norms; it does not directly negatively affect the relationship between ascription of responsibility and personal norms.

However, the debilitating effect of perceived life inconvenience from the overuse of DCTAs on consequence awareness and personal norms was not confirmed in this study. This inspires us to suggest that the inconvenient effects of a DCTA may need to be combined with precise accountability. The inconvenience caused by the overuse of a DCTA is magnified in the process of precise accountability, such as privacy concerns. When there is no accountability process, privacy is only restricted to a very few managers, and once the accountability process is involved, it can lead to stigmatization due to privacy breaches. Therefore, people will always carry this psychological pressure when using a DCTA. If the process of precise accountability is missing, a DCTA brings about only inconvenience to life, and people only need to measure the relationship between the inconvenience to mobility and prevention of COVID-19; they will naturally think that cooperating with the government to prevent COVID-19 is only a moral responsibility and will have no sense of responsibility for preventing COVID-19.

6.2. Theoretical Contributions

This study offers several theoretical contributions. First, this study evaluated the effect of DCTA overuse on promoting continuous cooperation with the government for COVID-19 prevention in a large-scale, mandatory-use environment and clarified the mechanism by which DCTA overuse promotes people's cooperation with the government for disease prevention from a psychological perspective, thus enriching the literature on the effectiveness of DCTAs in disease prevention. Second, this study validated the issue of the feasibility of the NAM model in explaining people's pro-social behavior in epidemics; it also verified that the overuse of digital health technology is an antecedent that triggers users to develop awareness of the consequences and ascription of responsibility, which expands the field of the use of the NAM and helps subsequent studies to apply it to investigate the impact of digital health technology on user psychology. Finally, this study verified the existence of a moderating effect of negative factors on the intrinsic mechanisms of the NAM model by analyzing the effects of perceived life inconvenience, which enriches the connotation of the NAM.

6.3. Practical Contributions

Our study also carries some practical implications for the prevention of COVID-19.

First, the government should not only develop functionally advanced DCTAs, but it should also have the ability to enforce their strict use nationwide. Meanwhile, the digital divide in the pandemic is worsening [23], making it difficult for many elderly and low-income people to use DCTAs because of accessibility issues. This requires the government to find ways to solve the tracking problem for this sector of the population, and the Chinese government's practice in this regard is worth learning from and promoting. In many areas of China, people who do not have electronic devices and have difficulty using DCTAs only need to bring their ID cards when they travel, while public transportation drivers and staff in public places can use the "register and present for others" feature of a DCTA to help them overcome the digital divide caused by the adoption of digital technology.

Second, the tracking accuracy of a DCTA must be further optimized. This helps reduce the perceived inconvenience of life caused by DCTA. People worry about being wrongly isolated or pursued because of wrong tracking by a DCTA, which may be due to the layout of wireless base stations. When different areas are covered by the same base station, the tracking may be confused, resulting in location misclassification. Therefore, the developers of DCTAs should work with wireless network providers to optimize the location algorithm and base station distribution to reduce their chances of being mislocated.

Third, a special governmental supervision department should be established to implement confidential supervision of people's tracking information through legislation and relevant technical measures and make timely adjustments to the errors that occur.

Finally, DCTA is only a precautionary measure that had to be taken to prevent COVID-19. The government should continuously adjust the level of DCTA use according to the changing situation of the COVID-19 pandemic. The inconvenience caused by the overuse of DCTAs to people's daily lives and their negative emotions should be reduced.

6.4. Limitations and Future Directions

There are some shortcomings in this study. First, the survey in this study is limited to Shanghai, while the sample size is insufficient; thus, there may be some problems with representativeness, and future studies are encouraged to adopt more representative research methods, such as big data analysis of the epidemic. Second, the "perceived inconvenience" in this study is a broad concept, which can be further subdivided into travel inconvenience, privacy concerns, and so on, in subsequent studies. Finally, the number of elderly respondents in this study was small, while the elderly are considered to be a very important group in the prevention of COVID-19; it is expected that future studies on the use of DCTAs among the elderly can be conducted.

Author Contributions: Conceptualization, J.C. and D.L.; methodology, J.C.; software, M.S.; validation, G.Z. and M.S.; formal analysis, G.Z. and M.S.; investigation, M.S.; data curation, D.L.; writing—original draft preparation, J.C.; writing—review and editing, J.C. and D.L.; supervision, M.S.; project administration, D.L. and M.S. All authors have read and agreed to the published version of the manuscript.

Funding: This research received no external funding.

Institutional Review Board Statement: Not applicable.

Informed Consent Statement: Not applicable.

Data Availability Statement: The data presented in this study are available upon request from the corresponding author. The data are not publicly available for ethical reasons.

Conflicts of Interest: The authors declare no conflict of interest.

Appendix A. Scale

Table A1. Scale.

Construct	No.	Item	References
Contact Tracing App Overuse (CTO)	CTO 1	I must use a DCTA to enter any place.	Lee, Kim, Fava, Mischoulon, Park, Shim, Lee, Lee, and Jeon [28]
	CTO 2	Even though using a DCTA to enter some sites caused queues, I had to use it.	
	CTO 3	I have to use a DCTA every time I travel.	
	CTO 4	Many places force me to use a DCTA.	

Table A1. *Cont.*

Construct	No.	Item	References
Awareness of Consequences (ACS)	ACS1	The overuse of DCTAs allowed me to understand the consequences of COVID-19 proliferation.	Sang, Yao, Zhang, Wang, Wang, and Liu [36] Kim, Woo, and Nam [38]
	ACS2	The overuse of DCTAs did not raise my awareness of preventing COVID-19 (Reverse).	
	ACS3	The overuse of DCTAs reminds me to avoid risky behaviors as much as possible.	
Ascription of Responsibility (ARE)	ARE1	The overuse of DCTAs has taught me that negative behaviors resulting in the spread of COVID-19 will be precisely pursued.	Kim, Woo, and Nam [38]
	ARE2	The overuse of DCTAs reminds me of my responsibility to cooperate in the prevention of COVID-19.	
	ARE3	The overuse of DCTAs makes me think that everyone must take responsibility for slowing the spread of COVID-19.	
Personal Norms (PNO)	PNO1	The overuse of DCTAs makes me feel obliged to cooperate in the prevention of COVID-19.	Kim, Woo, and Nam [38] Sang, Yao, Zhang, Wang, Wang, and Liu [36]
	PNO2	The overuse of DCTAs has forced me to cooperate in the prevention of COVID-19.	
	PNO3	The overuse of DCTAs makes me think it is morally responsible to cooperate in the prevention of COVID-19.	
Perceived Life Inconvenience (PLI)	PIC1	The overuse of DCTAs has caused inconvenience to my travel.	Seiders, Voss, Godfrey, and Grewal [51]
	PIC2	The overuse of DCTAs makes it inconvenient for me to get in and out of some places.	
	PIC3	The overuse of a DCTA adds inconveniences to my life, such as concerns about privacy leaks and being misplaced.	
	PIC4	The overuse of a DCTA forces me to spend a lot of time planning my life.	
Continuous Cooperation Against COVID-19 Intention (CAI)	CAI1	I intend to continue working with the relevant departments to prevent COIVD-19.	Kim, Woo, and Nam [38] He and Zhan [34]
	CAI2	I am willing to follow the guidance of the relevant department to continuously prevent COVID-19.	
	CAI3	Even if it costs me time and money, I am willing to keep working with the relevant departments to prevent COVID-19.	
	CAI4	I look forward to continuing to work with the relevant authorities to prevent COVID-19.	

References

1. Cao, J.; Zhang, G.; Liu, D. The Impact of Using mHealth Apps on Improving Public Health Satisfaction during the COVID-19 Pandemic: A Digital Content Value Chain Perspective. *Healthcare* **2022**, *10*, 479. [CrossRef]
2. Ish, P.; Agrawal, S.; Goel, A.D.; Gupta, N. Contact tracing: Unearthing key epidemiological features of COVID-19. *SAGE Open Med. Case Rep.* **2020**, *8*, 2050313X–2093348X. [CrossRef]
3. Kawakami, N.; Sasaki, N.; Kuroda, R.; Tsuno, K.; Imamura, K. The Effects of Downloading a Government-Issued COVID-19 Contact Tracing App on Psychological Distress During the Pandemic Among Employed Adults: Prospective Study. *JMIR Ment. Health* **2021**, *8*, e23699. [CrossRef]

4. Scholl, A.; Sassenberg, K. How Identification with the Social Environment and With the Government Guide the Use of the Official COVID-19 Contact Tracing App: Three Quantitative Survey Studies. *Jmir. Mhealth Uhealth* **2021**, *9*, e28146. [CrossRef]
5. Montagni, I.; Roussel, N.; Thiébaut, R.; Tzourio, C. Health Care Students' Knowledge of and Attitudes, Beliefs, and Practices Toward the French COVID-19 App: Cross-sectional Questionnaire Study. *J. Med. Internet Res.* **2021**, *23*, e26399. [CrossRef]
6. Grill, E.; Eitze, S.; De Bock, F.; Dragano, N.; Huebl, L.; Schmich, P.; Wieler, L.H.; Betsch, C. Sociodemographic characteristics determine download and use of a Corona contact tracing app in Germany—Results of the COSMO surveys. *PLoS ONE* **2021**, *16*, e256660. [CrossRef]
7. Ishimaru, T.; Ibayashi, K.; Nagata, M.; Hino, A.; Tateishi, S.; Tsuji, M.; Ogami, A.; Matsuda, S.; Fujino, Y.; Fujino, Y.; et al. Industry and workplace characteristics associated with the downloading of a COVID-19 contact tracing app in Japan: A nation-wide cross-sectional study. *Environ. Health Prev.* **2021**, *26*, 94. [CrossRef]
8. Joo, J.; Shin, M.M. Resolving the tension between full utilization of contact tracing app services and user stress as an effort to control the COVID-19 pandemic. *Serv. Bus.* **2020**, *14*, 461–478. [CrossRef]
9. Xiong, B.; Lin, F. How to Balance Governance Efficiency and Privacy Protection? A Textual Analysis of the Privacy Policies of the COVID-19 Contact-Tracing App in China and Singapore. *Int. J. Chin. Comp. Philos. Med.* **2020**, *18*, 113–143. [CrossRef]
10. Bianconi, A.; Marcelli, A.; Campi, G.; Perali, A. Efficiency of COVID-19 mobile contact tracing containment by measuring time-dependent doubling time. *Phys. Biol.* **2020**, *17*, 65006. [CrossRef]
11. Samuel, G.; Sims, R. The UK COVID-19 contact tracing app as both an emerging technology and public health intervention: The need to consider promissory discourses. *Health: Interdiscip. J. Soc. Study Health Illn. Med.* **2021**, *107*, 2411–2502. [CrossRef]
12. Ferrari, A.; Santus, E.; Cirillo, D.; Ponce-de-Leon, M.; Marino, N.; Ferretti, M.T.; Santuccione Chadha, A.; Mavridis, N.; Valencia, A. Simulating SARS-CoV-2 epidemics by region-specific variables and modeling contact tracing app containment. *npj Digit. Med.* **2021**, *4*, 9. [CrossRef]
13. Blom, A.G.; Wenz, A.; Cornesse, C.; Rettig, T.; Fikel, M.; Friedel, S.; Möhring, K.; Naumann, E.; Reifenscheid, M.; Krieger, U. Barriers to the Large-Scale Adoption of a COVID-19 Contact Tracing App in Germany: Survey Study. *J. Med. Internet Res.* **2021**, *23*, e23362. [CrossRef]
14. Maccari, L.; Cagno, V. Do we need a contact tracing app? *Comput Commun.* **2021**, *166*, 9–18. [CrossRef]
15. Gasteiger, N.; Gasteiger, C.; Vedhara, K.; Broadbent, E. The more the merrier! Barriers and facilitators to the general public's use of a COVID-19 contact tracing app in New Zealand. *Inform. Health Soc. Care* **2021**, *47*, 132–143. [CrossRef]
16. O'Callaghan, M.E.; Buckley, J.; Fitzgerald, B.; Johnson, K.; Laffey, J.; McNicholas, B.; Nuseibeh, B.; O'Keeffe, D.; O'Keeffe, I.; Razzaq, A.; et al. A national survey of attitudes to COVID-19 digital contact tracing in the Republic of Ireland. *Ir. J. Med. Sci.* **2020**, *190*, 863–887. [CrossRef]
17. Geber, S.; Ho, S.S. Examining the cultural dimension of contact-tracing app adoption during the COVID-19 pandemic: A cross-country study in Singapore and Switzerland. *Inf. Commun. Soc.* **2022**, 1–21. [CrossRef]
18. Williams, S.N.; Armitage, C.J.; Tampe, T.; Dienes, K. Public attitudes towards COVID-19 contact tracing apps: A UK-based focus group study. *Health Expect.* **2021**, *24*, 377–385. [CrossRef]
19. O'Callaghan, M.E.; Abbas, M.; Buckley, J.; Fitzgerald, B.; Johnson, K.; Laffey, J.; McNicholas, B.; Nuseibeh, B.; O'Keeffe, D.; Beecham, S.; et al. Public opinion of the Irish "COVID Tracker" digital contact tracing App: A national survey. *Digit. Health* **2022**, *8*, 2012837342. [CrossRef]
20. Albertus, R.W.; Makoza, F. An analysis of the COVID-19 contact tracing App in South Africa: Challenges experienced by users. *Afr. J. Sci. Technol. Innov. Dev.* 2022; *in press*. [CrossRef]
21. Schultz, É.; Touzani, R.; Mancini, J.; Ward, J.K. From contact tracing to COVID-19 pass holder; the tortured journey of the French TousAntiCovid contact tracing app. *Public Health* **2022**, *206*, 5–7. [CrossRef]
22. Al-Kuwari, M.G.; Ali Al Nuaimi, A.; Semaan, S.; Gibb, J.M.; AbdulMajeed, J.; Al Romaihi, H.E. Effectiveness of Ehteraz digital contact tracing app versus conventional contact tracing in managing the outbreak of COVID-19 in the State of Qatar. *BMJ Innov.* **2022**, *2021*. [CrossRef]
23. Giansanti, D.; Veltro, G. The Digital Divide in the Era of COVID-19: An Investigation into an Important Obstacle to the Access to the mHealth by the Citizen. *Healthcare* **2021**, *9*, 371. [CrossRef] [PubMed]
24. Jonker, M.; de Bekker-Grob, E.; Veldwijk, J.; Goossens, L.; Bour, S.; Rutten-Van Mölken, M. COVID-19 Contact Tracing Apps: Predicted Uptake in the Netherlands Based on a Discrete Choice Experiment. *Jmir. Mhealth Uhealth* **2020**, *8*, e20741. [CrossRef] [PubMed]
25. Dowthwaite, L.; Fischer, J.; Perez Vallejos, E.; Portillo, V.; Nichele, E.; Goulden, M.; McAuley, D. Public Adoption of and Trust in the NHS COVID-19 Contact Tracing App in the United Kingdom: Quantitative Online Survey Study. *J. Med. Internet Res.* **2021**, *23*, e29085. [CrossRef] [PubMed]
26. Nguyen, T.T.; Tran Hoang, M.T.; Phung, M.T. "To our health!" Perceived benefits offset privacy concerns in using national contact-tracing apps. *Libr. Hi Tech.* 2022; *ahead-of-print*. [CrossRef]
27. Isonne, C.; De Blasiis, M.R.; Turatto, F.; Mazzalai, E.; Marzuillo, C.; De Vito, C.; Villari, P.; Baccolini, V. What Went Wrong with the IMMUNI Contact-Tracing App in Italy? A Cross-Sectional Survey on the Attitudes and Experiences among Healthcare University Students. *Life* **2022**, *12*, 871. [CrossRef]

28. Lee, H.; Kim, J.; Fava, M.; Mischoulon, D.; Park, J.; Shim, E.; Lee, E.; Lee, J.H.; Jeon, H.J. Development and validation study of the Smartphone Overuse Screening Questionnaire. *Psychiat Res.* **2017**, *257*, 352–357. [CrossRef]
29. Burki, T. China's successful control of COVID-19. *Lancet Infect. Dis.* **2020**, *20*, 1240–1241. [CrossRef]
30. Radic, A.; Koo, B.; Gil-Cordero, E.; Cabrera-Sánchez, J.P.; Han, H. Intention to Take COVID-19 Vaccine as a Precondition for International Travel: Application of Extended Norm-Activation Model. *Int. J. Environ. Res. Public Health* **2021**, *18*, 3104. [CrossRef]
31. Si, H.; Shen, L.; Liu, W.; Wu, G. Uncovering people's mask-saving intentions and behaviors in the post-COVID-19 period: Evidence from China. *Sustain. Cities Soc.* **2021**, *65*, 102626. [CrossRef]
32. Schwartz, S.H. Normative explanations of helping behavior: A critique, proposal, and empirical test. *J. Exp. Soc. Psychol.* **1973**, *9*, 349–364. [CrossRef]
33. Song, Y.; Zhao, C.; Zhang, M. Does haze pollution promote the consumption of energy-saving appliances in China? An empirical study based on norm activation model. *Resour. Conserv. Recycl.* **2019**, *145*, 220–229. [CrossRef]
34. He, X.; Zhan, W. How to activate moral norm to adopt electric vehicles in China? An empirical study based on extended norm activation theory. *J. Clean Prod.* **2018**, *172*, 3546–3556. [CrossRef]
35. Ashraf Javid, M.; Ali, N.; Abdullah, M.; Campisi, T.; Shah, S.A.H. Travelers' Adoption Behavior towards Electric Vehicles in Lahore, Pakistan: An Extension of Norm Activation Model (NAM) Theory. *J. Adv. Transp.* **2021**, *2021*, 1–14. [CrossRef]
36. Sang, P.; Yao, H.; Zhang, L.; Wang, S.; Wang, Y.; Liu, J. Influencing factors of consumers' willingness to purchase green housing: A survey from Shandong Province, China. *Environ. Dev. Sustain.* **2019**, *22*, 4267–4287. [CrossRef]
37. Ataei, P.; Karimi, H.; Moradhaseli, S.; Babaei, M.H. Analysis of farmers' environmental sustainability behavior: The use of norm activation theory (a sample from Iran). *Arab. J. Geosci.* **2022**, *15*, 8396. [CrossRef]
38. Kim, Y.G.; Woo, E.; Nam, J. Sharing economy perspective on an integrative framework of the NAM and TPB. *Int. J. Hosp. Manag.* **2018**, *72*, 109–117. [CrossRef]
39. Hay, D.F. Prosocial Development. *J. Child. Psychol Psyc.* **1994**, *35*, 29–71. [CrossRef]
40. De Groot, J.I.M.; Steg, L. Morality and Prosocial Behavior: The Role of Awareness, Responsibility, and Norms in the Norm Activation Model. *J. Soc. Psychol.* **2009**, *149*, 425–449. [CrossRef]
41. Schwartz, S.H. Normative Influences on Altruism. *Adv. Exp. Soc. Psychol.* **1977**, *10*, 221–279. [CrossRef]
42. Wang, S.; Wang, J.; Zhao, S.; Yang, S. Information publicity and resident's waste separation behavior: An empirical study based on the norm activation model. *Waste Manag.* **2019**, *87*, 33–42. [CrossRef]
43. Richards, C.S.; McReynolds, W.T.; Holt, S.; Sexton, T. Effects of information feedback and self-administered consequences on self-monitoring study behavior. *J. Couns Psychol.* **1976**, *23*, 316–321. [CrossRef]
44. Van Quaquebeke, N. Paranoia as an Antecedent and Consequence of Getting Ahead in Organizations: Time-Lagged Effects Between Paranoid Cognitions, Self-Monitoring, and Changes in Span of Control. *Front. Psychol.* **2016**, *7*, 1446. [CrossRef] [PubMed]
45. Spencer, M.B.; Dupree, D.; Swanson, D.P. Parental Monitoring and Adolescents' Sense of Responsibility for Their Own Learning: An Examination of Sex Differences. *J. Negro Educ.* **1996**, *65*, 30. [CrossRef]
46. Valentine, S.R. Men and Women Supervisors' Job Responsibility, Job Satisfaction, and Employee Monitoring. *Sex. Roles.* **2001**, *45*, 179–197. [CrossRef]
47. Chi, X.; Cai, G.; Han, H. Festival travellers' pro-social and protective behaviours against COVID-19 in the time of pandemic. *Curr Issues Tour.* **2021**, *24*, 3256–3270. [CrossRef]
48. Arkorful, V.E.; Lugu, B.K.; Shuliang, Z. Unearthing mask waste separation behavior in COVID-19 pandemic period: An empirical evidence from Ghana using an integrated theory of planned behavior and norm activation model. *Curr. Psychol.* 2021; *Online ahead of print*. [CrossRef]
49. Kahneman, D.; Tversky, A. Prospect Theory: An Analysis of Decision under Risk. *Econometrica* **1979**, *47*, 263. [CrossRef]
50. Tran, C.D.; Nguyen, T.T. Health vs. privacy? The risk-risk tradeoff in using COVID-19 contact-tracing apps. *Technol. Soc.* **2021**, *67*, 101755. [CrossRef]
51. Seiders, K.; Voss, G.B.; Godfrey, A.L.; Grewal, D. SERVCON: Development and validation of a multidimensional service convenience scale. *J. Acad Market. Sci.* **2007**, *35*, 144–156. [CrossRef]
52. PudongReleases. Shanghai Fully Implement the "Place Code" "Digital Sentry" Service. 2022. Available online: https://www.pudong.gov.cn/006012/20220404/672292.html (accessed on 28 June 2022).
53. Hair, J.; Hollingsworth, C.L.; Randolph, A.B.; Chong, A.Y.L. An updated and expanded assessment of PLS-SEM in information systems research. *Ind. Manag. Data Syst.* **2017**, *117*, 442–458. [CrossRef]
54. Sharma, A.; Dwivedi, Y.K.; Arya, V.; Siddiqui, M.Q. Does SMS advertising still have relevance to increase consumer purchase intention? A hybrid PLS-SEM-neural network modelling approach. *Comput Hum. Behav.* **2021**, *124*, 106919. [CrossRef]
55. Cao, J.; Liu, F.; Shang, M.; Zhou, X. Toward street vending in post COVID-19 China: Social networking services information overload and switching intention. *Technol. Soc.* **2021**, *66*, 101669. [CrossRef] [PubMed]
56. Salehan, M.; Kim, D.; Kim, C. Use of Online Social Networking Services from a Theoretical Perspective of the Motivation-Participation-Performance Framework. *J. Assoc. Inf. Syst.* **2017**, *18*, 141–172. [CrossRef]
57. Podsakoff, P.M.; MacKenzie, S.B.; Lee, J.; Podsakoff, N.P. Common method biases in behavioral research: A critical review of the literature and recommended remedies. *J. Appl. Psychol.* **2003**, *88*, 879–903. [CrossRef] [PubMed]
58. Kock, N. Common Method Bias in PLS-SEM. *Int. J. E-Collab.* **2015**, *11*, 1–10. [CrossRef]

59. Benitez, J.; Henseler, J.; Castillo, A.; Schuberth, F. How to perform and report an impactful analysis using partial least squares: Guidelines for confirmatory and explanatory IS research. *Inf. Manag.-Amster.* **2020**, *57*, 103168. [CrossRef]
60. Chin, W.W.; Marcolin, B.L.; Newsted, P.R. A Partial Least Squares Latent Variable Modeling Approach for Measuring Interaction Effects: Results from a Monte Carlo Simulation Study and an Electronic-Mail Emotion/Adoption Study. *Inform. Syst Res.* **2003**, *14*, 189–217. [CrossRef]

life

MDPI

Comment

Why Has Digital Contact Tracing Worked Differently in Different Countries? Comment on Cao et al. The Impact of Digital Contact Tracing Apps Overuse on Prevention of COVID-19: A Normative Activation Model Perspective. *Life* 2022, *12*, 1371

Daniele Giansanti

Centre Tisp, Istituto Superiore di Sanità, 00131 Roma, Italy; daniele.giansanti@iss.it; Tel.: +39-06-49902701

Citation: Giansanti, D. Why Has Digital Contact Tracing Worked Differently in Different Countries? Comment on Cao et al. The Impact of Digital Contact Tracing Apps Overuse on Prevention of COVID-19: A Normative Activation Model Perspective. *Life* 2022, *12*, 1371. *Life* **2022**, *12*, 1592. https://doi.org/10.3390/life12101592

Academic Editor: Ying Chen

Received: 15 September 2022
Accepted: 10 October 2022
Published: 13 October 2022

I am writing you regarding your interesting article "The Impact of Digital Contact Tracing Apps Overuse on Prevention of COVID-19: A Normative Activation Model Perspective" [1] published in the Special Issue "The Digital Health in the Pandemic Era" [2].

I found that this work particularly stimulating and gives a great added value to the field.

Specifically, I believe that this study has great merit, drawing the attention of scholars to the usefulness of digital contact tracing in the fight against COVID-19.

During the COVID-19 pandemic, many countries have used digital contact tracing apps (DCTAs) to implement contact tracing. Although the use of DCTAs has contributed to the prevention and control of COVID-19, there are doubts in academia about their actual effectiveness [1]. In your study [1], the role of DCTAs in the prevention of COVID-19 was analysed in terms of both the responsibility and inconvenience to life in a large-scale DCTA overuse environment, based on the normative activation model. I completely agree with your findings suggesting that: (a) the overuse of a DCTA activates people's personal norms by triggering awareness of the consequences and ascription of responsibility, leading people to consistently cooperate with the government to prevent COVID-19. (b) However, the inconvenience of living with DCTA overuse weakens the effect of the awareness of consequences, ascription of responsibility and its role in influencing personal norms. (c) These effects may bear on people's willingness to consistently cooperate with the government to prevent COVID-19. The results from your study confirmed the effectiveness of DCTAs in counteracting pandemics from a social responsibility perspective in a large-scale environment where a DCTA is used, enriching the literature on DCTA research in the COVID-19 pandemic. I am convinced that the results of your study can also help governments to design, develop and improve policies to prevent COVID-19, as well as improve DCTAs' operating patterns.

I also, with some co-authors, faced DCTA use during the pandemic [3]. Specifically, I investigated the effectiveness of DCTA use in Italy [3]. We found that in Italy the DCTA showed both a low diffusion and a lack of capacity in the fight against COVID-19. Several factors have been identified affecting the use and diffusion of the DCTAs, for example, the strong impact of privacy issues and the digital divide [3–5]

Another study, based on a survey and published in the Special Issue confirmed this problem in Italy [6].

Ferretti et al. [7] had explored in the first phases of the pandemic the feasibility of protecting the population (that is, achieving transmission below the basic reproduction number) using isolation coupled with classical contact tracing by questionnaires versus algorithmic instantaneous contact tracing assisted by a mobile phone application [8]. The authors had concluded that although SARS-CoV-2 was spreading too fast to be contained by manual contact tracing, it could be controlled if this process was faster, more efficient, and increased in scale. According to the authors, the DCTAs that build a memory of proximity contacts and immediately notifies contacts of positive cases could achieve epidemic control if used by enough people. Therefore, according to these authors, DCTAs could be useful in controlling the COVID-19 epidemic.

However, the approach based on the DCTAs [8] did not run with the same effectiveness in the world. It showed positive experiences for containing the pandemic in some countries, such as for example in China and South Korea [8] and negative experiences in other countries, such as in Italy [3,5,6].

I understand that there are a lot of factors affecting this in a positive [1] or negative way [3–5], and that also the national disaster culture has an important role. Based on your experience [1] and field of study I would like to open a discussion with you and have your reply.

Funding: This research received no external funding.

Data Availability Statement: The data presented in this study are available upon request from the corresponding author. The data are not publicly available for ethical reasons.

Conflicts of Interest: The author declares no conflict of interest.

References

1. Cao, J.; Liu, D.; Zhang, G.; Shang, M. The Impact of Digital Contact Tracing Apps Overuse on Prevention of COVID-19: A Normative Activation Model Perspective. *Life* **2022**, *12*, 1371. [CrossRef] [PubMed]
2. Giansanti, D. The Digital Health: From the Experience of the COVID-19 Pandemic Onwards. *Life* **2022**, *12*, 78. [CrossRef] [PubMed]
3. Scrivano, N.; Gulino, R.A.; Giansanti, D. Digital Contact Tracing and COVID-19: Design, Deployment, and Current Use in Italy. *Healthcare* **2022**, *10*, 67. [CrossRef] [PubMed]
4. Kolasa, K.; Mazzi, F.; Leszczuk-Czubkowska, E.; Zrubka, Z.; Péntek, M. State of the Art in Adoption of Contact Tracing Apps and Recommendations Regarding Privacy Protection and Public Health: Systematic Review. *JMIR Mhealth Uhealth* **2021**, *9*, e23250. [CrossRef] [PubMed]
5. Giansanti, D.; Veltro, G. The Digital Divide in the Era of COVID-19: An Investigation into an Important Obstacle to the Access to the mHealth by the Citizen. *Healthcare* **2021**, *9*, 371. [CrossRef] [PubMed]
6. Isonne, C.; De Blasiis, M.R.; Turatto, F.; Mazzalai, E.; Marzuillo, C.; De Vito, C.; Villari, P.; Baccolini, V. What Went Wrong with the IMMUNI Contact-Tracing App in Italy? A Cross-Sectional Survey on the Attitudes and Experiences among Healthcare University Students. *Life* **2022**, *12*, 871. [CrossRef] [PubMed]
7. Ferretti, L.; Wymant, C.; Kendall, M.; Zhao, L.; Nurtay, A.; Abeler-Dörner, L.; Parker, M.; Bonsall, D.G.; Fraser, C. Quantifying SARS-CoV-2 transmission suggests epidemic control with digital contact tracing. *Science* **2020**, *368*, eabb6936. [CrossRef] [PubMed]
8. Giansanti, D.; Scrivano, N.; Gulino, R.A. A map point on the role of the telemedicine and e-Health in the digital contact tracing during the COVID-19 pandemic. *J. Public Health Emerg.* **2021**, *5*, 39. [CrossRef]

life

MDPI

Reply

Reply to Giansanti, D. Why Has Digital Contact Tracing Worked Differently in Different Countries? Comment on "Cao et al. The Impact of Digital Contact Tracing Apps Overuse on Prevention of COVID-19: A Normative Activation Model Perspective. *Life* 2022, *12*, 1371"

Junwei Cao [1], Dong Liu [2], Guihua Zhang [3] and Meng Shang [4,*]

[1] School of Business, Yangzhou University, Yangzhou 225127, China
[2] Department of Global Business, Yeungnam University, Gyeongsan 38541, Korea
[3] Department of Business, Yeungnam University, Gyeongsan 38541, Korea
[4] School of Flight, Anyang Institute of Technology, Anyang 455008, China
* Correspondence: shangmengdr@163.com

Citation: Cao, J.; Liu, D.; Zhang, G.; Shang, M. Reply to Giansanti, D. Why Has Digital Contact Tracing Worked Differently in Different Countries? Comment on "Cao et al. The Impact of Digital Contact Tracing Apps Overuse on Prevention of COVID-19: A Normative Activation Model Perspective. *Life* 2022, *12*, 1371". *Life* **2022**, *12*, 1593. https://doi.org/10.3390/life12101593

Academic Editor: Ying Chen

Received: 30 September 2022
Accepted: 10 October 2022
Published: 13 October 2022

Thank you for your comments [1] for our Manuscript. We would like to discuss with you the question, "Why has digital contact tracing worked differently in different countries?" from the perspective of national disaster cultures.

Many studies in Western countries have shown that digital contact tracking APPs (DCTAs) cause a lot of inconvenience, leading to their effectiveness in preventing COVID-19, such as digital divide issues, diffusion issues, and data security issues [2–4]. In particular, one review noted that compared to Western DCTAs, the biggest problem with Chinese DCTAs (such as the Alipay Health Code app) is the difficulty in achieving an optimal balance between data protection standards and public health interests [3]. However, our study shows that the massive overuse of DCTA in China, while causing inconvenience, also helps to stimulate a sense of consequence and responsibility, in which people tend to ignore the inconvenience and consistently prevent COVID-19 [5]. I think it helps to understand such differences from the perspective of national disaster culture.

There are three main types of disaster management culture: state-oriented, individualistic, and fatalistic. In a state-oriented risk management culture, people believe that the direction of disasters is determined by both the environment and people, and that action must be taken to manage disasters. However, people do not know much about coping mechanisms, so they show a high level of trust in national disaster management authorities and have high expectations of the role of government. They believe that national-level involvement can be effective in managing disasters. They will trust and cooperate with the disaster management measures of the national disaster management authority and will be highly compliant with these measures [6,7]. In a culture of individualistic-centered disaster management, people believe that risk prevention is possible and that negative consequences should be reduced through action. However, they tend to have a high level of awareness of how they can respond to disasters, are not trusting of the role of government disaster management, and prefer to take action on their own to protect themselves [6,7]. In a fatalistic risk management culture, people will believe that disasters are forces of nature that cannot be denied, are unpredictable, and are inevitable. These people lack confidence in crisis resolution, and they expect the nation's disaster management authorities to take action, but do not take nationally issued information about disasters seriously, and do not seriously cooperate with disaster management [6,7].

The national culture of disaster management in Chinese society is state oriented. Chinese people are highly subservient to the government's instructions and arrangements to resolve their crisis. Chinese people have a high level of trust in the national disaster management authorities, and they believe that state involvement helps to manage disasters efficiently. As the Chinese government takes strong measures to prevent COVID-19, it inspires the public to cooperate with the government by creating a sense of collective responsibility and mission. Chinese people like to obtain knowledge and policies on epidemic prevention released by the government from social media, and also tend to express their views on responsibility and concerns about the epidemic through social media. Therefore, when the Chinese government overused DCTA nationwide, although it caused inconvenience to people, they were more inclined to sacrifice their own interests to cooperate with the state's epidemic prevention when weighing the pros and cons in a state-oriented disaster response culture. Chinese also tend to show compassion for patients, blame those who caused the consequences and give appreciation to COVID-19 prevention heroes on social media [8,9]; in addition, with DCTAs' tracking feature, the troublemakers who caused the spread of COVID-19 can be better located. Thus, although the Chinese perceived a lack of transparency in the operation of DCTAs and an unclear scope of data storage, they were also willing to overlook the potential loss of personal benefits from DCTAs [10] and cooperate with the national disaster authorities for effective use; ultimately, DCTAs have been shown to be extremely effective in China. Moreover, the sense of responsibility is very strong not only in China, but also in countries with Eastern cultures, such as South Korea, which has been very successful in preventing COVID-19. A study analyzed Korean posts on Twitter about COVID-19 prevention and found that the most content was about "attribution of responsibility" [11]. Korea is also a country where DCTAs are used on a large scale, and where the epidemic prevention department gives each place a phone number for people to register their tracks. If someone is infected with COVID-19, his or her track will likely be made public on the disease administration's website.

In contrast, in some western countries, their disaster management culture may be centered on individualism. For example, in Europe, many British people refuse to use such apps because of privacy concerns [12], Irish people refuse to use it because they fear that tech companies or the government will use the app to monitor users even after the COVID-19 pandemic is over [13]. In the individualistic-centered disaster management culture, they are distrustful of government or company-led preventive measures. They are worried about their own interests and do not use DCTAs, and they prefer to face COVID-19 through their own perceptions rather than using DCTAs promoted by the relevant authorities or the government. Therefore, we can also infer that this part of the population can be individualistic in disaster management issues when they are capable of coping with COVID-19, and once they fail to cope, then they are likely to turn to a fatalistic disaster management culture, believing that COVID-19 is a force of nature, unpredictable and unavoidable, lacking confidence in the prevention of COVID-19, hoping to rely on the help of national disaster management authorities, but not taking national control measures, such as the strict use of DCTAs seriously.

Therefore, I think that future research can empirically study the effectiveness of DCTAs in different countries from the perspective of national disaster management culture, which will help each country to construct its own disaster management plan to cope with possible future international public health emergencies.

Author Contributions: Conceptualization, J.C. and D.L.; methodology, J.C.; software, M.S.; validation, G.Z. and M.S.; formal analysis, G.Z. and M.S.; investigation, M.S.; data curation, D.L.; writing—original draft preparation, J.C.; writing—review and editing, J.C. and D.L.; supervision, M.S.; project administration, D.L. and M.S. All authors have read and agreed to the published version of the manuscript.

Funding: This research received no external funding.

Data Availability Statement: The data presented in this study are available upon request from the corresponding author. The data are not publicly available for ethical reasons.

Conflicts of Interest: The authors declare no conflict of interest.

References

1. Giansanti, D. Why Has Digital Contact Tracing Worked Differently in Different Countries? Comment on Cao et al. The Impact of Digital Contact Tracing Apps Overuse on Prevention of COVID-19: A Normative Activation Model Perspective. *Life* 2022, *12*, 1371. *Life* **2022**, *12*, 1592. [CrossRef]
2. Giansanti, D.; Veltro, G. The digital divide in the era of COVID-19: An investigation into an important obstacle to the access to the mHealth by the citizen. *Healthcare* **2021**, *9*, 371. [CrossRef] [PubMed]
3. Kolasa, K.; Mazzi, F.; Leszczuk-Czubkowska, E.; Zrubka, Z.; Péntek, M. State of the art in adoption of contact tracing apps and recommendations regarding privacy protection and public health: Systematic review. *JMIR mHealth uHealth* **2021**, *9*, e23250. [CrossRef] [PubMed]
4. Scrivano, N.; Gulino, R.A.; Giansanti, D. Digital contact tracing and COVID-19: Design, deployment, and current use in italy. *Healthcare* **2021**, *10*, 67. [CrossRef] [PubMed]
5. Cao, J.; Liu, D.; Zhang, G.; Shang, M. The impact of digital contact tracing apps overuse on prevention of COVID-19: A normative activation model perspective. *Life* **2022**, *12*, 1371. [CrossRef] [PubMed]
6. Reuter, C.; Kaufhold, M.; Schmid, S.; Spielhofer, T.; Hahne, A.S. The impact of risk cultures: Citizens' perception of social media use in emergencies across Europe. *Technol. Forecast. Soc.* **2019**, *148*, 119724. [CrossRef]
7. Cornia, A.; Dressel, K.; Pfeil, P. Risk cultures and dominant approaches towards disasters in seven European countries. *J. Risk Res.* **2014**, *19*, 288–304. [CrossRef]
8. Liao, Q.; Yuan, J.; Dong, M.; Yang, L.; Fielding, R.; Lam, W.W.T. Public Engagement and Government Responsiveness in the Communications About COVID-19 During the Early Epidemic Stage in China: Infodemiology Study on Social Media Data. *J. Med. Internet Res.* **2020**, *22*, e18796. [CrossRef] [PubMed]
9. Su, Y.; Wu, P.; Li, S.; Xue, J.; Zhu, T. Public emotion responses during COVID-19 in China on social media: An observational study. *Hum. Behav. Emerg. Technol.* **2020**, *3*, 127–136. [CrossRef]
10. Joo, J.; Shin, M.M. Resolving the tension between full utilization of contact tracing app services and user stress as an effort to control the COVID-19 pandemic. *Serv. Bus.* **2020**, *14*, 461–478. [CrossRef]
11. Park, H.W.; Park, S.; Chong, M. Conversations and Medical News Frames on Twitter: Infodemiological Study on COVID-19 in South Korea. *J. Med. Internet Res.* **2020**, *22*, e18897. [CrossRef] [PubMed]
12. Williams, S.N.; Armitage, C.J.; Tampe, T.; Dienes, K. Public attitudes towards COVID-19 contact tracing apps: A UK-based focus group study. *Health Expect.* **2021**, *24*, 377–385. [CrossRef] [PubMed]
13. Callaghan, M.E.O.; Abbas, M.; Buckley, J.; Fitzgerald, B.; Johnson, K.; Laffey, J.; McNicholas, B.; Nuseibeh, B.; Keeffe, D.O.; Beecham, S.; et al. Public opinion of the Irish "COVID Tracker" digital contact tracing App: A national survey. *Digit. Health* **2022**, *8*, 2012837342. [CrossRef]

life

MDPI

Article

IoT Framework for a Decision-Making System of Obesity and Overweight Extrapolation among Children, Youths, and Adults

Saeed Ali Alsareii [1,*,†], Ahmad Shaf [2,†], Tariq Ali [2], Maryam Zafar [2], Abdulrahman Manaa Alamri [1], Mansour Yousef AlAsmari [1], Muhammad Irfan [3] and Muhammad Awais [4]

1 Department of Surgery, College of Medicine, Najran University Saudi Arabia, Najran 11001, Saudi Arabia
2 Department of Computer Science, COMSATS University Islamabad, Sahiwal Campus, Sahiwal 57000, Pakistan
3 Electrical Engineering Department, College of Engineering, Najran University Saudi Arabia, Najran 11001, Saudi Arabia
4 Department of Computer Science, Edge Hill University, St Helens Rd, Ormskirk L39 4QP, UK
* Correspondence: alsareii@nu.edu.sa
† These authors contributed equally to this work.

Citation: Alsareii, S.A.; Shaf, A.; Ali, T.; Zafar, M.; Alamri, A.M.; AlAsmari, M.Y.; Irfan, M.; Awais, M. IoT Framework for a Decision-Making System of Obesity and Overweight Extrapolation among Children, Youths, and Adults. *Life* **2022**, *12*, 1414. https://doi.org/10.3390/life12091414

Academic Editors: Tao Huang and Daniele Giansanti

Received: 8 August 2022
Accepted: 5 September 2022
Published: 10 September 2022

Abstract: Approximately 30% of the global population is suffering from obesity and being overweight, which is approximately 2.1 billion people worldwide. The ratio is expected to surpass 40% by 2030 if the current balance continues to grow. The global pandemic due to COVID-19 will also impact the predicted obesity rates. It will cause a significant increase in morbidity and mortality worldwide. Multiple chronic diseases are associated with obesity and several threat elements are associated with obesity. Various challenges are involved in the understanding of risk factors and the ratio of obesity. Therefore, diagnosing obesity in its initial stages might significantly increase the patient's chances of effective treatment. The Internet of Things (IoT) has attained an evolving stage in the development of the contemporary environment of healthcare thanks to advancements in information and communication technologies. Therefore, in this paper, we thoroughly investigated machine learning techniques for making an IoT-enabled system. In the first phase, the proposed system analyzed the performances of random forest (RF), K-nearest neighbor (KNN), support vector machine (SVM), decision tree (DT), logistic regression (LR), and naïve Bayes (NB) algorithms on the obesity dataset. The second phase, on the other hand, introduced an IoT-based framework that adopts a multi-user request system by uploading the data to the cloud for the early diagnosis of obesity. The IoT framework makes the system available to anyone (and everywhere) for precise obesity categorization. This research will help the reader understand the relationships among risk factors with weight changes and their visualizations. Furthermore, it also focuses on how existing datasets can help one study the obesity nature and which classification and regression models perform well in correspondence to others.

Keywords: IoT; pandemic; obesity; classification; regression; real-time system

1. Introduction

Obesity refers to excessive amounts of body fat. Obesity is not only caused by food genetics, the environment could also be a cause. The intake of energy and not consuming this energy through physical activity could also be a primary reason for obesity [1]. Obesity is the relationship between calorie intake and energy expenditure. It is a significant health issue associated with chronic illness and has a negative impact and long-term effects on patients and their families. As obesity is a risk factor for a number of diseases worldwide, it can be a threat to the world in the future. The Asia region is already dealing with malnutrition (as many cases have reported). Therefore, the number of obesity cases is increasing significantly with time [2].

Since 1975, the global obesity rate has increased thrice according to the World Health Organization (WHO) [3]. In 2013, the Indonesian Basic Health Research national survey (RISEKDAS) noted that obesity cases were rapidly increasing in Indonesia. Obesity can affect both men and women. The rate of obesity in adult men was 13.9%, 7.8%, and 19.7% in 2007, 2010, and 2013, respectively. In contrast, the rate of obesity in adult women was 14.8%, 15.5%, and 32.9% in 2007, 2010, and 2013, respectively [4]. However, in 2018, according to RISEKDAS (the same survey), the rates decreased to 14.5% in men and 29.3% in women [5].

The 2016 data show that the obesity rate has hit over 650 million people globally [6]. From age 18 and older, the ratio of people who are overweight increased to 39% [7]. Obesity and being overweight lead to other dangerous consequences that could lead to health anxiety. Obesity is the prime reason for significant lifestyle diseases, such as cancer, type II diabetes, lung disease, chronic pulmonary disease, and asthma.

Underdeveloped countries and populations are high victims of these diseases. NCDs (non-communicable diseases) and lifestyle diseases caused 36 million (63%) global deaths in 2008. Of these 36 million people, 80% were affected in underdeveloped countries and the middle class, 13% affected the upper class, and 29% affected those under age 60. Selected literature studies showed an annual increase of 10 million deaths due to NCDs. A survey from 2016 showed an increase of 71% (56.9 million), predicting 75% to 88.5% of deaths (until 2030) from NCDs in emerging countries, while the ratio predicted in developing countries is 65% [8]. Body mass index (BMI) is a primary risk element for the rise in diseases linked to sedentary lifestyles [9]. BMI helps in assessing body composition by calculating "weight/height".[2] However, BMI is considered a lousy sign of the proportion of body fat because BMI is dependent on age and does not count the fat on different body sites. According to the Institute of Medicine's 2012 report, there are population-based obesity prevention initiatives that address obesity and being overweight, such as a balanced diet, regular exercise, context- and setting-specific advice, and sound social norms [10].

There are several risk factors associated with obesity. In general, these factors are divided into categories, such as lifestyle factors (e.g., consuming junk food, alcohol, stress, and low physical activity), as well as demographic and socioeconomic elements (e.g., age, gender, marital status, place of residence, and genetic elements) [11]. Some risk factors can be avoided while others cannot. To implement an effective risk reduction strategy, the individual and population levels need to understand the factors that can be avoided [12]. The available data have helped numerous studies in exploring better approaches.

Epidemiological data modeling techniques (using machine learning) are popular in scholarly publications. These techniques can contribute to a better understanding of illness distribution, general health, risk identification, and health risk factors. There are several methods and algorithms available for this purpose [13]. The techniques require exact data classifications to assist in identifying risk detection from the information to lessen the danger signs and morbidity and mortality caused by obesity. Based on data showing compliance with dietary guidelines for obesity prevention, machine learning is applied to predict the likelihood of obesity [14]. Electronic health records are used in machine learning for predicting obesity in children, predicting obesogenic environments for children, and aggregating clinical data, such as metabolomic lipidomics and model drug dose responses [15].

In an online study conducted in Bangladesh (November 2020), 338 adults were examined. Sociodemographic statistics, health-related information, physical activity-related details, and nutrition measurements were all covered in the questionnaire. With two scenarios ('before' and 'during' the pandemic commencement) taken into consideration, inferential statistics (i.e., chi-square test, McNemar test) were employed to analyze the relationships between BMI and examined variables [16]. P0.05 was regarded as statistically significant. Results revealed that 30.5% of people were overweight "before" the COVID-19 pandemic and 34.9% of people were overweight "during" the pandemic. This suggests that 4.4% of the participants experienced significant weight gain after the pandemic started.

A recent report from Riyadh shows that 24.5% of women and 19% of men are suffering from obesity [17]. In 2021, the United Kingdom of Saudi Arabia showed significant increases in obesity rates. The ration varied in men and women but overall statistics showed that there were increases of 26.8%, 24%, 23.5%, 23.3%, 20.6%, 20.2%, 19.8%, 19.7%, and 14.2% in Riyadh, Makkah region, Hail (23.5 %), Al-Jouf, the northern border region, Madinah, Jazan, Tabuk, and Al-Baha, respectively, as depicted in Figure 1. This increase in people being overweight has led to an increase in diseases. The report further revealed that diseases such as obesity, diabetes, high BP, heart illness, stroke, and cancer are rising at ratios of (19.0%, 24.5%), (13.5%, 10.6%), (13.7%, 12.7%), (5.5%, 3.9%), (1.5%, 1.3%), and (1.3%, 1.8%) in both males and females, respectively, as shown in Figure 2.

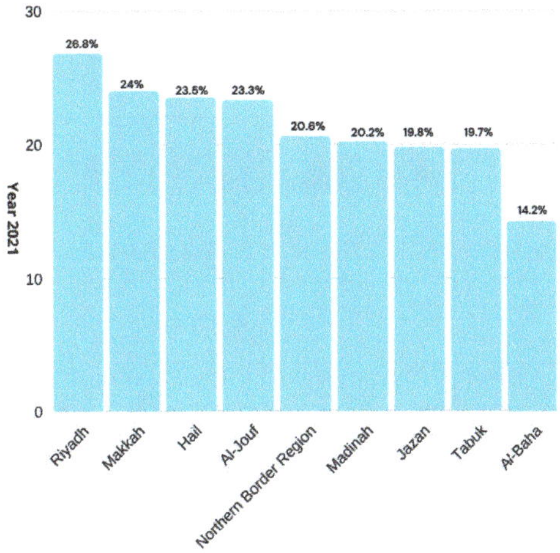

Figure 1. Obesity rate in the provinces of the Saudi Kingdom.

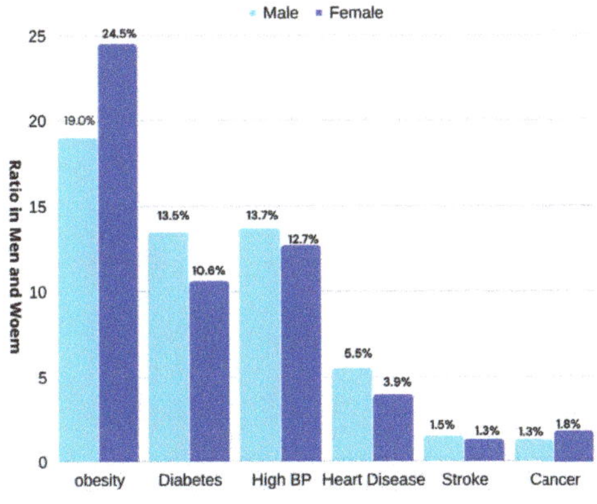

Figure 2. Increased specific disease rate in males and females of the Saudi Kingdom.

Several machine learning algorithms are applied with several features to predict specific health conditions. A branch of machine learning known as ANN (artificial neural networks) correlates input parameters and corresponds to output data. ANN has reported several applications in engineering and medicine with variable success rates. Dugan et al. [18] employed artificial intelligence to predict childhood obesity. Six models were used in this research for the study. These models were naïve Bayes, random tree, ID3, j48, random forest, and Bayes net-trained. These models were applied to the clinical decision support system on CHICA. The results showed that ID3 performed well, giving a high ratio of accurate results at 85% and a sensitivity rate of 90%. Jindal et al. used techniques for collective machine learning for obesity prediction. The prediction accuracy proposed for ensemble machine learning approaches was 89.68%. The generalized linear model, partial least squares, and random forest were used in the ensemble prediction through a Python interface.

Hammond et al. [15] used public records and electronic health records for the prediction of obesity in childhood. Several machine learning algorithms were trained for regression and binary classification. The results showed considerable accuracy in the first two years of data collection. The results showed that children at age five could become obese. To distinguish between low, medium, and high obesity, they used logistic regression (using a separate random forest classifier). They employed LASSO regression to 'prophesy' their continuous BMI values. The bootstrap was run 100 times to obtain a better performance of the model.

Obesity at the national level was predicted using data on food sales; see Dunstan et al. [19]. Three machine learning models were applied to the data obtained from seventy-nine countries. The authors researched basic information from the synergic nature of categories by analyzing food sales. They used five categories. The research considered 60% of countries for 10% (concerning the prevalence range). Moreover, 87% of countries projected the prevalence of obesity with an absolute error of less than 20%. The research showed that baked goods and flour were the most appropriate food categories for the prediction of obesity. Extreme gradient boosting, RF, and SVM were utilized for this model.

Singh and Tawfik [20] presented a machine learning model that might predict adolescent weight gain and obesity. In this study, seven machine learning methods were employed. J48 pruned tree, K-NN, bagging, and other algorithms were used, such as multi-layer perception and random forest. An unaltered and unbalanced dataset was used to vote on the effectiveness of all of the proposed algorithms. The MLP algorithm resulted in a 96% precision value. While the F1-score gave results of 93.96%. Gerl et al. [21] exhibited the use of large population cohorts for the prediction of different measures of obesity. A perplexing lipidomic signature was identified for BFP. A total of 73% of BFP variants were predicted based on age, gender, and lipidome, with the complete range of BFP having mistakes.

Montanezet al. [22] used publicly available genetic profiles and studied machine learning algorithms for predicting obesity. Many machine learning models were involved in this study, such as the SVM algorithm, decision tree, K-NN algorithm, and the decision rule for predicting susceptibility to chronic hepatitis with the help of SNP data. Of all the techniques, SVM produced the best results for the prediction model. According to the simulation findings, the SVM area was below the curve value of 90.5%.

Borrel and Samuel [23] worked on risk mortality and the US adult body mass index category. The effects of obesity and excess weight on the Cox proportional hazard regression were looked at to obtain the death prevalence. They calculated the rate of progress through time for all causes and the mortality rate dependent on peers at a normal weight. They also looked into the mortality rate of persons with obesity/were overweight and had cardiovascular disease. Their proposed results showed CVD caused death in obese adults (over 20% compared to normal-weight adults).

During the pandemic, the obesity rate increased due to lockdowns, and it become extremely important to have digital methods to monitor physical activities and the obeseness of people. Various challenges were involved in the understanding of risk factors and the obesity ratio. Traditionally, statistical analyses were used for understanding obesity, imposing independent linearity and a limited number of prediction sets. Therefore, this study focused on the different machine learning models for the risk identification of obesity. It evaluated the effectiveness of machine learning techniques, such as regression and classification on accessible data in order to compile a list of criteria that could be used to diagnose obesity and being overweight. These results helped us to design an IoT-enabled decision system that might be accessible worldwide where internet facilities are available. Thus, the paper provides the following contributions:

- A novel IoT framework was designed that could be accessible from anywhere and any time for the early prediction of obesity, from the given link http://mlobesity.herokuapp.com/ (accessed on: 22 August 2022)
- A decision-making system was developed with the assistance of state-of-the-art machine learning algorithms.
- The proposed expert system involves both classification and regression models for clear visualization of given data.
- This system could help doctors in making early decisions that might significantly increase the prediction of a patient's current obesity condition

The remainder of the paper is organized as follows: the proposed machine learning algorithms and IoT system architecture are explained in Section 2. In Sections 3 and 4, the results and discussion are covered; the conclusion and future work are discussed in Section 5.

2. Materials and Methods

Classification and regression are supervised machine learning algorithms used for accurate assessments and instructions. For classification and regression, the process includes the following steps: data collection, preprocessing, data visualization, model training, testing, and evaluating. The research discusses the target population, study sample, and at-risk population. The study does not predict any new risk factors. The sample data focus on the population from the ages >20 to <60, excluding pregnancy and genetic factors.

2.1. Dataset Explanation

This study analyzes the data on eating habits and health to estimate the prevalence of obesity among persons from Mexico, Peru, and Colombia. The data were categorized using the values of Insufficient weight, normal weight, overweight level I, overweight level II, obesity type I, obesity type II, and obesity type III, thanks to the class variable NObesity (obesity level) assigned to the records. The dataset consisted of 2111 records and 17 attributes. The SMOTE filter and the Weka tool were used to artificially produce 77% of the data, while a website platform collected 23% of the data directly from users. The dataset is categorized into three parts: **Food intake indicators:** FAVC (frequent consumption of high-calorie foods), FCVC (frequent consumption of vegetables), NCP (number of meals), CAEC (intake of food between meals), CH_2O (daily water intake), CALC (alcohol intake). **Body attribute**: TUE (time utilizing technological devices), FAF (regular exercise frequency), SCC (calorie-ingestion tracking), MTRANS (utilized for transportation). **Other attributes**: gender, age, height, weight, smoke, and family history.

$$BMI = \frac{Weight}{height^2} \tag{1}$$

Dataset attributes were categorized according to the mass body index as shown in Equation (1) for each individual; the results were compared with the data provided by the WHO and the Mexican normativity.

- Underweight Less than 18.5;
- Normal 18.5 to 24.9;
- Overweight 25.0 to 29.9;
- Obesity I 30.0 to 34.9;
- Obesity II 35.0 to 39.9;
- Obesity III higher than 40.

BMI is considered a 'lousy' sign relating to the proportion of body fat because BMI is dependent on age and does not count the fat on different body sites. Therefore, a detailed analysis of individual eating habits, physical activities, and other attributes is needed to understand obesity in a better way.

2.2. Dataset Preprocessing

Categorical and continuous data were separated into two groups. Classification and regression are considered supervised machine learning algorithms used for accuracy assessment and instruction. The selected dataset had noise—some values were small and some had a considerable enough amount of data for the supervised, trained machine learning model. Data samples containing outliers were discarded; the remaining data were filtered with data mining. Data mining involves clustering, classification, feature selection, association, calculation, outlier analysis, and pattern discovery. Incomplete data were removed during the data cleaning process. Similarly, several steps were involved in the data post-processing, such as pattern interpretation, pattern evolution, pattern visualization, and pattern selection. 1. K-fold assists in the accuracy of the ML (machine learning) model after training. 2. Spyder IDE helps establish a Python environment data science application using anaconda distribution.

2.3. Decision Tree

A classification model that recursively divides the datasets into sub-parts is known as a decision tree. There are root nodes, internal nodes in the decision tree, and terminal nodes developed by the subdivision of the tree. Each node was derived from a single parent and could have many child nodes. A decision tree helps in the decision-making process. The context of the decision tree decides the probability of sets. The simple structure of the decision tree has nodes and terminal nodes, which is a supervised approach to classification. Nodes represent the dataset's properties, and their results are displayed by terminal nodes. C4.5 and random forest are examples of algorithms used to implement the decision tree [24].

2.4. Random Forest

Several applications rely on decision tree architecture during training, such as regression and classification; random decision forest is also an ensemble learning technique. Random forest utilizes several decision trees (CART) and then gives the most accurate outcome with the combination of these trees. The decision tree algorithm uses the Gini index technique, which measures the probability that a selected element from the set will be erroneously categorized. The total squared possibility for each class is decreased by 1 from the Gini index calculations. This technique increases the predictive power of the system. Removing the bias created by the decision tree model adds to the system. Additionally, using the "random Forest" R package, random forest can naturally order the relevance of variables in regression or classification tasks [25,26].

2.5. Support Vector Machine

SVM offers excellent empirical findings and a strong theoretical base. Several agents have used SVM to complete tasks, including digit recognition, object identification in text categorization, and human activity recognition [27–31]. Based on the 'A' mathematical model for problems involving regression; classification was supplied by the statistical learning systems. A key benefit of SVM involves the availability of trustworthy tools and techniques for solving issues swiftly and efficiently.

2.6. K-Nearest Neighbor

One data mining technique is the K-nearest neighbor (KNN) approach used for categorization, which assigns a batch of data based on learning previously labeled or categorized data. The outcomes of newly categorized query instances based on the majority of the proximity to existing categories in KNN fall under the category of supervised learning, including KNN. The following are processes involved in categorizing using the K-nearest neighbor (KNN) algorithm: 1. Establishes the k parameter; 2. Determines the separation between training and test data using the Euclidean distance calculation; 3. Arranges the formed distances; 4. Establishes the distance closest to the sequence K; 5. Matches the proper class; 6. Assigns the class as the data class to be assessed by counting the number of classes from the nearest neighbors [32].

2.7. Naïve Bayes

Naïve Bayes data mining techniques help make predictions in many fields and are used by many researchers. The framework for a hybrid strategy that uses naïve Bayes for parameter optimization and genetic algorithms for prediction is presented in this research. According to the naïve Bayes model, parameters with zero values show weaknesses in the results. This problem can be resolved by applying genetic algorithm optimization. The problem 'suggested' optimizing genetic algorithms for the study. The study was initialized with an analysis of the literature on the subject of child obesity and adequate data mining models for the prediction of childhood obesity. Following the review, 19 attributes were chosen, and the NB approach was used to predict child obesity. A 75% increase in accuracy was seen in the first test to gauge the utility of the proposed approach [33].

2.8. Logistical Regression

Using prior observations from a dataset, a statistical analysis technique called logistic regression predicts a binary outcome, such as yes or no. Using a logistic regression model, a dependent data variable is predicted by looking at the correlation between the independent variables that are already present. For instance, logistic regression may be used to foretell a candidate's outcome in a political election or whether a high school student will be accepted into a particular college. These simple choices between two options allow for binary outcomes. Thirty input variables were gathered from the patient records, including clinical information (gender, age, body mass index, and concurrent disorders), laboratory testing, and histopathologic results of the gallbladder. The identical database was used to produce a logistic regression model, and similar data were compared to the outcome [34].

2.9. IoT Enabled System Architecture

This system is regarded as a multiple-user access system, allowing numerous users to connect to the cloud simultaneously, as shown in Figure 3. There is only a single universal receiver shared by all users. An IoT system with cloud administration was created to classify obesity. Because it is a distributed system, the cloud is the best solution for a healthcare system that enables doctors to obtain data more easily. Our suggested IoT system comprises four key phases: (1) data collecting, (2) textual data classification, (3) diagnosis, and (4) user interface. Its goal is to lower disease rates through early detection of obesity.

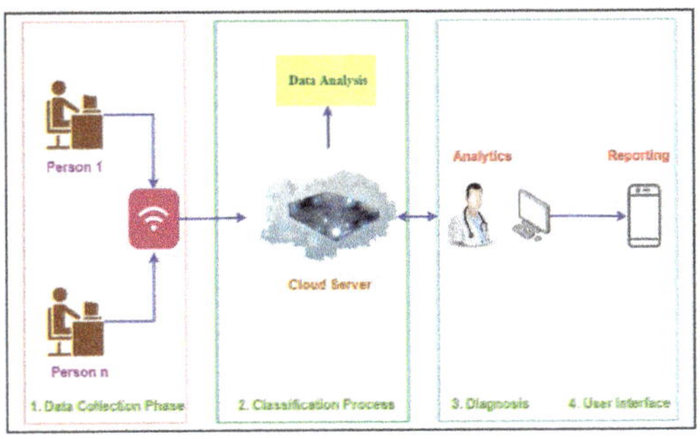

Figure 3. Proposed IoT system architecture.

This figure demonstrates that the user is the origin of the entire process. With the web application interface, users engaging with the server and application interfaces are directly coupled. Therefore, when a user interacts with the web interface, a specific request is sent to the server. Upon receiving a request, the server examines it to determine what is the need of the user (obesity prediction, check his/her history, download report, or doctor's advice).

Then the server will decide where to transmit the user's request after considering the needs of the user. Therefore, the server looks for an expert system that can handle the user's request and deliver the results. The server assigns the user's duties after identifying the optimal expert system. The user's task is inputted into the expert system as a string because the entire model is reliant on textual information, which is utilized to identify obesity in its early stages. After receiving a string input, the algorithm eliminates any extraneous words that are found during the prediction stage. After eliminating superfluous words, the user-provided data are used by the prediction engine to make predictions. Following the calculation of the outcome, the results are sent to the expert system. The server receives the results that the expert system collects. After obtaining the expert system's results, the server sends it to the web interface, where the user can access his/her results and move forward in light of the report.

3. Results

The following metrics help in evaluating machine learning models for classification and regression. **Regression Metrics**: MBE (mean bias error), RMSE (root mean square error), MABE (mean absolute bias error), and R^2 (determination coefficients). **Classification metrics**: discuss the classification report and confusion matrix. F1-score, recall, accuracy, and precision are included in the classification report and their equations are shown in Equations (2)–(5). Two dimensions, "actual" and "predicted," are included in the confusion matrix. For each dimension, there are values for true positive (TruePos), true negative (TrueNeg), false positive (FalsePos), and false negative (FalseNeg).

- True positive: The difference between the actual and anticipated classes is 1.
- True negative: The difference between the actual and projected classes is 0.
- False positive: The predicted class is 1, while the actual class is 0.
- False negative: The predicted class is 0, whereas the actual class is 1.

The following class labels were used for regression and classification purposes: 'Normal_Weight', 'Insufficient_Weight', 'Overweight_Level_I', 'Overweight_Level_II', 'Obesity_Type_I', 'Obesity_Type_II', 'Obesity_Type_III' with the indexes of '0', '1', '2', '3', '4', '5', and '6', respectively.

The following formulas help in the calculation of classification metrics:

$$Accuracy = \frac{(TruePos + TrueNeg)}{TruePos + FalsePos + FalseNeg + TrueNeg} \tag{2}$$

$$Precision(P) = \frac{TruePos}{TruePos + FalsePos} \tag{3}$$

$$Recall(R)\ or\ Sensitivity(S) = \frac{TruePos}{TruePos + FalseNeg} \tag{4}$$

$$F1\text{-}score = 2 \times \frac{Precision \times Recall}{Precision + Recall} \tag{5}$$

The precision determines how closely the real value resembles the measured value, while accuracy assesses how closely the measured value resembles the actual value. Recall and sensitivity indicate a machine learning model's overall usefulness. MBE, RMSE, MABE, and R^2 are used for regression problems as represented in Equations (6)–(9). If the MBE is low and close to zero, the prediction model performs well. Furthermore, zero represents the optimal situation. The prediction model effectiveness (in the short term) is assessed by the RMSE metric. It always has a positive value, which ought to be close to zero. MABE evaluates the severity of an association. The objective is to come as close to zero. The R^2 approach shows how well a method can forecast a set of quantifiable facts. Its value is a number between 0 and 1.

$$MBE = \frac{1}{q}\sum_{n=1}^{q}(b_n - c_n)^2 \tag{6}$$

$$RMSE = \sqrt{\frac{1}{q}\sum_{n=1}^{q}(b_n - c_n)^2} \tag{7}$$

$$MABE = \frac{1}{q}\sum_{n=1}^{q}|b_n - c_n| \tag{8}$$

$$R^2 = 1 - \frac{\sum(b_n - c_n)^2}{\sum(b_n - \overline{b_n})^2} \tag{9}$$

3.1. Confusion Matrix

The confusion matrix clarifies the performance of the classification algorithm. The accuracy value can be misled if the number of classes in a dataset is more than one or the dataset has unequal observations. A confusion matrix gives a clear idea of the results of the classification model and highlights the errors. It contains the summary of the predicted results applied to a classified problem [35]. The percentage of accurate classification in all of the predictions is indicated by accuracy. The matrix contains several values, but the confusion matrix tells precisely where the process went wrong. There are two axes in the confusion matrix. The Y-axis shows the test values of the dataset, while the x-axis represents the prediction results of the test values. There are seven classes in the dataset predicted by machine learning algorithms. The confusion matrix of the decision tree, regression logistic, KNN, naïve Bayes, SVM, and random forest are shown in Figures 4–9. The colorful boxes represent the actual scores of the classes, while the values in other boxes show the mistaken values.

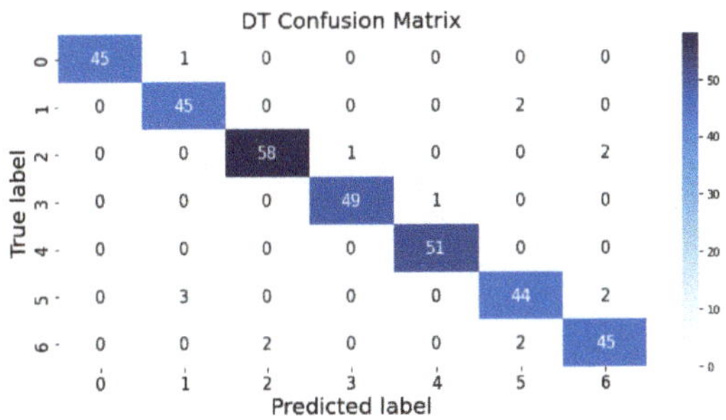

Figure 4. Decision tree prediction on each class testing sample.

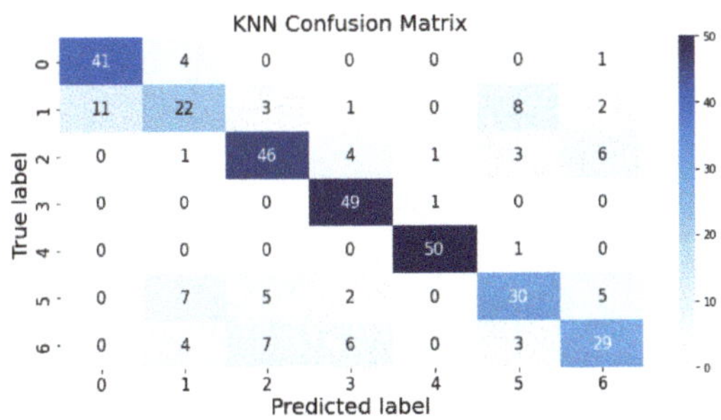

Figure 5. KNN prediction on each class testing sample.

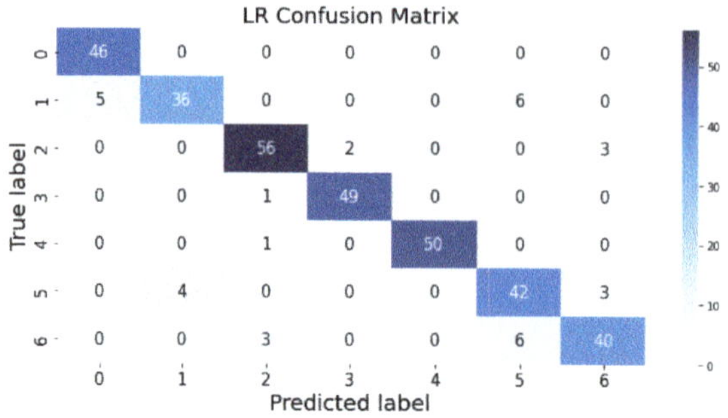

Figure 6. Logistic regression prediction on each class testing sample.

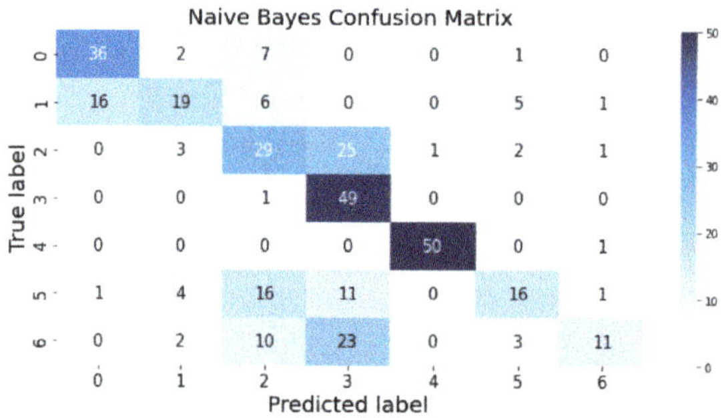

Figure 7. Naïve Bayes prediction on each class testing sample.

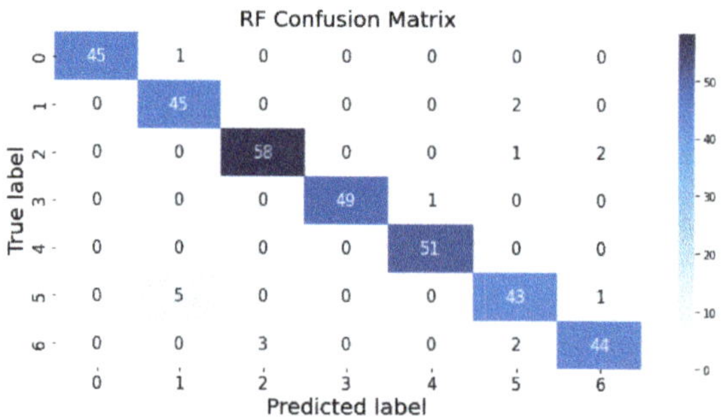

Figure 8. Random Forest prediction on each class testing sample.

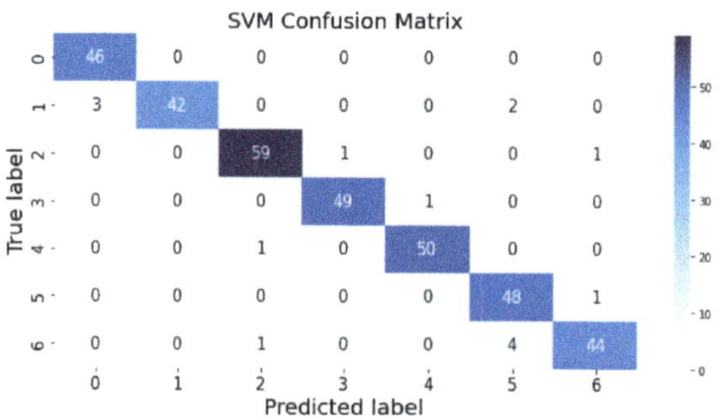

Figure 9. Support vector machine prediction on each class testing sample.

3.2. Real-Time Analysis

Figures 10–15 represent the predicted and real values of different algorithms. The dotted black line shows the real value that we obtained during the real-time analysis and the colored lines represent the predicted values of the algorithms. These figures map 353 samples of obesity, with a total of 16 columns; each sample value is the sum of 16 columns.

Figure 10 represents the predicted value of the decision tree, which shows that real values match with the predicted values most of the time and provide good results as already described in Table 1 with an accuracy of 95%. This algorithm is able to validate the model by using statistical data, which makes it more reliable.

Table 1. Performance analysis of the decision tree, KNN, and logistic regression.

	Decision Tree			KNN			Logistic Regression		
Classes	Precision	Recall	F1-Score	Precision	Recall	F1-Score	Precision	Recall	F1-Score
0	1	0.98	0.99	0.79	0.89	0.84	0.9	1	0.95
1	0.92	0.96	0.94	0.58	0.47	0.52	0.9	0.77	0.83
2	0.97	0.95	0.96	0.75	0.75	0.75	0.92	0.92	0.92
3	0.98	0.98	0.98	0.79	0.98	0.87	0.96	0.98	0.97
4	0.98	1	0.99	0.96	0.98	0.97	1	0.98	0.99
5	0.92	0.9	0.91	0.67	0.61	0.64	0.78	0.96	0.82
6	0.92	0.92	0.92	0.67	0.59	0.63	0.87	0.82	0.84
accuracy	0.95			0.76			0.9		

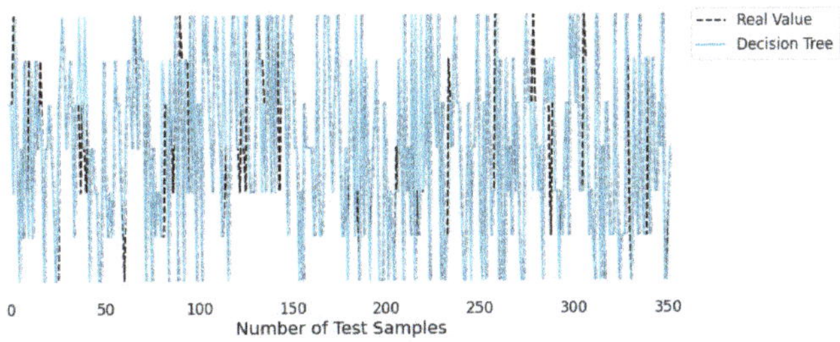

Figure 10. Real-time analysis of each testing sample against the predicted values of the decision tree.

Figure 11 represents the predicted value of naïve Bayes, which shows that real values did not match with the predicted values most of the time and provided very bad results, as shown in Table 2, with an accuracy of 59%. This algorithm assumes that all predicates are independent and very rarely occur in real life.

Table 2. Performance analysis of random forest, naïve Bayes, and support vector machine.

	Random Forest			Naïve Bayes			Support Vector Machine		
Classes	Precision	Recall	F1-Score	Precision	Recall	F1-Score	Precision	Recall	F1-Score
0	1	0.98	0.99	0.68	0.78	0.73	0.94	1	0.97
1	0.88	0.96	0.92	0.63	0.4	0.49	1	0.89	0.94
2	0.95	0.95	0.95	0.42	0.48	0.45	0.97	0.97	0.97
3	1	0.98	0.99	0.45	0.98	0.62	0.98	0.98	0.98
4	0.98	1	0.99	0.98	0.98	0.98	0.98	0.98	0.98
5	0.9	0.88	0.89	0.59	0.33	0.42	0.89	0.98	0.93
6	0.94	0.9	0.92	0.73	0.22	0.34	0.96	0.9	0.93
accuracy	0.95			0.59			0.96		

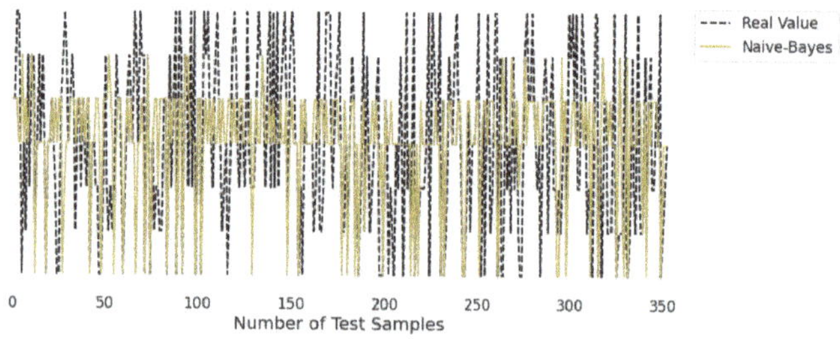

Figure 11. Real-time analysis of each testing sample against the predicted values of naïve Bayes.

Figure 12 presents the predicted value of SVM, which shows that there were very few values where the predicted values did not match the real values and, thus, it provided very good results, as shown in Table 2, with an accuracy of 96%. This algorithm even works with unstructured and semi-structured data.

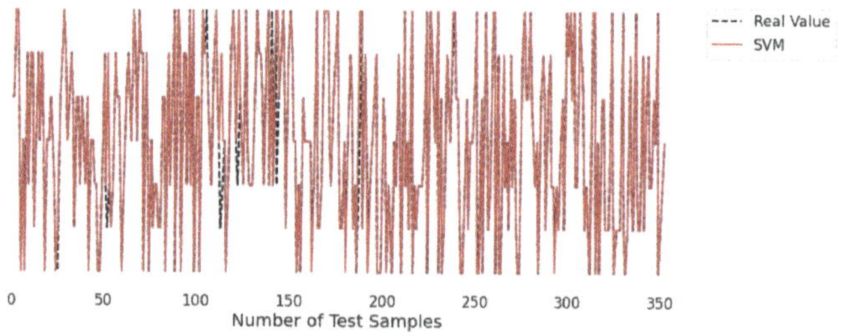

Figure 12. Real-time analysis of each testing sample against the predicted values of SVM.

Figure 13 presents the predicted value of KNN, which shows that there were few values where the predicted values matched the real values and some values where the predicted values did not match the real values; thus, it provided average results, as described in Table 2, with an accuracy of 76%. This algorithm does not perform well on a small dataset.

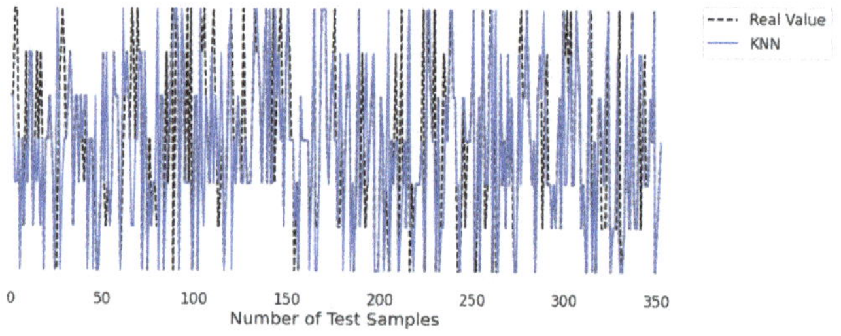

Figure 13. Real-time analysis of each testing sample against the predicted values of KNN.

Figure 14 presents the predicted value of logistic regression, which shows that there were very few values where the predicted values did not match the real values; thus, it provided good results, as shown in Table 1, with an accuracy of 90%. This algorithm is very fast at classifying unknown records.

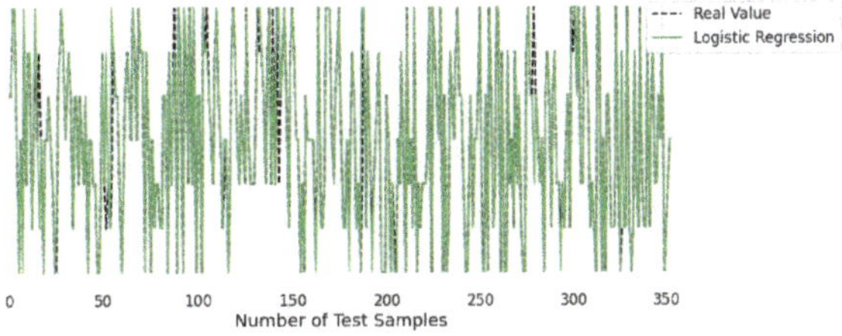

Figure 14. Real-time analysis of each testing sample against the predicted values of logistic regression.

Figure 15 presents the predicted value of the random forest, which shows that there were very few values where the predicted values did not match the real values and, thus, it provided very good results, as shown in Table 2, with an accuracy of 95%. This algorithm can be used to solve classification and regression problems.

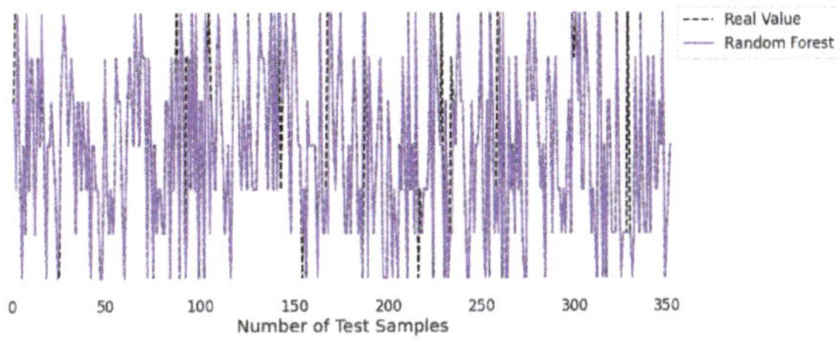

Figure 15. Real-time analysis of each testing sample against the predicted values of random forest.

3.3. Comparison with Existing Schemes

Table 3 shows a fair comparison between the proposed and existing work. There is a lack of studies on the number of machine learning models and statistical matrices for classification reports. It is not clear how the existing work will perform when the number of algorithms increases.

Table 3. Comparison with existing work.

References	Models	Precision	Recall	F1-Score	Accuracy
[24]	SVM	0.62	0.64	-	-
	DT	0.97	0.97	-	-
[36]	NB	0.90	0.91	-	-
	LR	0.90	0.91	-	-
	J48	0.97	0.97	-	-
[37]	DL	-	-	-	0.82
Proposed work	SVM	0.96	0.95	0.95	0.96
	KNN	0.74	0.75	0.74	0.76
	LR	0.90	0.91	0.90	0.90
	DT	0.95	0.95	0.95	0.95
	NB	0.64	0.59	0.57	0.59
	RF	0.95	0.95	0.95	0.95

In [24], two machine learning algorithms—SVM and DT—were used for obesity detection, with predicted precision and recall values, but not the F1-score or accuracy value. Similarly, in [36], there were two machine learning algorithms with prediction values of precision and recall only. In [37], the deep learning approach was adopted for classification purposes. In terms of statistical matrices, only accuracy was considered. These studies do not offer a complete classification report. Furthermore, the existing work does not discuss the error rate in the predicted values. The error rate helps in determining whether the prediction can be considered for further use or not.

However, the proposed system calculates the error rate of the predicted value against each machine learning model. In the proposed work, all machine learning algorithm error rate values were calculated in the form of MBE, RMSE, MABE, and R^2. The proposed system utilizes six machine learning models with optimized configuration settings of SVM, KNN, LR, DT, RF, and NB. It also shows the complete analysis of precision, recall, F1 score, and accuracy as shown in Table 4 .

Table 4. Error rates of the predicted values.

	MBE (MJ/m^2)	RMSE (MJ/m^2)	MABE (MJ/m^2)	R^2
Decision Tree	−0.006	0.374	0.119	0.901
Naïve-Bayes	0.074	3.62	1.337	−1.057
SVM	0.025	0.235	0.082	0.939
KNN	−0.074	2.125	0.589	0.415
Logistic Regression	0.037	0.643	0.201	0.834
Random Forest	−0.008	0.156	0.054	0.959

4. Discussion

This study focused on several machine learning algorithms for early obesity diagnosis. In order to create a list of criteria that could be used to diagnose obesity and being overweight, we assessed the usefulness of machine learning algorithms on accessible data. Our study showed that the SVM performed the best, with the DT and RF classifiers coming in second for early obesity detection. The SVM's remarkable performance across all experiments may be explained by the fact that it employed an adaptive weighting strategy during training. All selected machine learning models were employed for accuracy, precision, F1-score, and recall. Accuracy evaluates how closely the measured value resembles the actual value; precision measures how closely the real value resembles the measured value. A machine learning model's recall or sensitivity reveals its usefulness.

The research focused on obtaining maximum outputs from the classifiers by using true positive, false positive, true negative, false negative, the confusion matrix, and classification report, which resulted in precision, F1-score, recall, and accuracy ratios. The metrics listed aid in assessing machine learning models for regression: MABE (mean absolute bias error), RMSE (root mean square error), MBE (mean bias error), and R^2 (determination coefficients), whereas classification metrics include the confusion matrix and classification report (F1-score, recall, accuracy, and precision). The dataset for this study included 2111 records. A website platform assisted in collecting 23% of the data directly from users while the SMOTE filter and the Weka tool were utilized to artificially construct 77% of the data.

In terms of MBE, RMSE, MABE, and R^2—naïve Bayes predicted the results with a higher error rate and lower determination coefficient value. In terms of MBE, KNN predicted the results with the lowest error rate while the SVM secured the second lowest value. Similarly, the SVM achieved the second lowest value compared to random forest when RMSE, MABE, and R^2 results were calculated.

Furthermore, the proposed study utilized maximum machine learning models to obtain a detailed overview of the predicted values as compared to [24,36,37]. The SVM showed the highest value for precision, F1-score, recall, and accuracy, and 'suggested' the best prediction after taking a close look at the end values. Only accuracy, such as in [37], was not enough to obtain a finer-grained idea of the classification performance. Classifier working was identified by analyzing the value of the precision, F1-score, and recall.

The analysis of seven classes in six machine learning models showed that naïve Bayes had less value for precision, F1-score, and recall. Nonetheless, it is important to point out that the results of this analysis are positive and imply that the suggested SVM can achieve very high performances above 96%.

The proposed IoT system gathers textual data on obesity using data collection tools. The textual data are then communicated to the cloud via the WIFI module, where it goes through preprocessing and classification phases before being scaled to fit the suggested machine learning model, which employs a classifier to detect obesity and extract features from the processed data. The patient can access his/her database to find the classification results during the analytic phase. By submitting the data and receiving the classification report in a couple of seconds, the patient can quickly identify obesity (if there is any). The report is sent to the patient's doctor in the final step, who will choose the best course of action.

There are certain limitations to the proposed research, despite the fact that it provides considerable potential to address situations in real-life. Due to the lack of accessibility of the datasets gathered from overweight patients, one of these drawbacks is that it exclusively uses datasets that are publicly available. Therefore, in order to categorize the activity patterns, future studies should concentrate on gathering and analyzing the datasets of exclusively fat or overweight persons.

5. Conclusions

Obesity is a major public health problem worldwide. The prevalence of obesity has increased dramatically in the past few decades, especially during the COVID-19 pandemic. It is now considered a global epidemic. This is problematic for several reasons, e.g., there is an increased risk of developing serious health conditions, such as heart disease and diabetes. Therefore, we proposed a real-time expert system that successfully determines the possible threat factors related to obesity and being overweight. Several statistical, machine learning, and data visualization methods have been applied to publicly accessible obesity datasets. We performed a fair comparison of machine learning algorithms in terms of precision, recall, F1 score, and accuracy. From the list of proposed algorithms, the SVM outperforms its counterpart schemes. In case of error rates, the following statistical measurements were considered: MBE (MJ/m^2), RMSE (MJ/m^2), MABE (MJ/m^2), and R^2. In MBE (MJ/m^2), SVM has the lowest error rate nearer to zero, while for RMSE (MJ/m^2), MABE (MJ/m^2), and R^2—random forest has a better performance compared to the others. Our expert system takes input from users via a web interface and passes the data to multiple algorithms to make a classification report. This report will be sent to the patient's doctor for necessary actions. In this way, we can easily entertain obesity cases in the initial stages.

Author Contributions: Conceptualization, A.S. and S.A.A.; methodology, A.S., T.A. and A.M.A.; software, A.S. and M.I.; validation, M.I., M.Y.A. and T.A.; investigation, M.Z. and S.A.A.; resources, M.I.; data curation, A.S. and M.A.; writing—original draft preparation, M.Z. and A.M.A.; writing—review and editing, T.A., A.S. and M.Y.A.; visualization, A.S. and M.I.; supervision, M.I. and T.A.; project administration, S.A.A.; funding acquisition, S.A.A. and M.I. All authors have read and agreed to the published version of the manuscript.

Funding: The authors acknowledge the support from the Deputy for Research and Innovation-Ministry of Education, Kingdom of Saudi Arabia (grant NU/IFC/ENT/01/020) under the institutional funding committee at Najran University, Kingdom of Saudi Arabia.

Institutional Review Board Statement: Not applicable.

Informed Consent Statement: Not applicable.

Data Availability Statement: The dataset can be downloaded from the following link: https://archive.ics.uci.edu/ml/datasets/Estimation+of+obesity+levels+based+on+eating+habits+and+physical+condition+ (accessed on 1 June 2022).

Acknowledgments: The authors acknowledge the support from the Deputy for Research and Innovation-Ministry of Education, Kingdom of Saudi Arabia (grant NU/IFC/ENT/01/020) under the institutional funding committee at Najran University, Kingdom of Saudi Arabia.

Conflicts of Interest: The authors declare no conflict of interest.

References

1. Ferdowsy, F.; Rahi, K.S.A.; Jabiullah, M.I.; Habib, M.T. A machine learning approach for obesity risk prediction. *Curr. Res. Behav. Sci.* **2021**, *2*, 100053. [CrossRef]
2. Katz, D.A.; McHorney, C.A.; Atkinson, R.L. Impact of obesity on health-related quality of life in patients with chronic illness. *J. Gen. Intern. Med.* **2000**, *15*, 789–796. [CrossRef]
3. WHO. Obesity and Overweight (n.d.). Available online:https://www.who.int/news-room/fact-sheets/detail/obesity-and-overweight (accessed on 5 May 2022).
4. Oddo, V.M.; Maehara, M.; Rah, J.H. Overweight in Indonesia: An observational study of trends and risk factors among adults and children. *BMJ Open* **2019**, *9*, e031198. [CrossRef] [PubMed]
5. Fruh, S.M. Obesity: Risk factors, complications, and strategies for sustainable long-term weight management. *J. Am. Assoc. Nurse Pract.* **2017**, *29*, S3–S14. [CrossRef]
6. WHO. Obesity (n.d.). Available online: https://www.who.int/health-topics/obesity (accessed on 5 May 2022).
7. Nuertey, B.D.; Alhassan, A.I.; Nuertey, A.D.; Mensah, I.A.; Adongo, V.; Kabutey, C.; Addai, J.; Biritwum, R.B. Prevalence of obesity and overweight and its associated factors among registered pensioners in Ghana; A cross sectional studies. *BMC Obes.* **2017**, *4*, 26. [CrossRef] [PubMed]
8. Available online: https://jamanetwork.com/journals/jamainternalmedicine/article-abstract/2323411 (accessed on 10 May 2022).
9. Chatterjee, A.; Gerdes, M.W.; Martinez, S.G. Identification of risk factors associated with obesity and overweight—A machine learning overview. *Sensors* **2020**, *20*, 2734. [CrossRef] [PubMed]
10. Safaei, M.; Sundararajan, E.A.; Driss, M.; Boulila, W.; Shapi'i, A. A systematic literature review on obesity: Understanding the causes & consequences of obesity and reviewing various machine learning approaches used to predict obesity. *Comput. Biol. Med.* **2021**, *136*, 104754. [CrossRef]
11. Lee, A.; Cardel, M.; Donahoo, W.T. Social and Environmental Factors Influencing Obesity. Endotext. 2019. Available online: MDText.com (accessed on 12 May 2022).
12. Institute of Medicine (US) Committee on an Evidence Framework for Obesity Prevention Decision Making; Kumanyika, S.K.; Parker, L.; Sim, L.J. *Obesity Prevention Strategies in Concept and Practice*; National Academies Press: Washington, DC, USA, 2010.
13. Available online: https://academic.oup.com/ije/article/49/6/1763/5814327?login=true (accessed on 12 May 2022).
14. Thamrin, S.A.; Arsyad, D.S.; Kuswanto, H.; Lawi, A.; Nasir, S. Predicting obesity in adults using machine Learning techniques: An analysis of Indonesian Basic Health Research 2018. *Front. Nutr.* **2021**, *8*, 669155. [CrossRef]
15. Hammond, R.; Athanasiadou, R.; Curado, S.; Aphinyanaphongs, Y.; Abrams, C.; Messito, M.J.; Gross, R.; Katzow, M.; Jay, M.; Razavian, N.;et al. Predicting childhood obesity using electronic health records and publicly available data. *PLoS ONE* **2019**, *14*, e0215571. Erratum in *PLoS ONE* **2019**, *14*, e0223796. [CrossRef]
16. Akter, T.; Zeba, Z.; Hosen, I.; Al-Mamun, F.; Mamun, M.A. Impact of the COVID-19 pandemic on BMI: Its changes in relation to socio-demographic and physical activity patterns based on a short period. *PLoS ONE* **2022**, *17*, e0266024. [CrossRef]
17. Survey: Among Saudis, More Women Are Obese than Men. *Saudi Gazette*, 23 July 2022. Available online: https://saudigazette.com.sa/article/623202 (accessed on 5 May 2022).
18. Dugan, T.M.; Mukhopadhyay, S.; Carroll, A.; Downs, S. Machine learning techniques for prediction of early childhood obesity. *Appl. Clin. Inform.* **2015**, *6*, 506–520. [CrossRef] [PubMed]
19. Dunstan, J.; Aguirre, M.; Bastías, M.; Nau, C.; Glass, T.A.; Tobar, F. Predicting nationwide obesity from food sales using machine learning. *Health Inform. J.* **2020**, *26*, 652–663. [CrossRef] [PubMed]
20. Singh, B.; Tawfik, H. Machine learning approach for the early prediction of the risk of overweight and obesity in young people. In *Lecture Notes in Computer Science*; Springer International Publishing: Berlin/Heidelberg, Germany, 2020; pp. 523–535.
21. Gerl, M.J.; Klose, C.; Surma, M.A.; Fernandez, C.; Melander, O.; Männistö, S.; Borodulin, K.; Havulinna, A.S.; Salomaa, V.; Ikonen, E.; et al. Machine learning of human plasma lipidomes for obesity estimation in a large population cohort. *PLoS Biol.* **2019**, *17*, e3000443. [CrossRef] [PubMed]
22. Montanez, C.A.C.; Fergus, P.; Hussain, A.; Al-Jumeily, D.; Abdulaimma, B.; Hind, J.; Radi, N. Machine learning approaches for the prediction of obesity using publicly available genetic profiles. In Proceedings of the 2017 International Joint Conference on Neural Networks (IJCNN), Anchorage, AK, USA, 14–19 May 2017; pp. 2743–2750.
23. Borrell, L.N.; Samuel, L. Body mass index categories and mortality risk in US adults: The effect of overweight and obesity on advancing death. *Am. J. Public Health* **2014**, *104*, 512–519. [CrossRef] [PubMed]
24. Cervantes, R.C.; Palacio, U.M. Estimation of obesity levels based on computational intelligence. *Inform. Med. Unlocked* **2020**, *21*, 100472. [CrossRef]
25. Yu, C.-S.; Lin, Y.-J.; Lin, C.-H.; Wang, S.-T.; Lin, S.-Y.; Lin, S.H.; Wu, J.L.; Chang, S.-S. Predicting metabolic syndrome with machine learning models using a decision tree algorithm: Retrospective cohort study. *JMIR Med. Inform.* **2020**, *8*, e17110. [CrossRef]
26. Breiman L. Random forests. *Mach. Learn.* **2001**, *45*, 5–32. [CrossRef]
27. Joachims, T. Text categorization with support vector machines: Learning with many relevant features. In Proceedings of the European Conference on Machine Learning, Chemnitz, Germany, 21–23 April 1998; Springer: Berlin/Heidelberg, Germany, 1998; pp. 137–142.
28. Kim, Y.; Ling, H. Human activity classification based on micro-Doppler signatures using a support vector machine. *IEEE Trans. Geosci. Remote Sens.* **2009**, *47*, 1328–1337.

29. Niedermeyer, E.; da Silva, F.L. (Eds.) *Electroencephalography: Basic Principles, Clinical Applications, and Related Fields*; Lippincott Williams & Wilkins: Philadelphia, PA, USA, 2005.
30. Parsons, T.D.; Rizzo, A.A. Affective outcomes of virtual reality exposure therapy for anxiety and specific phobias: A meta-analysis. *J. Behav. Ther. Exp. Psychiatry* **2008**, *39*, 250–261. [CrossRef]
31. De la Hoz, E.; De la Hoz, E.; Ortiz, A.; Ortega, J.; Martínez-Álvarez, A. Feature selection by multi-objective optimisation: Application to network anomaly detection by hierarchical self-organising maps. *Knowl.-Based Syst.* **2014**, *71*, 322–338. [CrossRef]
32. Bekele, E.; Wade, J.; Bian, D.; Fan, J.; Swanson, A.; Warren, Z.; Sarkar, N. Multimodal adaptive social interaction in virtual environment (MASI-VR) for children with Autism spectrum disorders (ASD). In Proceedings of the 2016 IEEE Virtual Reality (VR), Greenville, SC, USA, 19–23 March 2016; pp. 121–130.
33. Wanto, A.; Siregar, M.N.H.; Windarto, A.P.; Hartama, D.; Ginantra, N.L.W.S.R.; Napitupulu, D.; Negara, E.S.; Lubis, M.R.; Dewi, S.V.; Prianto, C. *Data Mining: Algoritma Dan Implementasi*; Yayasan Kita Menulis: Medan, Indonesia, 2020.
34. Adnan, M.H.B.M.; Husain, W. A hybrid approach using Naïve Bayes and Genetic Algorithm for childhood obesity prediction. In Proceedings of the 2012 International Conference on Computer Information Science (ICCIS), Kuala Lumpur, Malaysia, 12–14 June 2012; Volume 1, pp. 281–285.
35. Breiman, L.; Friedman, J.H.; Olshen, R.A.; Stone, C.J. *Classification and Regression Trees*; Routledge: Abingdon, UK, 2017.
36. De-La-Hoz-Correa, E.; Mendoza Palechor, F.; De-La-Hoz-Manotas, A.; Morales Ortega, R.; Sánchez Hernández, A.B. Obesity Level Estimation Software Based on Decision Trees. 2019. Available online: https://repositorio.cuc.edu.co/handle/11323/4176 (accessed on 6 August 2022)
37. Kivrak, M. Deep learning-based prediction of obesity levels according to eating habits and physical condition. *J. Cogn. Syst.* **2021**, *6*, 24–27. [CrossRef]

life

MDPI

Article

A Turf-Based Feature Selection Technique for Predicting Factors Affecting Human Health during Pandemic

Alqahtani Saeed [1], Maryam Zaffar [2,*], Mohammed Ali Abbas [2], Khurrum Shehzad Quraishi [3], Abdullah Shahrose [4], Muhammad Irfan [5], Mohammed Ayed Huneif [6], Alqahtani Abdulwahab [6], Sharifa Khalid Alduraibi [7], Fahad Alshehri [7], Alaa Khalid Alduraibi [7] and Ziyad Almushayti [7]

1 Department of Surgery, Faculty of Medicine, Najran University, Najran 61441, Saudi Arabia
2 Faculty of Computer Sciences, IBADAT International University, Islamabad 44000, Pakistan
3 Department of Chemical Engineering, Pakistan Institute of Engineering and Applied Sciences (PIEAS), Islamabad 44000, Pakistan
4 Department of Computer Science, HITEC University, Taxila 47080, Pakistan
5 Electrical Engineering Department, College of Engineering, Najran University Saudi Arabia, Najran 61441, Saudi Arabia
6 Department of Pediatrics, College of Medicine, Najran University, Najran 61441, Saudi Arabia
7 Department of Radiology, College of Medicine, Qassim University, Buraidah 52571, Saudi Arabia
* Correspondence: maryam.zaffar82@gmail.com

Abstract: Worldwide, COVID-19 is a highly contagious epidemic that has affected various fields. Using Artificial Intelligence (AI) and particular feature selection approaches, this study evaluates the aspects affecting the health of students throughout the COVID-19 lockdown time. The research presented in this paper plays a vital role in indicating the factor affecting the health of students during the lockdown in the COVID-19 pandemic. The research presented in this article investigates COVID-19's impact on student health using feature selections. The Filter feature selection technique is used in the presented work to statistically analyze all the features in the dataset, and for better accuracy. ReliefF (TuRF) filter feature selection is tuned and utilized in such a way that it helps to identify the factors affecting students' health from a benchmark dataset of students studying during COVID-19. Random Forest (RF), Gradient Boosted Decision Trees (GBDT), Support Vector Machine (SVM), and 2- layer Neural Network (NN), helps in identifying the most critical indicators for rapid intervention. Results of the approach presented in the paper identified that the students who maintained their weight and kept themselves busy in health activities in the pandemic, such student's remained healthy through this pandemic and study from home in a positive manner. The results suggest that the 2- layer NN machine-learning algorithm showed better accuracy (90%) to predict the factors affecting on health issues of students during COVID-19 lockdown time.

Keywords: mental stress; COVID-19; feature selection; artificial intelligence; human health; pandemic; lock down

Citation: Saeed, A.; Zaffar, M.; Abbas, M.A.; Quraishi, K.S.; Shahrose, A.; Irfan, M.; Huneif, M.A.; Abdulwahab, A.; Alduraibi, S.K.; Alshehri, F.; et al. A Turf-Based Feature Selection Technique for Predicting Factors Affecting Human Health during Pandemic. *Life* **2022**, *12*, 1367. https://doi.org/10.3390/life12091367

Academic Editor: Tao Huang

Received: 18 July 2022
Accepted: 24 August 2022
Published: 1 September 2022

1. Introduction

The COVID-19 was triggered by Sars-Cov-2 coronavirus, which was initially identified in Wuhan, China in December 2019 [1,2]. This disease spread in the whole world rapidly and has significantly affected many aspects of life including mental health, social life, supply chain, energy consumption, education, etc., [3,4]. Lockdown measures were taken by governments all over the world to impede the disperse of the disease. People all over the world were restricted to quarantine and keep social distancing to determine the number of people who have become infected [5]. Studies have shown that the lockdown during this COVID-19 had different physiological effects including anxiety, stress, confusion [6], and anger [7]. Similar effects were observed in the education domain and various educational stakeholders were affected by lockdown in COVID-19. According to a report of UNSECO [8], about 1.6 billion students faced school closure issues. Face-to-face education was replaced by e-learning. This transformed the lives of the students reducing them to their homes. Students' mental and physical health is affected by the COVID-19 lockout situation. Different studies are taking part to figure out the reasons causing the disturbance in the students' mental health. During COVID-19, a variety of statistical tools were used to investigate the elements that influence the students' mental health. The results of different existing studies reported that lockdown causes depression, anxiety, mental stress and health issues in quarantined populations during COVID-19 [9]. Furthermore, social distancing and different lockdown measures during COVID-19 negatively affect the health of student [10]. The efficient execution of education depends on the health of students. As the students are the main pillar of society and the nation's leadership and control will rely on them in the future. Therefore, it is necessary to put maximum possible effort to maintain the health of students [11].

Artificial intelligence (AI) plays a vital role in predicting coronavirus effects in the future by analyzing the covid data [12]. Different Al-based supervised and unsupervised algorithms are being employed in studies for COVID-19 predictions and analysis [13]. As the education sector is the base of every country's development, so AI techniques help the educational stakeholders and government officials of countries all over the world to plan strategies and techniques to maintain the health of students. The focus of this study is to employ the AI-based technique for the identification of factors affecting the students' mental health in the pandemic of COVID-19. The following are the primary contributions of the proposed work:

- Identifying the factors affecting the health of students in the lockdown phase of COVID-19;
- To assist the educational stakeholders in taking proactive measures for maintaining the student's health for the duration of COVID-19;
- Proposing AI-based identification of features affecting the student's health during the lockdown period of COVID-19;
- Explore AI approaches namely RF, GBDT, SVM, and NN along with Turf feature selection for selecting the optimal feature set affecting the health of students.

In this study, COVID-19 related student data will be evaluated in order to determine factors affecting students' health during COVID-19 lockdown. The paper is structured in the subsequent pattern: a summary of related literature is provided in Section 2; the suggested AI-based strategy is described in detail in Section 3; Section 4 presents the analysis evaluation of the proposed technique; and the research and future work is summarized in Section 5.

2. Related Work

Different studies are conducted to illustrate the effects of lockdown during COVID-19. In this section, an overview of existing approaches is presented, focusing on the students' mental health during the lockdown period in COVID-19. Different studies in different countries are conducted all over the world to analyze the health of students during the lockdown in this pandemic situation. Some of the studies are selected from the related work on the mental and physical health of the students, available on Google Scholar. Table 1 shows the reference of the papers, country in which the study was conducted, also presents the different variations in sizes of datasets collected for analyzing the factors affecting the health of students in a pandemic situation. Furthermore, Table 1 analyzes that whether the existing studies are utilizing AI techniques or not. In the end, Table 1 also presents what is the conclusion of the recent studies regarding the factors affecting the mental health of students.

Table 1. Analysis of studies on student's health during COVID-19 in different countries.

Reference	Country	Level of Students	Size of Dataset	Machine Learning	Effect on Mental and Physical Health in COVID-19
[6]	Malaysia	Postgraduate	-	No	Loneliness, anxiety, stress, and depression.
[14]	Asian students in Poland	Medical students	85	No	Feeling of isolation to students who live abroad.
[15]	Bangladesh	College Students	400	No	In COVID-19, perceptions of e-Learning failure and fear about academic year failure were connected with psychological distress.
[16]	Bulgaria	Graduate and undergraduate students	134	Yes	Availability of separate rooms for students affects their education.
[17]	China	Non-graduating undergraduate students	1172	YES XGBOOST	School closure, Social distancing or Isolation, and Online learning are the reason for anxiety.
[18]	China	Secondary vocational students	5783	No	Good family functioning can positively affect the mental health of students.
[19]	Philippines	College Students	952	No	Socioeconomic gaps and the digital divide affect the mental health of students.
[20]	India	Undergraduate and post-graduate	516	Yes	Uncertainty regarding examination affects the mental health of students.
[21]	Jordan	Medical Student	1404	No	Students focus on strategies to prevent covid.
[22]	Pakistan	Higher Educational Institutions	494	No	Unaffordability of digital devices and the internet.
[23]	USA (United States of America)	University Students	195	No	Fear of own health and dear one's health affects the mental health of students.
[24]	UAE (United Arab Emirates)	Medical and non-medical students	1485	No	Fear of the unknown might affect the mental health of students so that students must be aware of the COVID-19.
[25]	Saudi Arabia	University students	400	Yes	Females and fourth-semester students face anxiety during COVID-19.
[26]	New Zealand	Mater level Graduate-level Teaching degree (Mathematics education learning)	3	No	Teachers help in the transition of a new way of learning that affects students.
[27]	Greece	Undergraduate forestry students	181	No	Students must be counseled properly to control negative emotions during the lockdown.
[7]	Iran	Public school students	20,697	No	Behavioral and socializing changes during COVID-19 affects mental health.

The paper presents 16 most relevant literature on student's health in COVID-19. Recent studies are evaluated on 5 different parameters: the country through which dataset is taken, study level of student's understudy, number of students in the dataset, utilization of machine learning technique for identification of factors affecting student's health, and, lastly, the factor identified by the existing studies that may affect the health of students all through lockdown phase of COVID-19. Different the different levels and sizes of students with varying datasets sizes. The studies focus on graduate, undergraduate, college, public schools, and medical and forestry students of different countries. The recent literature indicates that there is so much gap in studies regarding machine learning utilization for analyzing the mental health of students. Different factors come across while analyzing the existing literature on the students' mental health. Mainly, the following factors were found to be very crucial in association with the students' mental health in the COVID-19 pandemic.

- Loneliness [6];
- The feeling of isolation [13];
- Fear of academic year loss;
- Availability of space for studies;
- Family functioning;
- Females have more mental health issues than male students during COVID-19;
- Fear of own health;
- Fear of dear one's health;
- Poverty;
- Student Counseling.

Different factors are found in the literature that has an association with the mental health of students. These factors will help the educational admiration to take measures for maintaining the health of students during COVID-19. Different remote techniques and activities should be planned by educational stakeholders to minimize the anxiety of students during the lockdown period of the pandemic. However, as the health of students is an important concern so there is a need for deep insight into the data of students during such pandemic situations. However, the need for AI algorithms is still there for a better insight into data and its analysis. Main shortcomings in recent studies regarding the health of students in COVID-19 are still required to address, some of the shortcomings found in the literature that may help the educational stakeholders to build educational strategies. Firstly, there is a need to the utilization of feature selection techniques to identify the features affection health of students during the lockdown in COVID-19. To our knowledge, there has never been a study that conducted a comprehensive literature analysis and identified factors affecting the health of kids during COVID-19's lockdown period based on feature selection, whereas [28] has presented and utilized AI, but did not consider feature selection. In the coming sections of this article, we will discuss our novel proposed approach for the analysis of factors affecting the health of students in COVID-19.

3. Methods and Materials

In this section proposed approach for identifying the factors affecting the health of students is presented. As it is very important to figure out that what are factors affecting the health of Figure 1 presents the main flow of the proposed approach main steps of the proposed approach is as follows:

- Dataset Selection;
- Dataset Cleaning;
- Feature Selection;
- Machine learning algorithm.

Each of the steps is explained further in detail in the coming subsections.

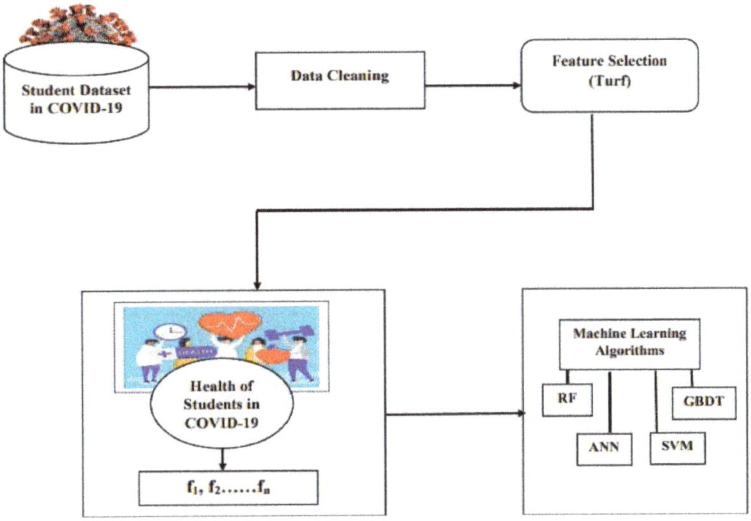

Figure 1. Proposed flow AI-Based feature selection of factors affecting the health of students in COVID-19.

3.1. Student Dataset in COVID-19

A benchmark dataset of 1182 students in COVID-19 [29] is utilized to analyze the factors affecting the health of students in the lockdown period of COVID-19. The dataset is freely available and, hence, utilized easily for research purpose. Table 2 describes the main properties of the student dataset.

Table 2. Student COVID-19 dataset description.

Number of Students	1182
Number of features	19
Features	Id of the student, home location of students, Student_age, Time consumed _online Class, Rating of Online Class experience, Instruction medium for an online class, Time consumed_ self-study, Time consumed_ fitness, sleeping_ time, Time consumed_ social media, preferred social media platform, Time consumed_ TV, meals _per day, changes _weight, Health issue_ lockdown, Stressbusters, Utilization _time, what you miss the most
Target feature	Health issue during lockdown
Number of classes	2

3.2. Data Preprocessing

Python programming language platform is utilized for coding the proposed approach, and its various libraries like NumPy, pandas for better insight of data [30]. Different steps are taken to preprocess the imbalanced dataset, firstly by scaling and data cleaning by deleting ids, dropping duplicating rows, and filling all NA values. Moreover, categorical features are mapped to numbers. Furthermore, to convert the text features like (stress buster, what you miss most), pretrained bert is utilized for generating word vectors. Then words are mapped to a single feature by following the normalization formula as:

$$x = \frac{sum(vector)}{max(vector) - min(vector)}. \quad (1)$$

Figures 2 and 3 represents variable count after and before sampling, whereas SMOTE (Synthetic minority oversampling technique) addresses imbalance class issues very effectively in various domains of research [31]. SMOTE oversampling technique is applied to resample student's datasets for COVID-19. Based on feature space similarity, the SMOTE approach combines extra minority samples [32]. Let k = nearest neighbor for xi using Euclidean distance.

Random Selection of k nearest neighbor

Feature vector difference between k and x_i

Adding M in x_i

Equation (2) presents the formula for calculating SMOTE

This is example 2 of an equation:

$$x_{new} = x_i\left(x_i^k - x_i\right) \times \delta. \quad (2)$$

x_i^k = A nearest neighbors of x_i, and δ is an arbitrary value belongs to (0, 1).

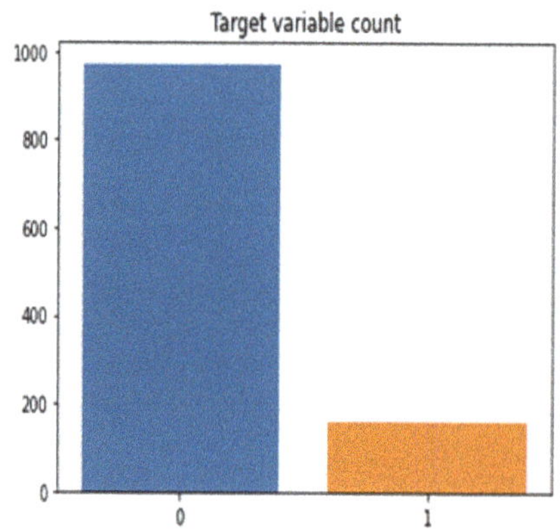

Figure 2. Variable count before sampling.

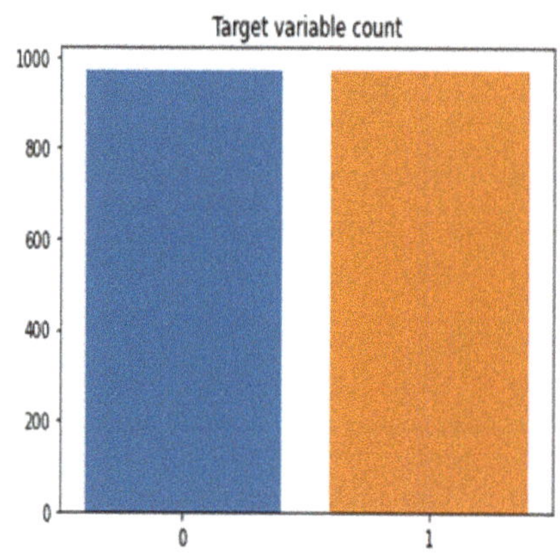

Figure 3. Variable count after sampling.

3.3. Feature Selection

Feature selection is a process to obtain an optimal set of features, to obtain better classification accuracy. There are different types of feature selection algorithm filter and wrapper feature selection. Filter feature selection is high in speed [33] and consumes less time, and is the main reason for selecting filter feature selection in our proposed approach.

Filter feature selection is further divided into two types, univariant and multivariant filter feature selection methods. The univariant filter feature ignores the features dependencies and that leads to a poor selection of feature set [34], whereas multivariant feature selection takes consideration of feature dependencies while selecting the feature set [35]. Turf is the tuned form of Relief multivariant filter feature selection. When selecting relief features, feature dependencies are taken utilizing the full feature vector, which may ignore the noisy features, so that Turf feature selection step by step low-quality features, hence, generating optimal feature set [36]. The Turf algorithm is presented in Algorithm 1.

Algorithm 1. TuRF algorithm [36].

a = features in dataset

Let p = iterations
For i:= 1 to p do
Estimation of feature weights through ReliefF
Features sorting through weight
+
+ remove p/a of outstanding features with smallest weights
end for
return final ReliefF weight estimations for outstanding features

3.4. Machine Learning Algorithms

After the selection of features, classification is performed. SVM (Support Vector Machine) is a classifier for binary classification of data. The hyperplane is used to solve the learning problem in SVM. A robust method with different kernel values is considered one of the best classifiers for classification [37]. RF (Random Forest) utilized various trees to predict. It is being utilized by different research areas of research with remarkable results. RF produces high classification accuracy with an even dataset with a large number of features. It handles unbalanced data by accessing important features. Whereas GBDT (Gradient Boosting Decision Tree) is selected due to its property of selecting fewer parameters as compared to the other classification algorithms. In existing research, in machine learning, GBDT shows tremendous results. It is based on the CART algorithm. GBDT merges the concept of regression and boosting tree and intends the use of residual gradient to optimize the assimilation process of regression tree [38]. ANN (Artificial Neural Network) is a popular classification technique utilized in different areas of research like agriculture, medical, security, education, business, art, etc. It is very easy to use and can manage complex data [39]. Moreover, the performance of the proposed approach presented in this paper is evaluated through accuracy, precision, recall, and f-measure, whereas accuracy is defined as the predicted observations over a total number of observations [40–42]. Precision is the fraction of the recovered instances that belong to the target class, whereas F-measure is the harmonic mean of precision and recall. Equations (3)–(6) presents the formula of evaluation parameters, whereas *TP*, *FN*, and *FP* stand for true positive, false negative, and false-positive respectively.

$$\text{Accuracy} = \frac{TP + FN}{TP + FN + FP + FN} \tag{3}$$

$$\text{Precision} = \frac{TP}{TP + FP} \tag{4}$$

$$\text{Recall} = \frac{TP}{TP + FN} \tag{5}$$

$$\text{F} - \text{Measure} = \frac{2(Precision \times Recall)}{Precision + Recall} \tag{6}$$

whereas Table 3 presents the Parameters of classification algorithm utilized in proposed work.

Table 3. Parameters of classification algorithm utilized in proposed work.

SR	Name	Parameters
1	GBDT	n_estimators = 19, learning_rate = 0.3, max_depth = 7, random_state = 0
2	SVM	SVC(C = 5, break_ties = False, cache_size = 200, class_weight = Balanced, degree = 3, gamma = 11, kernel = 'rbf')
3	Random Forest	bootstrap = True, criterion = 'gini', max_depth = 15, max_features = 'auto', min_samples_leaf = 1, min_samples_split = 2, n_estimators = 20
4	NN-2 Layers	Momentum0.9, learning rate = 0.003, layers = 2, drop_out = 0.1, optimizer = adam, Loss Binary Class

4. Results

Results of the proposed approach for the identification of factors affecting the health of students in COVID-19 will be discussed in detail in this section. Figure 4 explains the proposed method in detail with results. The results show that the dataset of a feature vector of 16 features is balanced through applying SMOTE technique. The health of students is taken as a target feature, and the Turf feature selection technique is utilized to detect the factors influencing the health of students. Different classification algorithms are applied to the selected feature datasets of student's health during COVID-19. The performance of the suggested method was assessed using accuracy, precision, recall, and f-measure assessment metrics.

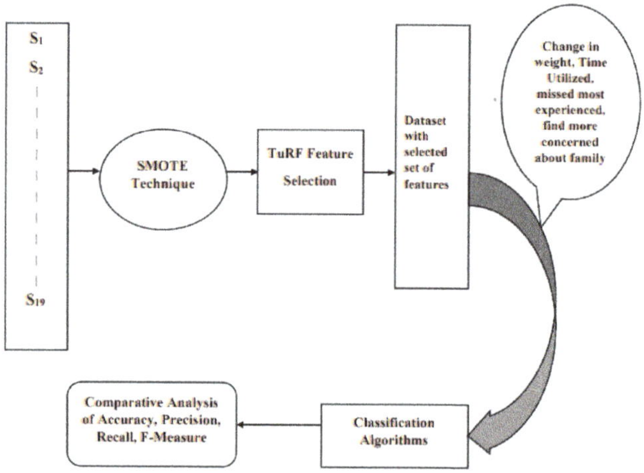

Figure 4. Some feature selection process of factors affecting the student's health during COVID-19.

The result s shows that the student who utilized their time during lockdown period in COVID-19 in different activities remain healthy. Utilization of time appears as the main factor affecting the health of students. The academic organization may keep that factor in front and must plan activities, guide, and motivate students to participate in some indoor actives in such a way that maintains their health. Emotional attachment of students with family members also affects the health of students, as the fear of any family loss due to COVID-19 affects the health of students. Moreover, change in the weight of students during COVID-19 also affects the health of students. Figure 5 presents the results of four classifiers, GBDT, RF, SVM, and NN, on students COVID-19 dataset, whereas the accuracy describes the number of healthy students correctly classified by proposed work over a total number of students. Results show that a Neural network (NN) outperforms other existing classification algorithms in terms of accuracy. However, GBDT also performs well on students COVID-19 dataset and showed around 87% of accuracy. Equation (7) presents the accuracy formula for the student COVID-19 dataset.

$$Accuracy = \frac{Number\ of students correcly\ classified}{Total\ number\ of\ students} \tag{7}$$

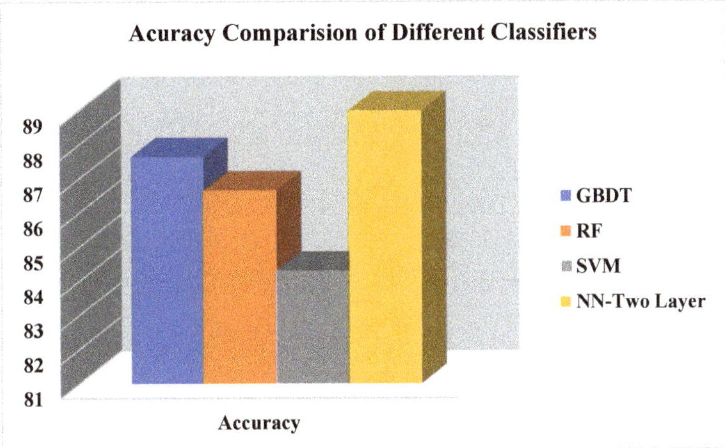

Figure 5. Comparison of accuracy of proposed COVID-19 approach.

Figure 6 presents the performance evaluation of the proposed work in terms of precision, whereas precision calculates the number of healthy students in the COVID-19 student dataset correctly classified by proposed work divided by the total number of healthy students in the COVID-19 dataset, classified by the proposed approach. Results show that neural network performs better than other classification algorithms. Equation (8) presents the formula of precision for calculation precision of proposed approach on student COVID-19 dataset.

$$Precision = \frac{Number\ of\ healthy\ students\ identified\ by\ propsed\ approach}{Total\ number\ of\ health\ and\ unhealthy\ students\ \ classified\ by\ proposed\ approach} \tag{8}$$

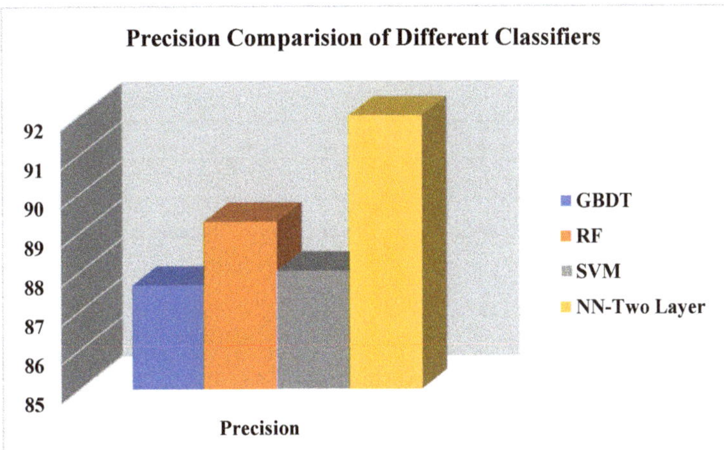

Figure 6. Comparison of precision of proposed COVID-19 approach.

The results in Figure 7 show the performance evaluation of the proposed work in terms of recall. The recall is the calculation of a total number of healthy students in the COVID-19 student dataset classified by the proposed approach divided by the total number of healthy students in the COVID-19 student dataset. The results show that the GBDT classifier outperforms other classifiers in recall performance evaluation measures. Furthermore, RF and NN also show better performance. Equation (9) presents the formula for calculating the recall for evaluating proposed approach on students COVID-19 students.

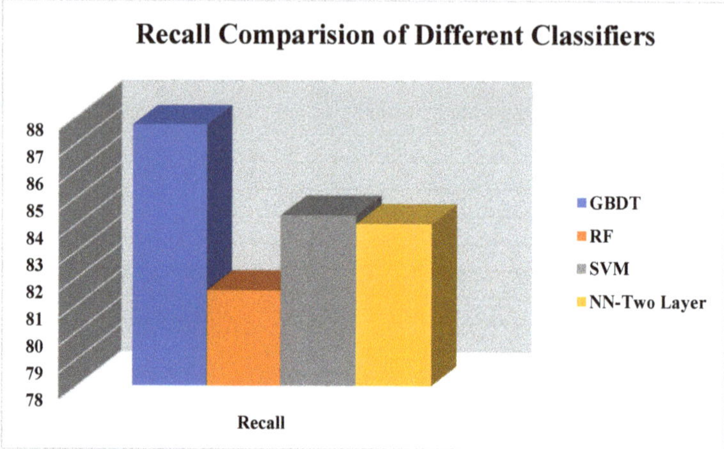

Figure 7. This Comparison of recall of proposed COVID-19 approach.

This is example 2 of an equation:

$$Recall = \frac{Number\ of\ healthy\ studentsclassfied\ by\ the\ proposed\ approach}{Total\ number\ of\ healthy\ students} \quad (9)$$

Figure 8 presents a comparison of the performance of four classifiers in terms F-measure performance evaluation measure, whereas the f-measure of the proposed approach considers precision and recall both, presented already in Equation (6). Results show that GBDT and NN give better performance on the proposed work on the COVID-19 student dataset in terms of F-measure.

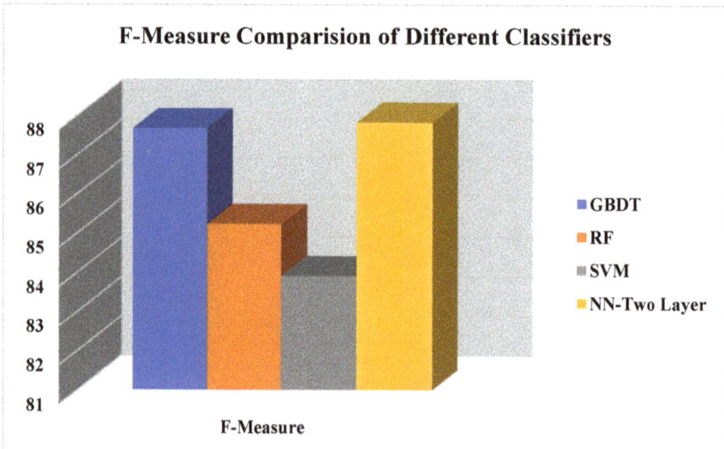

Figure 8. This Comparison of f-measure of proposed COVID-19 approach.

5. Conclusions

COVID-19 affects every field of life, the educational sector all over the world faces different issues. During the lockdown, students face a lot of issues, whereas health issue becomes the main issue. Results presented in the proposed approach identifies the main factors affecting the health of students during the lockdown. Results show that the health of students affects the factors that how they utilized their time during the lockdown in COVID-19, whereas weight and family concerns also appear as factors affecting the health of students during a lockdown of COVID-19. Henceforth, there is a need to take proactive measures to discover the approaches to sustain the health of students, either by guiding them in health time utilization activities or by counseling them about family matters. These well-timed taken measures may reduce the health issues in students caused by pandemic situation in COVID-19. Moreover, reported results in this paper 'show that neural network outperforms and shows 90% accuracy on the proposed approach as compared to GBDT, RF, and SVM.

Author Contributions: M.Z., A.S. (Alqahtani Saeed), M.A.A. and K.S.Q. performed data curation, methodology, visualization, software, investigation, simulations, analysis, and writing the manuscript. A.S. (Abdullah Shahrose), M.I., A.K.A., S.K.A., Z.A., M.A.H., A.A. and F.A. were responsible for conceptualization, editing, investigation, resource management, project management, analysis, comparisons, and visualization. All authors have read and agreed to the published version of the manuscript.

Funding: The authors would like to express their gratitude to the Ministry of Education and the Deanship of Scientific Research, Najran University, Kingdom of Saudi Arabia, for their financial and technical support under code number NU/MID/18/031.

Institutional Review Board Statement: Not applicable.

Informed Consent Statement: Not applicable.

Acknowledgments: The authors would like to express their gratitude to the Ministry of Education and the Deanship of Scientific Research, Najran University. Kingdom of Saudi Arabia, for their financial and technical support under code number NU/MID/18/031.

Conflicts of Interest: The authors declare no conflict of interest.

References

1. Li, Q.; Med, M.; Guan, X.; Wu, P.; Wang, X. Early transmission dynamics in Wuhan, China, of novel coronavirus–infected pneumonia. *N. Engl. J. Med.* **2020**, *382*, 1199–1209. [CrossRef] [PubMed]
2. Wei, J.T.; Liu, Y.X.; Zhu, Y.C.; Qian, J.; Ye, R.Z. Impacts of transportation and meteorological factors on the transmission of COVID-19. *Int. J. Hyg. Environ. Health* **2020**, *230*, 113610. [CrossRef] [PubMed]
3. Rahman, M.; Paul, K.C.; Hossian, A.; Ali, G.G.N.; Rahman, S. Machine Learning on the COVID-19 Pandemic, Human Mobility and Air Quality: A Review. *IEEE Access* **2021**, *9*, 72420–72450. [CrossRef] [PubMed]
4. Silva, P.C.L.; Batista, P.V.C.; Lima, H.S.; Alves, M.A.; Guimarães, F.G. COVID-ABS: An agent-based model of COVID-19 epidemic to simulate health and economic effects of social distancing interventions. *Chaos Solitons Fractals* **2020**, *139*, 110088. [CrossRef]
5. Shafiq, S.; Nipa, S.N.; Sultana, S.; Rahman, M.R.U.; Rahman, M. Exploring the triggering factors for mental stress of university students amid COVID-19 in Bangladesh: A perception-based study. *Child. Youth Serv. Rev.* **2021**, *120*, 105789. [CrossRef]
6. Azlan, C.A.; Wong, J.H.D.; Tan, L.K.; Huri, M.H.N.A.D.; Ung, N.M. Teaching and learning of postgraduate medical physics using Internet-based e-learning during the COVID-19 pandemic–A case study from Malaysia. *Phys. Med.* **2020**, *80*, 10–16. [CrossRef]
7. Karasmanaki, E.; Tsantopoulos, G. Impacts of social distancing during COVID-19 pandemic on the daily life of forestry students. *Child. Youth Serv. Rev.* **2021**, *120*, 105781. [CrossRef]
8. d'Orville, H. COVID-19 causes unprecedented educational disruption: Is there a road towards a new normal? *Prospects* **2020**, *49*, 11–15. [CrossRef]
9. Amerio, A.; Brambilla, A.; Morganti, A.; Aguglia, A.; Bianchi, D. COVID-19 lockdown: Housing built environment's effects on mental health. *Int. J. Environ. Res. Public Health* **2020**, *17*, 5973. [CrossRef]
10. Elmer, T.; Mepham, K.; Stadfeld, C. Students under lockdown: Comparisons of students' social networks and mental health before and during the COVID-19 crisis in Switzerland. *PLoS ONE* **2020**, *15*, e0236337.
11. Andriningrum, H.; Gunawan, I. Cultivatation of Healthy Life for Students in School: A Literature Review. In Proceedings of the International Conference on Education and Technology (ICET 2018), Malang, Indonesia, 26–27 October 2018.
12. Vaishya, R.; Javiad, M.; Khan, I.H.; Haleem, A. Artificial Intelligence (AI) applications for COVID-19 pandemic. *Diabetes Metab. Syndr. Clin. Res. Rev.* **2020**, *14*, 337–339. [CrossRef] [PubMed]

13. Hussain, A.A.; Bouachir, O.; Turjman, F.A.; Aloqaily, M. AI techniques for COVID-19. *IEEE Access* **2020**, *8*, 128776–128795. [CrossRef]
14. Rzymski, P.; Nowicki, M. COVID-19-related prejudice toward Asian medical students: A consequence of SARS-CoV-2 fears in Poland. *J. Infect. Public Health* **2020**, *13*, 873–876. [CrossRef]
15. Hasan, N.; Bao, Y. Impact of "e-Learning crack-up" perception on psychological distress among college students during COVID-19 pandemic: A mediating role of "fear of academic year loss". *Child. Youth Serv. Rev.* **2020**, *118*, 105355. [CrossRef] [PubMed]
16. Llieva, G.; Yankova, T.; Belcheva, S.K.; Lvanova, S. Effects of COVID-19 Pandemic on University Students' Learning. *Information* **2021**, *12*, 163.
17. Wang, C.; Zhao, H.; Zhang, H. Chinese College Students Have Higher Anxiety in New Semester of Online Learning During COVID-19: A Machine Learning Approach. *Front. Psychol.* **2020**, *11*, 3465. [CrossRef]
18. Pan, Y.; Yang, Z.; Han, H.; Qi, S. Family functioning and mental health among secondary vocational students during the COVID-19 epidemic: A moderated mediation model. *Personal. Individ. Differ.* **2021**, *171*, 110490. [CrossRef]
19. Cleofas, J.V.; Rocha, I.C.N. Demographic, gadget and internet profiles as determinants of disease and consequence related COVID-19 anxiety among Filipino college students. *Educ. Inf. Technol.* **2021**, *26*, 6771–6786. [CrossRef]
20. Khattar, A.; Jain, P.R.; Quadri, S.M.K. Effects of the disastrous pandemic COVID 19 on learning styles, activities and mental health of young Indian students-a machine learning approach. In Proceedings of the 2020 4th International Conference on Intelligent Computing and Control Systems (ICICCS), Madurai, India, 13–15 May 2020.
21. Khasawneh, A.I.; Humeidan, A.A.; Alsulaiman, J.W.; Bloukh, S.; Ramadan, M. Medical students and COVID-19: Knowledge, attitudes, and precautionary measures. A descriptive study from Jordan. *Front. Public Health* **2020**, *8*, 258. [CrossRef]
22. Baloch, G.M.; Sundarasen, S.; Chinna, K.; Nurunnabi, M.; Kamaludin, K. COVID-19: Exploring impacts of the pandemic and lockdown on mental health of Pakistani students. *PeerJ* **2021**, *9*, e10612. [CrossRef]
23. Son, C.; Hegde, S.; Smith, A.; Wang, X.; Sasangohar, F. Effects of COVID-19 on college students' mental health in the United States: Interview survey study. *J. Med. Internet Res.* **2020**, *22*, e21279. [CrossRef] [PubMed]
24. Saddik, B.; Hussein, A.; Askari, F.S.S.; Kheder, W.; Temsah, M.H. Increased levels of anxiety among medical and non-medical university students during the COVID-19 pandemic in the United Arab Emirates. *Risk Manag. Healthc. Policy* **2020**, *13*, 2395. [CrossRef] [PubMed]
25. Khoshaim, H.B.; Sukayt, A.A.; Chinna, K.; Nurunnabi, M.; Sundarasen, S. Anxiety Level of University Students During COVID-19 in Saudi Arabia. *Front. Psychiatry* **2020**, *11*, 1397. [CrossRef] [PubMed]
26. Calder, N.; Jafri, M.; Guo, L. Mathematics Education Students' Experiences during Lockdown: Managing Collaboration in eLearning. *Educ. Sci.* **2021**, *11*, 191. [CrossRef]
27. Ranjbar, K.; Hosseinpour, H.; Shahriarirad, R.; Ghaem, H.; Jafari, K. Students' attitude and sleep pattern during school closure following COVID-19 pandemic quarantine: A web-based survey in south of Iran. *Environ. Health Prev. Med.* **2021**, *26*, 1–10. [CrossRef]
28. Yuduang, N.; Ong, A.K.; Vista, N.B.; Prasetyo, Y.T.; Nadlifatin, R.; Persada, S.F.; Gumasing, M.J.; German, J.D.; Robas, K.P.; Chuenyindee, T.; et al. Utilizing structural equation modeling–artificial neural network hybrid approach in determining factors affecting perceived usability of mobile mental health application in the Philippines. *Int. J. Environ. Res. Public Health* **2022**, *19*, 6732. [CrossRef]
29. Chaturvedi, K.; Vishwakarma, K.D.; Singh, N. COVID-19 and its impact on education, social life and mental health of students: A survey. *Child. Youth Serv. Rev.* **2021**, *121*, 105866. [CrossRef]
30. McKinney. W. Why python for data analysis? In *Book Python for Data Analysis: Data Wrangling with Pandas, NumPy, and Ipython*; O' Reilly Media, Inc.: Sebastopol, CA, USA, 2012; Volume 1, pp. 1–17.
31. Byeon, H. Predicting the depression of the South Korean elderly using SMOTE and an imbalanced binary dataset. *Int. J. Adv. Comput. Sci. Appl.* **2021**, *12*, 74–79. [CrossRef]
32. Shen, F.; Zhao, X.; Kou, G.; Alsaadi, F.E. A new deep learning ensemble credit risk evaluation model with an improved synthetic minority oversampling technique. *Appl. Soft Comput.* **2021**, *98*, 106852. [CrossRef]
33. Wah, Y.B.; Ibrahim, N.; Hamid, H.A.; Rahman, S.A.; Fong, S. Feature Selection Methods: Case of Filter and Wrapper Approaches for Maximising Classification Accuracy. *Pertanika J. Sci. Technol.* **2018**, *26*, 329–340.
34. Yusta, S.C. Different metaheuristic strategies to solve the feature selection problem. *Pattern Recognit. Lett.* **2009**, *30*, 525–534. [CrossRef]
35. Moore, J.H.; White, B.C. Tuning Relief, F for genome-wide genetic analysis. In Proceedings of the European Conference on Evolutionary Computation, Machine Learning and Data Mining in Bioinformatics, Valencia, Spain, 11–13 April 2007.
36. Urbanowicz, R.J.; Meeker, M.; Cava, W.L.; Olson, R.S.; Moore, J.H. Relief-based feature selection: Introduction and review. *J. Biomed. Inform.* **2018**, *85*, 189–203. [CrossRef] [PubMed]
37. Sun, J.; Zhang, R.; Chen, M.; Chen, B.; Wang, X. Identification of Porosity and Permeability While Drilling Based on Machine Learning. *Arab. J. Sci. Eng.* **2021**, *46*, 7031–7045. [CrossRef]
38. Kostopoulos, G.; Kotsiantis, S.; Pintelas, P. Estimating student dropout in distance higher education using semi-supervised techniques. In Proceedings of the 19th Panhellenic Conference on Informatic, Athens, Greece, 1–3 October 2015.
39. Chuenyindee, T.; Buaphiban, T. Utilization of random forest classifier and artificial neural network for predicting factors influencing the perceived usability of COVID-19 contact tracing "Morchana" in Thailand. *Int. J. Environ. Res. Public Health* **2022**, *19*, 7979.

40. Velazquez, R.M.; Tobon, D.P.; Sanchez, A.; Saddik, A.E.; Petriu, E. A Machine Learning Approach as an Aid for Early COVID-19 Detection. *Sensor* **2021**, *21*, 4202. [CrossRef]
41. Almalki, Y.E.; Qayyum, A.; Irfan, M.; Haider, N.; Glowacz, A. A Novel Method for COVID-19 Diagnosis Using Artificial Intelligence in Chest X-ray Images of the article. *Healthc. J.* **2021**, *9*, 522. [CrossRef]
42. Irfan, M.; Iftikhar, M.A.; Yasin, S.; Draz, U.; Ali, T.; Hussain, S.; Bukhari, S.; Alwadie, A.S.; Rahman, S.; Glowacz, A.; et al. Role of Hybrid Deep Neural Networks (HDNNs), Computed Tomography, and Chest X-rays for the Detection of COVID-19. *Int. J. Environ. Res. Public Health* **2021**, *18*, 3056. [CrossRef]

life

Article

Did Usage of Mental Health Apps Change during COVID-19? A Comparative Study Based on an Objective Recording of Usage Data and Demographics

Maryam Aziz [1,*], Aiman Erbad [1], Mohamed Basel Almourad [2], Majid Altuwairiqi [3], John McAlaney [4] and Raian Ali [1,*]

[1] College of Science and Engineering, Hamad Bin Khalifa University, Doha P.O. Box 5825, Qatar
[2] College of Technological Innovation, Zayed University, Dubai P.O. Box 144534, United Arab Emirates
[3] College of Computer and Information Technology, University of Taif, Taif 21974, Saudi Arabia
[4] Faculty of Science and Technology, Bournemouth University, Bournemouth BH12 5BB, UK
* Correspondence: maur33838@hbku.edu.qa (M.A.); raali2@hbku.edu.qa (R.A.)

Abstract: This paper aims to objectively compare the use of mental health apps between the pre-COVID-19 and during COVID-19 periods and to study differences amongst the users of these apps based on age and gender. The study utilizes a dataset collected through a smartphone app that objectively records the users' sessions. The dataset was analyzed to identify users of mental health apps (38 users of mental health apps pre-COVID-19 and 81 users during COVID-19) and to calculate the following usage metrics; the daily average use time, the average session time, the average number of launches, and the number of usage days. The mental health apps were classified into two categories: guidance-based and tracking-based apps. The results include the increased number of users of mental health apps during the COVID-19 period as compared to pre-COVID-19. Adults (aged 24 and above), compared to emerging adults (aged 15–24 years), were found to have a higher usage of overall mental health apps and guidance-based mental health apps. Furthermore, during the COVID-19 pandemic, males were found to be more likely to launch overall mental health apps and guidance-based mental health apps compared to females. The findings from this paper suggest that despite the increased usage of mental health apps amongst males and adults, user engagement with mental health apps remained minimal. This suggests the need for these apps to work towards improved user engagement and retention.

Keywords: mental health; COVID-19; mindfulness; digital health; mobile health; social isolation

Citation: Aziz, M.; Erbad, A.; Almourad, M.B.; Altuwairiqi, M.; McAlaney, J.; Ali, R. Did Usage of Mental Health Apps Change during COVID-19? A Comparative Study Based on an Objective Recording of Usage Data and Demographics. *Life* **2022**, *12*, 1266. https://doi.org/10.3390/life12081266

Academic Editor: Daniele Giansanti

Received: 11 July 2022
Accepted: 27 July 2022
Published: 19 August 2022

Publisher's Note: MDPI stays neutral with regard to jurisdictional claims in published maps and institutional affiliations.

1. Introduction

The World Health Organization (WHO) reports that as of 2017, there has been a 13% increase in mental health issues worldwide [1]. Mental health has been further affected since restrictions on social gatherings were placed to counter COVID-19 outbreaks [2]. The ramifications of COVID-19 include social isolation, economic crises, and unemployment, which are among the known risk factors for mental health illnesses [3]. Statista, a major provider of market and consumer data, surveyed around 23,000 participants worldwide, within the age range of 16–74 years, on mental health during the pandemic. The organization found that as of early 2021, 40% of the participants had reported a negative impact on their mental well-being during the last six months [4].

With the pandemic spreading, the interest in digital health for mental illnesses has accelerated [5,6]. Several studies, based on the populations of China [7], Spain [8], Canada [9], and Australia [10], discussed the adoption of mental health apps as a way to provide mental healthcare to their population. Along with this, there has been an increase in the number of apps claiming to provide mental health care in the market, as well as a boost in the number of downloads of these mental health apps [10,11]. The growing number of mental health app downloads could mean that more people are seeking mental health support or are receptive to it. Several demographic factors may need to be considered to identify these people, such as whether a certain age group or gender is predominantly downloading these apps. Additionally, downloading apps alone does not mean that a user is committing to use them; it would, however, show their receptiveness to help.

Jaworski et al. [12] investigated the daily usage of COVID Coach, a publicly available mental health app, with respect to the number of days the app was used. The study found that almost 50,000 people used the COVID Coach app from March 2020 to October 2020, and the app had a consistent daily active usage. Kozlov et al. [13] studied the usage of Mindfulness Coach based on the number of downloads and number of days. Their study found that the app is used infrequently and for short sessions. Almost 40% of returning users would open the app but not use it. Research conducted during COVID-19 from March to April 2020 focused on estimating the usage of popular mental health apps [14]. The study used monthly active users as a base metric to estimate usage and focused on the during-COVID-19 times. Wang et al. [15] found the number of downloads of mental health apps to increase during the COVID-19 period as compared to pre-COVID-19. Their research focused on the popular mental health apps based on the number of downloads and classified apps as per the developer's choice.

With respect to demographics, Mackenzie et al. [16] studied the differences in help-seeking outlooks amongst the different age and gender groups using a sample size of 206 participants. Their study used questionnaires to measure the help-seeking attitudes and the mental health conditions of the participants. They found older adults above 60 and females to be more receptive to seeking help regarding their mental health. Additionally, Segal et al. [17] performed a cross-sectional study on the beliefs and help-seeking attitudes of people based on their ages. They used questionnaires to gather information about the beliefs and help-seeking attitudes of people. Their results showed that older adults aged 60–95 years reported willingness to seek help on the same level as the adults aged 17–26 years. Furthermore, Kern et al. [18] conducted a study at the university level to investigate the help-seeking attitudes of young adults aged around 18–22 years old. Their survey responses showed that most students have a positive attitude towards receiving mental health support through mental health apps. Forbes et al. [19], on the other hand, found that their survey responses showed older adults above 60 as less likely to recognize their need for mental health care.

A major limitation in the research literature relates to the utilization of self-reported data to quantify the usage of mental health apps. This may lead to reporting bias which means that participants may underestimate or overestimate their mental health behaviors. Additionally, with regard to the use of smartphone apps, such as average screen time and average launches, users tend to underestimate their smartphone usage behavior [20,21]. Furthermore, previous research studies have mostly utilized usage metrics that do not take into consideration the actual time spent on mental health apps and focus only on the number of downloads or number of days of use of mental health apps. Torous et al. [22] state that almost 70% of users leave a health app after 10 uses. The number of downloads and the number of days alone are, therefore, not representative of the usage of mental health app users. Additionally, past research investigated the popular mental health apps downloaded by users to study their usage, whereas the change in usage may only be to certain categories of these apps. For example, some apps require spending time on them, whether to meditate or do mindfulness exercises, while others are based on reminders or recording daily moods and habits.

In this paper, we compare mental health app usage before and during the COVID-19 pandemic, using objective data collected from a smartphone app that monitors smartphone usage. We analyze the users who are using these mental health apps based on age and gender. We also classify the mental health apps into two categories to study the change in usage that may exist due to the type of app. Hence, taking into consideration the pre-COVID-19 and during-COVID-19 use of mental health apps, our research questions (RQ) are:

RQ1. Is there a change in the number of users?

RQ2. Is there a change in the usage with respect to daily average time spent, the daily average number of launches, average session time, and the number of days of use?

RQ3. Is there a change in the usage across the two categories of mental health apps: the guidance-based and the tracking-based apps?

2. Methods

2.1. Dataset

The dataset for this research was collected through an app that helps users monitor their phone usage by tracking their usage, such as phone unlocks, launches, and sessions of each app they use. It also collects information related to user demographics, including age and gender. The app privacy policy, to which all users consent, includes that collected data can be shared with academic partners for research purposes. Nevertheless, the app required the explicit consent of users before collecting and utilizing their data for this research.

Data collection for this study was first conducted in 2019 and then repeated in 2020 during COVID-19. The pre-COVID-19 data collection took place between June 2019 and September 2019. The purpose of that data collection was to study the relationship between smartphone usage and certain psychometrics of the user collected through a questionnaire they answered. The data collection during the COVID-19 period was conducted from October 2020 to April 2021. Throughout the paper, pre-COVID-19 is referred to interchangeably with 2019, while during-COVID-19 is referred to as 2020. The data collection was open to new participants who installed the app throughout the study duration, and our participants joined and withdrew from the app after a different number of days. The study was conducted over a period of 21 days from the start date of each user. The 21 days period choice covered three weeks, including weekends, and included a considerable number of users since some left the study after that.

The 2019 dataset had 376 users, while for the 2020 dataset, 557 users participated. Users in the 2020 study came from ten countries: Sweden, Australia, Netherlands, Canada, Germany, India, the United Kingdom, Brazil, France, and the United States. We wanted to restrict the study to countries where restrictions on social gatherings were applied. The 2019 dataset had around 70% users from the same 10 countries as the during-COVID-19 dataset. The datasets included anyone who participated, even for one day. Furthermore, the dataset included only Android smartphone users.

2.2. Data Preparation

The pre-processing of the data was conducted using the programming language Python 3.0 [23]. The dataset received had a total of 1070 users from the pre-COVID-19 and during-COVID-19 periods. However, a few users had chosen not to enter their age and gender, bringing the number of users taken for this study to 933. The parameter "Age" was categorized into five different categories: 15–24, 25–34, 35–44, 45–54, and 55–64. Based on the UNICEF age categorization [24], age was categorized into two categories of emerging adults (15–24 years) and adults (above 24). This categorization also helped to balance the dataset, given that the number of participants in categories above 15–24 was comparatively less.

Furthermore, the third-party app records each app session for each user, i.e., the app name and the timestamps of the start and end of use. The records of app sessions were thoroughly checked and cleaned to remove duplicates and anomalies and concatenate broken down sessions. Table 1 presents a sample of the data collected from the third-party app. From the app sessions, the time spent on each app was calculated in seconds. Additionally, each session was considered as a launch of an app.

Table 1. Sample of data collection from the third-party app.

User	Website	Start Time	End Time
u1	Facebook	18-January, 1:40:53	18-January, 1:42:47
u1	WhatsApp	18-January, 9:59:35	18-January, 10:00:03
u1	Happify	18-January, 10:00:03	18-January, 10:00:08
u1	Happify	18-January, 10:07:37	18-January, 10:07:38
u1	Happify	18-January, 10:28:52	18-January, 10:28:54

The raw data collected included only the name of the apps used but not the category of these apps, that is, whether they are in the social media, communications, or gaming category. Therefore, we extracted the app categories and descriptions based on the Google Play Store, using software utilizing Google Play API. Table 2 shows a sample of the app descriptions we received from Google Play API. Google Play Store groups apps into 49 categories, including health and fitness and medical categories [25]. App categorization is based on the developer's choice; therefore, a few apps were categorized incorrectly. Due to this, we took the apps in the health and fitness and the medical app categories as well as the top apps used in both productivity and lifestyle categories. Additionally, to ensure all health-related apps were extracted from other categories, we searched for keywords such as "health", "fitness", "medical", and "disease" in the description of the apps. Health and fitness apps contain apps related to personal fitness, workout tracking, health, and safety, while medical apps contain apps related to clinical references, clinical apps, and medical journals, amongst others. For this study, after we extracted the health and medical apps, we then manually checked these extracted apps to find the mental health apps.

Table 2. Sample of app categories and descriptions based on Google Play Store.

Apps	Title	Category	Category Id
Gmail	Gmail	COMMUNICATION	C7
WhatsApp	WhatsApp Messenger	COMMUNICATION	C7
Tumblr	Tumblr	SOCIAL	C44
Wysa	Wysa: stress, depression & anxiety therapy chatbot	HEALTH_AND_FITNESS	C31
360 medics	360 medics	MEDICAL	C36
7 Cups	7 Cups: Anxiety & Stress Chat	HEALTH_AND_FITNESS	C31

Around 800 apps were identified using the above method, out of which 690 were health and medical-related. Out of these 690 apps, 115 mental health-related apps were found. The apps classified as mental health covered various areas, including online therapy, mindfulness, meditation, and well-being. For example, the apps that met the inclusion criteria included I am Sober [26], Wysa [27], Happify [28], Headspace [29], and Calm [30]. Apps such as those focused primarily on yoga and healthy living were removed since these were more directed towards lifestyle than mental health.

The National Institute of Mental Health categorizes mental health apps into six categories: self-management apps, apps for improving thinking skills, skill-training apps, social support apps, passive symptom tracking apps, and data collection apps [31]. These six categories of apps can be grouped into two categories based on the time and frequency of use, which are the focuses of this study. The classification of the subcategories of the mental health apps was achieved by first coding the mental health apps based on whether they were for online therapy, mental health support, meditation exercises, mindfulness, or tracking. We then marked the apps for online therapy, mental health support, meditation exercises, and mindfulness as time-based apps since users need to spend time on these apps to meet the purpose for which the app is developed. We marked the tracking apps as frequency-based apps since users typically need to launch these apps frequently to meet the purpose. As such, apps that require a user to spend some time on them can then be called guidance-based since they mostly guide users towards developing coping and cognitive skills through mindfulness exercises, meditation, online therapy, and host support groups, among other mental well-being tools. Examples of guidance-based apps are Headspace, Calm, and Wysa. On the other hand, apps that require the user to launch them occasionally but not spend a long time on them can be referred to as tracking-based apps since these mostly include mood-tracking, symptoms tracking, and addiction-tracking apps. Examples of such apps include I am Sober and Anxiety Tracker [32]. The mental health apps were categorized according to these two categories of guidance-based and tracking-based apps. Since a few mental health apps belonged to both subcategories, the apps' primary focus was used when categorizing them as either a guidance-based app or a tracking-based app. There was a total of 36 tracking-based apps, while the guidance-based apps totaled 79 apps.

2.3. Data Preparation

The usage of mental health apps for each participant was measured through four different metrics, which are the average daily time spent on mental health apps (DT), the average daily number of launches of mental health apps (DL), the average duration of daily sessions of mental health apps (DS), and the number of days of use of mental health apps (UD) during the 21 days period.

The average daily time spent was measured by finding the total daily time spent, in minutes with fractions of seconds, on the apps over the 21 days. For the average daily launches of the apps, the total daily count of sessions on the apps was taken over 21 days. Additionally, the average duration of daily sessions was calculated by taking the sum of the sessions throughout the 21 days and averaging them over 21. The number of days of use was calculated as a count of the unique days of the usage of mental health apps. The four usage metrics were calculated separately for the overall mental health apps, guidance-based mental health apps, and tracking-based mental health apps.

The study was designed to answer three different research questions and, therefore, required different criteria for the usage metrics. The first question (RQ1) was directed toward the number of users of mental health apps before and during COVID-19 and did not require the utilization of the four usage metrics (DT, DL, DS, and UD). The second question (RQ2) focused on answering the change in the usage of the overall mental health apps before and during COVID-19. In this case, all four usage metrics were measured for the participants. Additionally, we had 11 users in 2019 and 14 users in 2020 with fewer than 2 days of usage. We also had 327 users in 2019 and 452 users in 2020 with no usage. The users with no usage or number of days of use fewer than 2 days over the 21 days were not considered. This was undertaken to take into consideration that users with fewer than 2 days of usage may have installed mental health apps as a trial and hence, may not have returned to use them. Furthermore, they had a negligible time spent on mental health apps (below 2 min). The third question (RQ3) dealt with the change in usage of the two categories of mental health apps before and during COVID-19. We took all four usage metrics in this case as well and did not consider users with no usage or usage fewer than 2 days. We found 11 users in 2019 and 14 users in 2020 with fewer than 2 days of usage for guidance-based apps. Meanwhile, for tracking-based apps, we had 2 users in 2019 and 5 users in 2020. With regards to users having no usage, we had 335 users in 2019 and 478 users in 2020 for guidance-based apps. For tracking-based apps, we had 359 users with no usage in 2019 and 531 users in 2020. Similar to RQ2, the users with fewer than 2 days of usage were not considered since they had negligible time spent on mental health apps and may have installed the mental health apps as a trial.

2.4. Data Analysis

The statistical analysis was performed on JASP 0.14.1 [33]. Chi-square tests were used to determine the change in the number of users from pre-COVID-19 to during COVID-19. Chi-square tests were also further used to determine the relationship between using mental health apps and the demographic variables of their users. Phi-coefficients were used as well to determine the effect size of these relationships. The normality of the data was checked by conducting Shapiro–Wilk tests on the four usage metrics (DT, DL, DS, and UD) with respect to the years and demographics for the overall mental health, guidance-based, and tracking-based apps. The majority of the usage metrics (that is, 66 out of the 96 distributions) did not have a normal distribution; hence, non-parametric tests were considered. Median and interquartile ranges (IQR) were used for the descriptive statistics since the usage measures were not normally distributed. Since the usage metrics were continuous variables, the Mann–Whitney U test was further applied to compare the usage from pre-COVID-19 to during COVID-19. Mann-Whitney U test was first conducted separately against the usage and the time periods for overall mental health apps, guidance-based and tracking-based apps. Then, the usage was tested against age and gender for the overall mental health apps, guidance-based, and tracking-based apps. Cohen's d was used to determine the effect size of these relationships.

3. Results

3.1. Descriptive Statistics

The first question (RQ1) is concerned with whether the number of mental health app users before and during COVID-19 is significantly different. For RQ1, we had a sample size of 376 participants in 2019, of which around 47% were females and around 55% were emerging adults. For the 2020 dataset, the sample size had 557 participants, of which 58% were females and 50% were emerging adults.

The second question (RQ2) is concerned with the change in the usage of overall mental health apps before and during COVID-19. For RQ2, we had a sample size of 38 participants in 2019, of which around 32% were females and around 30% were emerging adults. The 2020 dataset for RQ2 had a sample size of 81 participants, of which around 63% were females and around 40% were emerging adults. The sample size was reduced compared to RQ1 because of the exclusion of users with no usage or usage less than 2 days, as mentioned previously. Table 3 shows the descriptive statistics of the usage of mental health apps in 2019 and 2020.

Table 3. Overall mental health apps descriptive statistics.

	2019 (N = 38)				2020 (N = 81)			
	Daily Average Time spent	**Daily Average Number of Launches**	**Average Session Time**	**Number of Days of Use**	**Daily Average Time Spent**	**Daily Average Number of Launches**	**Average Session Time**	**Number of Days of Use**
Median	1.22	1.58	1.01	9.50	1.39	0.86	1.16	8.00
IQR	2.16	1.98	0.95	11.00	2.30	1.81	2.08	12.00

The third question (RQ3) is concerned with the change in the usage of guidance-based and tracking-based apps before and during COVID-19. For RQ3, we had a sample size of 30 participants in 2019 for guidance-based apps. Of the 30 participants, around 37% were females and around 27% were emerging adults. For the 2020 dataset, the sample size was 65 participants, of which around 62% were females and around 38% were emerging adults for guidance-based apps. Table 4 shows the descriptive statistics of the usage of guidance-based mental health apps in 2019 and 2020. For the tracking-based mental health apps, the 2019 dataset had a sample size of 15 participants, of which around 13% were females and around 47% were emerging adults. For the 2020 dataset, the sample size had 21 participants, of which around 76% were females and around 52% were emerging adults. Table 5 shows the descriptive statistics of the usage of the tracking-based mental health apps in 2019 and 2020. The categories of guidance-based and tracking-based mental health apps were taken independently; hence some users had usage for both guidance-based and tracking-based mental health apps and were considered participants of both these categories.

Table 4. Guidance-based mental health apps descriptive statistics.

	2019 (N = 30)				2020 (N = 65)			
	Daily average Time Spent	Daily Average Number of Launches	Average Session Time	Number of Days of Use	Daily Average Time Spent	Daily Average Number of Launches	Average Session Time	Number of Days of Use
Median	1.21	1.12	0.98	9.00	1.49	0.76	1.23	7.00
IQR	1.97	1.84	1.06	12.00	2.28	1.81	2.31	12.00

Table 5. Tracking-based mental health apps descriptive statistics.

	2019 (N = 15)				2020 (N = 21)			
	Daily Average Time Spent	Daily Average Number of Launches	Average Session Time	Number of Days of Use	Daily Average Time Spent	Daily Average Number of Launches	Average Session Time	Number of Days of Use
Median	0.63	1.00	1.14	7.00	0.58	0.95	1.01	8.00
IQR	1.26	1.48	0.73	11.00	1.65	0.95	0.99	10.00

With respect to the days of use, the users did not use the mental health apps regularly and had sparse usage, showing that they left for some days before returning to use the mental health apps. In the 2020 dataset, around 49% of the users had a usage of less than or equal to 7 days, while around 31% had a usage of around 21 days. On the other hand, for the 2019 dataset, 39% of the users had a usage of less than or equal to 7 days, while 37% of users had a usage of around 21 days. Figure 1 represents this sparsity in the usage of the users in 2019 for a sample of 40 random users. Similarly, Figure 2 represents the sparsity in the usage of the users in 2020 for a sample of 40 random users. The days of use for the 40 users in both years are marked. The visualization shows a general lack of consistent and durable usage. For example, u4 and u11 in 2019 used such apps for only a few days in early July.

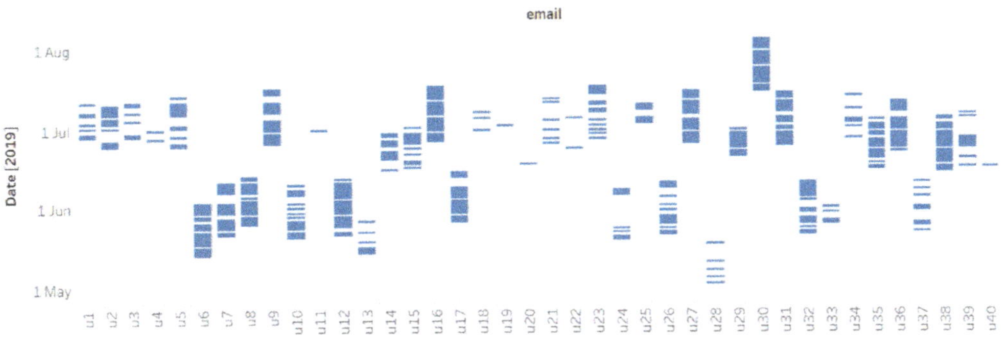

Figure 1. Usage sparsity for a sample of 40 users from 2019.

3.2. *App Usage*

3.2.1. RQ1: Changes in the Number of Mental Health Apps Users

We first compared the number of mental health app users between 2019 and 2020. The results, $\chi^2(\text{df} = 1, \text{N} = 933) = 4.77, p = 0.029$, show that the relationship is significant with an effect size, w = 0.071. Compared to 2019, the number of users increased in 2020. Furthermore, when comparing the years with age in 2019, the results, $\chi^2(\text{df} = 1, \text{N} = 376) = 15.86$, $p < 0.001$, showed the relationship to be significant, with an effect size of w = 0.21. Adults were more likely to be users of mental health apps compared to emerging adults. There was no significant difference between the change in the number of adult and emerging adult users in 2020. No significant relationship was found between gender and mental health app users, whether within 2019 or 2020.

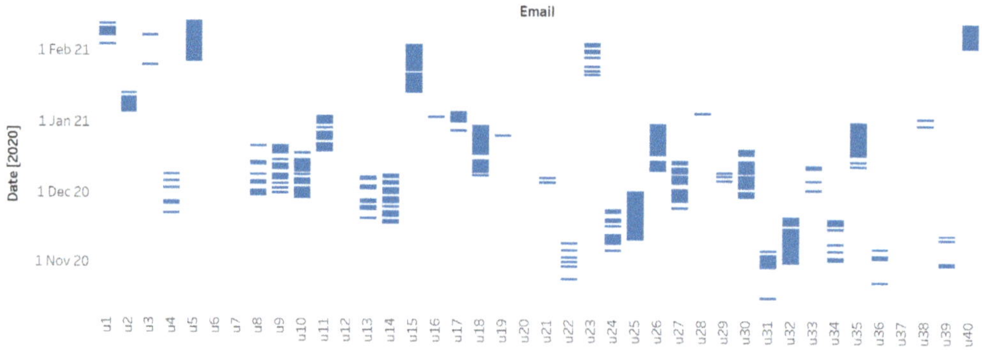

Figure 2. Usage sparsity for a sample of 40 users from 2020.

3.2.2. RQ2: Changes in Usage Time, Launches, Session Time and Number of Days

The results showed no statistically significant difference between the average daily usage time of mental health apps between 2019 and 2020. Additionally, no statistically significant relationship was found between mental health app usage and the demographics within 2019 and 2020. The results from comparing the change in mental health usage based on average session time also had no statistically significant relationship within and across 2019 and 2020.

The daily average number of launches of mental health apps was found to be statistically significant for gender in 2020. The average number of launches was greater for males (Mdn = 1.43, n = 30) compared to females (Mdn = 0.76, n = 51), $U = 982$, $p = 0.034$, $|r| = 0.28$.

For the number of days of use, age, and gender were both significant for 2020. The results showed that males (Mdn = 13.0 days, n = 30) had a higher number of days of use compared to females (Mdn = 7.0 days, n = 51), $U = 994.5$, $p = 0.025$, $|r| = 0.30$. On the other hand, in terms of days of use, adults (Mdn = 10.5 days, n = 48) were more likely to use mental health apps compared to emerging adults (Mdn = 5.0 days, n = 33), $U = 547$, $p = 0.018$, $|r| = 0.31$.

3.2.3. RQ3: Changes in Usage across the Two Categories of Mental Health Apps

The results for the daily average time spent on guidance-based mental health apps were found to have no significance with the period, age, and gender. Comparing the average session time for guidance-based apps with the year 2019 and 2020 showed a significant relationship where users in 2020 (Mdn = 1.34 min, n = 65) had a higher daily average session time compared to users in 2019 (Mdn = 1.05 min, n = 30), $U = 695$, $p = 0.025$, $|r| = 0.29$.

The results for the daily average number of launches of guidance-based mental health apps showed age to be significant for 2019 and both age and gender to be significant for 2020. For 2019, the results showed emerging adults (Mdn = 0.38, n = 8) as having lower daily average launches compared to adults (Mdn = 1.60, n = 22), $U = 39$, $p = 0.023$, $|r| = 0.56$. Similarly, in 2020, emerging adults (Mdn = 0.48, n = 25) were found to have lower daily average launches compared to adults (Mdn = 1.60, n = 40), $U = 288.5$, $p = 0.004$, $|r| = 0.42$. Additionally, males (Mdn = 1.48, n = 25) were found to have higher number of daily average launches compared to females (Mdn = 0.74, n = 40), $U = 649$, $p = 0.045$, $|r| = 0.30$.

The number of days of use of guidance-based apps showed significant results for age in both 2019 and 2020. In 2019, adults (Mdn = 12 days, n = 22) had a higher number of usage days of guidance-based mental health apps compared to emerging adults (Mdn = 3.5 days, n = 8), $U = 32$, $p = 0.009$, $|r| = 0.64$. Similarly, in 2020, adults (Mdn = 11.5 days, n = 40) again showed a higher number of usage days for guidance-based mental health apps compared to emerging adults (Mdn = 4 days, n = 25), $U = 247.5$, $p < 0.001$, $|r| = 0.51$.

For tracking-based mental health apps, a significant relationship for gender was found in 2020 where males (Mdn = 17 days, n = 5) had a higher usage in terms of number of days compared to females (Mdn = 6 days, n = 16), $U = 68.5$, $p = 0.020$, $|r| = 0.712$.

4. Discussion

This study compares the use of mental health apps before and during the pandemic amongst different age and gender groups. There are compelling reasons to directly focus on this area of study since, with remote lifestyle becoming the norm, digital tools are being used increasingly to provide guided and unguided mental health care remotely [34]. The findings from this paper are objective, as the data were used to accurately quantify the mental health app usage of users from 10 different countries, as compared to the use of self-reported data in the literature. Additionally, these findings can also be applied to a social isolation setting since limiting the spread of COVID-19 has resulted in prolonged periods of social isolation. Being socially isolated for even less than 10 days can cause long-term mental health problems [35]. Table 6 summarizes the findings related to RQ1, and Table 7 summarizes the findings related to RQ2 and RQ3.

Table 6. Summary of results of the analysis performed to answer RQ1.

	Number of Users					
	N (df)	χ^2	*p*			
Year	933 (1)	4.77	0.029 *			
		2019			**2020**	
	n (df)	χ^2	*p*	**n (df)**	χ^2	*p*
Age	376 (1)	15.86	<0.001 *	557 (1)	2.82	0.093
Gender	376 (1)	1.61	0.205	557 (1)	2.95	0.086

* *Significance level* < 0.05.

Table 7. Summary of results of the analysis performed to answer RQ2 and RQ3.

		Mental Health Apps					
Year		n	*U*	*p*			
		119					
Daily average time spent			1476.50	0.72			
Daily average number of launches			1741.50	0.25			
Average session time			1215.00	0.065			
Number of days of use			1711.00	0.33			
			2019			**2020**	
		n	*U*	*p*	n	*U*	*p*
		38			81		
Daily average time spent	Age		121.50	0.39		748.00	0.68
	Gender		119.50	0.26		910.00	0.16
Daily average number of launches	Age		111.50	0.24		625.00	0.11
	Gender		138.50	0.59		982.00	0.034 *
Average session time	Age		141.00	0.82		911.00	0.26
	Gender		125.00	0.34		769.00	0.97
Number of days of use	Age		108.00	0.20		547.00	0.018 *
	Gender		142.50	0.68		994.50	0.025 *

Table 7. *Cont.*

			Guidance-based mental health apps				
Year		n	*U*	*p*			
		95					
Daily average time spent			853.50	0.33			
Daily average number of launches			1037.00	0.62			
Average session time			695.00	0.025 *			
Number of days of use			1062.50	0.49			
			2019			**2020**	
		n	*U*	*p*	n	*U*	*p*
		30			65		
Daily average time spent	Age		46.50	0.054		382.50	0.12
	Gender		65.00	0.093		395.00	0.16
Daily average number of launches	Age		39.00	0.023 *		288.50	0.004 *
	Gender		83.00	0.37		351.00	0.045 *
Average session time	Age		78.00	0.66		581.00	0.28
	Gender		73.00	0.19		483.00	0.83
Number of days of use	Age		32.00	0.009 *		247.50	<0.001 *
	Gender		76.00	0.23		360.50	0.059
			Tracking-based mental health apps				
Year		n	*U*	*p*			
		36					
Daily average time spent			162.50	0.89			
Daily average number of launches			167.00	0.77			
Average session time			159.00	0.98			
Number of days of use			153.00	0.90			
			2019			**2020**	
		n	*U*	*p*	n	*U*	*p*
		15			21		
Daily average time spent	Age		24.00	0.69		79.00	0.098
	Gender		12.00	0.93		44.50	0.74
Daily average number of launches	Age		22.50	0.56		75.70	0.16
	Gender		10.00	0.67		63.00	0.063
Average session time	Age		32.00	0.69		73.00	0.22
	Gender		18.00	0.48		26.00	0.28
Number of days of use	Age		22.50	0.56		63.50	0.57
	Gender		11.00	0.79		68.5	0.02 *

* *Significance level* < 0.05.

4.1. Mental Health Apps and Subcategories with Respect to Number of Users

The findings from this study show that there was a significant increase in the number of mental health app users from 2019 to 2020. This is in line with the report from ORCHA [11], stating that the number of downloads of mental health apps has increased during the pandemic. This increase in downloads of mental health apps could be attributed to the social isolation that has come into play due to the pandemic. This is supported by a study conducted by Chan and Honey [36] to understand user perceptions of mental health apps. They stated that although face-to-face mental health support cannot be replaced, mental health apps have the potential to be an add-on source for some and an alternative option for others to receive mental health support. With respect to guidance-based apps, the average session time also increased in 2020 compared to 2019. With mindfulness, meditation, online therapy, and other mental well-being exercises being a part of guidance-based apps, the increased average session time could be explained by users shifting their mental health care and needs to digital platforms.

4.2. Mental Health Apps and Subcategories with Respect to Gender

Overall mental health app usage showed an interesting outcome of gender being significant in 2020 with respect to the daily average number of launches and number of days of use of mental health apps. Moreover, in studying the guidance-based apps, gender was also significant in 2020 with respect to the daily average number of launches. For tracking-based apps, gender was again significant in 2020 with respect to the number of days of use. In all cases, males were found to be higher users compared to females. The previous research that was undertaken in this domain show males as having a more negative attitude when it comes to mental health treatment and hence, not receiving the care they need [37]. However, previous research undertaken in this domain mostly used self-reporting, and the participants may have not properly assessed themselves. The current study is conducted using objective data to measure the usage of mental health apps and identify the users of these apps. Therefore, the current results show male and female users to have similar usage pre-pandemic but increased usage by males during the pandemic. When using a self-reporting data collection method, e.g., surveys and interviews, whether for research or counselling purposes, females are more open when declaring their mental health help-seeking behavior as compared to males [16]. However, when using the objective measures of usage of mental health apps, both males and females had similar usage in 2019 and, hence, similar help-seeking behaviors. Further, in 2020, males had comparatively higher usage of mental health apps, which may suggest they are more likely to seek support for their mental health using digital means. This may relate to the stigma associated with seeking help when it relates to mental health, which is more prominent in males [38], and the tendency of males to prefer more anonymous options than females. In other words, our results suggest that while males are less likely than females to seek help through traditional means of therapy and health institutions, they are more open to adopting digital means for that same purpose.

4.3. Mental Health Apps and Subcategories with Respect to Age

The results also showed age to be significant in 2019 with respect to the number of users; that is, adults are more likely to be users of mental health apps compared to emerging adults. The during-pandemic period also showed adults using mental health apps more as opposed to emerging adults with respect to average daily launches and number of days of use of mental health apps. In addition, when dividing mental health apps into categories of guidance-based and tracking-based, both the 2019 and 2020 datasets showed that the daily average number of launches and number of days of use for guidance-based apps was comparatively higher for adults than emerging adults. This means that whether the period was pre-pandemic or during-pandemic, adults' usage of mental health apps is higher than emerging adults. These findings align with the past research of Mackenzie et al. [16], which found that adults are highly likely to seek help regarding their mental health compared to emerging adults. Moreover, this also shows that, regardless of a crisis, emerging adults are not likely to seek mental health support, despite having access to them. The research undertaken by Kern et al. [18] showed that although young adults were interested in adopting mental health apps, their usage of these apps was limited. In general, mobile health apps attract young adults due to the immediate access provided by them; however, young adults tend to abandon them, citing costs and user experience as demotivating factors for continued use [39].

4.4. User Engagement and Retention of Mental Health Apps

The average time spent on overall, guidance-based, or tracking-based mental health apps showed no significant differences in the current study. This unchanged usage for the two time periods could be explained by the fact that the users require the mental health apps only when needed and may use them with extended breaks between usage [12]. This is also seen in the sparse usage of mental health apps in 2019 and 2020, with 37% of the users using for 21 days from their start date for 2019 and 31% of the users using for 21 days from their start date for 2020, resulting in their average time spent being unchanged during both time periods. For guidance-based apps, we would expect users to spend more time on them doing meditation exercises or mindfulness practices compared to tracking-based apps. However, the relatively similar time may suggest that the users do not follow through with the apps' objectives. According to Kozlov et al. [13], the mindfulness app called Mindfulness Coach had returning users who would only launch the app but not use it for the mindfulness exercises. This means that these apps fail to retain users for extended periods and may need to adopt a just-in-time intervention technique [40] to increase user engagement. The intervention method would provide mental health support to users when they need it or when they ask for it. Furthermore, user engagement and retention with mental health apps can be enhanced by adopting gamification techniques, i.e., the use of game-like design elements for a meaningful purpose [41,42]. Gamification uses gaming dynamics such as rewards and levels to keep the users motivated and engaged [43]. Additionally, Chiauzzi and Newell [44] found that 23% of users leave mental health apps after 1 week, with users using a mood-tracking app as intended for no more than 2 weeks. Based on a meta-analysis of mental health interventions [45], the length of an intervention for mental health last from 4 to 16 weeks, whereas users do not use mental health apps long enough for an intervention. Privacy concerns, lack of effectiveness of the apps, absence of user-centered design, and inadequate usability standards could be the reasons behind the low engagement with mental health apps for long-term periods [22]. Despite mental health apps having a low user engagement, they show the users' intent on seeking mental health support. The hesitation in adopting mental health apps as a steady medium of support is understandable since most mental health apps are developed without the presence of a mental health professional [46,47]. Additionally, a study on mental health apps based on clinical and scientific evidence found that only a small amount of mental health apps are based on clinical and evidence-based interventions. This means that a large number of mental health apps available on the Apple Store and Google Play Store do not go through rigorous testing within healthcare contexts to ensure the effectiveness and safety of mental health apps [48]. These mental health apps also do not remain in the market for long since they are mostly developed by small teams of developers with no future plans for support and upgrades [49].

4.5. Limitations

The study has its limitations with regard to the sample size used. The sample size was large enough to study mental health users, overall mental health app usage, and guidance-based mental health apps. However, for tracking-based apps, the number of users was considerably small since, despite the large sample size of 378 users in 2019 and 557 users in 2020, overall mental health apps were used by 38 users in 2019 and by 81 users in 2020. Although this number of users helped us to make a comparison between the users and non-users of mental health apps, they also limited the further analysis that could have been conducted on the users of mental health apps themselves. Another limitation in the study is the time span of 21 days taken for this study to include most users since some users left the study. As previously noted, mental health apps are utilized with extended periods of break, and as such, only when facing the need, the time span of the study may need to be lengthened to account for the periods of breaks in between. Furthermore, the third-party app collected the age of participants as an age range rather than an exact number. The utilization of an exact age could help divide age into further refined categories such as late adolescents and young adults. This would help to produce more accurate results with respect to the usage of mental health apps and age.

5. Conclusions

With the findings of this paper, the change in mental health app usage pre-pandemic and during-pandemic can be realized. The help-seeking behavior that was associated with females is found to be equally existing in males pre-pandemic and increased during-pandemic. In addition, it was also found that during the pandemic, adults are engaging more in mental health apps than emerging adults. This study helps in identifying the groups that are truly using these apps and those that are not since objective data were collected with regards to mental health app usage. This research shows how mental health apps are helping users through the pandemic, and hence, regulations on these apps should be enforced to ensure their safety of usage for the mental health of users. Despite the change in usage amongst users, user engagement and retention were not substantial for these apps; hence, the findings from this paper also suggest a need to adopt improved user engagement and retention techniques such as just-in-time intervention by developers of mental health apps.

Future work in this area may focus on studying the users who used mental health apps occasionally and whether they used them intensively during these times. Another future direction of study may investigate and identify the users who left mental health apps and whether they left due to improved conditions or disinterest in the mental health apps. The type of apps these users use can also be studied to identify the categories of mental health apps that have the most dropout. Furthermore, for this study, only two demographics of age and gender were considered, while future studies may also investigate the impact of culture and education as well as personality and type of mental health issues. Future studies in this domain could also look into the usage of guidance-based apps and tracking-based apps with respect to identifying user clusters that, in fact, do utilize these subcategories as per the apps' objectives and those that do not.

Author Contributions: Conceptualization, M.A. (Maryam Aziz), A.E. and R.A.; Data curation, M.B.A., M.A. (Majid Altuwairiqi) and R.A.; Formal analysis, M.A. (Maryam Aziz) and J.M.; Funding acquisition, M.B.A. and M.A. (Majid Altuwairiqi); Investigation, M.A. (Maryam Aziz); Methodology, M.A. (Maryam Aziz), A.E., M.B.A., M.A. (Majid Altuwairiqi), J.M. and R.A.; Supervision, A.E. and R.A.; Writing—original draft, M.A. (Maryam Aziz); Writing—review & editing, A.E. and R.A. All authors have read and agreed to the published version of the manuscript.

Funding: This work received support from Zayed University, UAE, under grant number R18053 and the Scientific Research Department at Taif University, grant number 1-441-79. Open Access funding has been provided by the Qatar National Library.

Institutional Review Board Statement: The study was conducted in accordance with the Declaration of Helsinki and approved by the Institutional Review Board of QATAR BIOMEDICAL RESEARCH INSTITUTE (protocol code QBRI-IRB 2021-08-102 and 12 August 2021).

Informed Consent Statement: Informed consent was obtained from all subjects involved in the study.

Data Availability Statement: Derived data supporting the findings of this study are available from the corresponding author on request.

Conflicts of Interest: The authors declare no conflict of interest.

References

1. World Health Organization. Mental Health, World Health Organization. (n.d.). Available online: https://www.who.int/health-topics/mental-health#tab=tab_2 (accessed on 29 November 2021).
2. Nobles, J.; Martin, F.; Dawson, S.; Moran, P.; Savović, J. The Potential Impact of COVID-19 on Mental Health Outcomes and the Implications for Service Solutions-ARC West, 2020. Available online: https://arc-w.nihr.ac.uk/covid-response/rapid-reports/potential-impact-of-covid-19-on-mental-health-outcomes-and-the-implications-for-service-solutions/ (accessed on 29 November 2021).
3. Longyear, R.L.; Kushlev, K. Can Mental Health Apps be Effective for Depression, Anxiety, and Stress during a Pandemic? *Pract. Innov.* **2021**, *6*, 131–137. [CrossRef]
4. Varella, S. Impact of COVID-19 on Mental Health Worldwide 2021 Statista, Statista. (2021). Available online: https://www.statista.com/statistics/1218053/impact-of-covid-19-on-mental-wellbeing-worldwide/ (accessed on 29 November 2021).
5. Figueroa, C.A.; Aguilera, A. The Need for a Mental Health Technology Revolution in the COVID-19 Pandemic. *Front. Psychiatry* **2020**, *11*, 523. [CrossRef] [PubMed]
6. Torous, J.; Keshavan, M. COVID-19, mobile health and serious mental illness. *Schizophr. Res.* **2020**, *218*, 36. [CrossRef] [PubMed]

7. Dong, L.; Bouey, J.; Bouey, J. Public Mental Health Crisis during COVID-19 Pandemic, China. *Emerg. Infect. Dis.* **2020**, *26*, 1616. [CrossRef] [PubMed]
8. Marques, G.; Drissi, N.; Díez, I.d.; de Abajo, B.S.; Ouhbi, S. Impact of COVID-19 on the psychological health of university students in Spain and their attitudes toward Mobile mental health solutions. *Int. J. Med. Inform.* **2021**, *147*, 104369. [CrossRef]
9. Strudwick, G.; Sockalingam, S.; Kassam, I.; Sequeira, L.; Bonato, S.; Youssef, A.; Mehta, R.; Green, N.; Agic, B.; Soklaridis, S.; et al. Digital Interventions to Support Population Mental Health in Canada During the COVID-19 Pandemic: Rapid Review. *JMIR Ment. Health* **2021**, *8*, E26550. Available online: https://Mental.Jmir.Org/2021/3/e26550 (accessed on 22 December 2021). [CrossRef]
10. Marshall, J.M.; Dunstan, D.A.; Bartik, W. The role of digital mental health resources to treat trauma symptoms in Australia during COVID-19. *Psychol. Trauma Theory Res. Pract. Policy* **2020**, *12*, S269–S271. [CrossRef] [PubMed]
11. ORCHA, COVID-19: Digital Health Trends & 2021 Opportunities Report-ORCHA. 2021. Available online: https://orchahealth.com/covid-19-digital-health-trends-2021-opportunities-report/ (accessed on 29 November 2021).
12. Jaworski, B.K.; Taylor, K.; Ramsey, K.M.; Heinz, A.; Steinmetz, S.; Pagano, I.; Moraja, G.; Owen, J.E. Exploring Usage of COVID Coach, a Public Mental Health App Designed for the COVID-19 Pandemic: Evaluation of Analytics Data. *J. Med. Internet Res.* **2021**, *23*, e26559. [CrossRef]
13. Kozlov, E.; Bantum, E.; Pagano, I.; Walser, R.; Ramsey, K.; Taylor, K.; Jaworski, B.; Owen, J. The Reach, Use, and Impact of a Free mHealth Mindfulness App in the General Population: Mobile Data Analysis. *JMIR Ment. Health* **2020**, *7*, e23377. [CrossRef] [PubMed]
14. Wasil, A.R.; Gillespie, S.; Schell, T.; Lorenzo-Luaces, L.; DeRubeis, R.J. Estimating the real-world usage of mobile apps for mental health: Development and application of two novel metrics. *World Psychiatry* **2021**, *20*, 137. [CrossRef] [PubMed]
15. Wang, X.; Markert, C.; Sasangohar, F. Investigating Popular Mental Health Mobile Application Downloads and Activity During the COVID-19 Pandemic. *Hum. Factors* **2021**, *00*, 18720821998110. [CrossRef] [PubMed]
16. Mackenzie, C.S.; Gekoski, W.L.; Knox, V.J. Age, gender, and the underutilization of mental health services: The influence of help-seeking attitudes. *Aging Ment. Health* **2006**, *10*, 574–582. [CrossRef] [PubMed]
17. Segal, D.L.; Coolidge, F.L.; Mincic, M.S.; O'Riley, A. Beliefs about mental illness and willingness to seek help: A cross-sectional study. *Aging Ment. Health* **2005**, *9*, 363–367. [CrossRef] [PubMed]
18. Kern, A.; Hong, V.; Song, J.; Lipson, S.K.; Eisenberg, D. Mental health apps in a college setting: Openness, usage, and attitudes. *Mhealth* **2018**, *4*, 20. [CrossRef] [PubMed]
19. Forbes, M.K.; Crome, E.; Sunderland, M.; Wuthrich, V.M. Perceived needs for mental health care and barriers to treatment across age groups. *Aging Ment. Health* **2017**, *21*, 1072–1078. [CrossRef] [PubMed]
20. Ohme, J.; Araujo, T.; de Vreese, C.H.; Piotrowski, J.T. Mobile data donations: Assessing self-report accuracy and sample biases with the iOS Screen Time function. *Mob. Media Commun.* **2020**, *9*, 293–313. [CrossRef]
21. McAlaney, J.; Almourad, M.B.; Powell, G.; Ali, R. Perceptions and Misperceptions of Smartphone Use: Applying the Social Norms Approach. *Information* **2020**, *11*, 513. [CrossRef]
22. Torous, J.; Nicholas, J.; Larsen, M.E.; Firth, J.; Christensen, H. Clinical review of user engagement with mental health smartphone apps: Evidence, theory and improvements. *Evid. Based Ment. Health* **2018**, *21*, 116–119. [CrossRef] [PubMed]
23. Python, Welcome to Python.org, (n.d.). Available online: https://www.python.org/ (accessed on 29 November 2021).
24. UNICEF. *UNICEF Programme Guidance for the Second Decade: Programming with and for Adolescents*; UNICEF: New York, NY, USA, 2018.
25. Google Play Store, Choose a Category and Tags for Your App or Game-Play Console Help, Google Play Store. (n.d.). Available online: https://support.google.com/googleplay/android-developer/answer/9859673?hl=en (accessed on 29 November 2021).
26. I Am Sober, I Am Sober Sobriety App for Android & iOS, I Am Sober. (n.d.). Available online: https://iamsober.com/ (accessed on 29 November 2021).
27. Wysa, Wysa-Everyday Mental Health, Wysa. (n.d.). Available online: https://www.wysa.io/ (accessed on 22 December 2021).
28. Happify, Happify: Science-Based Activities and Games, Happify. (n.d.). Available online: https://www.happify.com/ (accessed on 29 November 2021).
29. Headspace, Meditation and Sleep Made Simple-Headspace, Headspace. (n.d.). Available online: https://www.headspace.com/ (accessed on 29 November 2021).
30. Calm, Experience Calm, Calm. (n.d.). Available online: https://www.calm.com/ (accessed on 29 November 2021).
31. National Institute of Mental Health. NIMH Technology and the Future of Mental Health Treatment, National Institute of Mental Health. 2019. Available online: https://www.nimh.nih.gov/health/topics/technology-and-the-future-of-mental-health-treatment (accessed on 29 November 2021).
32. Appstronaut Studios. Mental health, Appstronaut Studios. 2020. Available online: http://www.appstronautstudios.com/mental-health/ (accessed on 29 November 2021).
33. JASP. JASP-A Fresh Way to Do Statistics, JASP. (n.d.). Available online: https://jasp-stats.org/ (accessed on 29 November 2021).
34. Torous, J.; Myrick, K.J.; Rauseo-Ricupero, N.; Firth, J. Digital Mental Health and COVID-19: Using Technology Today to Accelerate the Curve on Access and Quality Tomorrow. *JMIR Ment. Health* **2020**, *7*, E18848. Available online: https://Mental.Jmir.Org/2020/3/e18848 (accessed on 22 December 2021). [CrossRef]
35. Pietrabissa, G.; Simpson, S.G. Psychological Consequences of Social Isolation During COVID-19 Outbreak. *Front. Psychol.* **2020**, *11*, 2201. [CrossRef] [PubMed]

36. Chan, A.H.Y.; Honey, M.L.L. User perceptions of mobile digital apps for mental health: Acceptability and usability-An integrative review. *J. Psychiatr. Ment. Health Nurs.* **2021**, *29*, 147–168. [CrossRef]
37. Gonzalez, J.M.; Alegrid, M.; Prihoda, T.J. How do attitudes toward mental health treatment vary by age, gender, and ethnicity/race in young adults? *J. Community Psychol.* **2005**, *33*, 611–629. [CrossRef]
38. Clement, S.; Schauman, O.; Graham, T.; Maggioni, F.; Evans-Lacko, S.; Bezborodovs, N.; Morgan, C.; Rüsch, N.; Brown, J.S.L.; Thornicroft, G. What is the impact of mental health-related stigma on help-seeking? A systematic review of quantitative and qualitative studies. *Psychol. Med.* **2015**, *45*, 11–27. [CrossRef] [PubMed]
39. Hasan, N.; Bao, Y.; Chiong, R. A multi-method analytical approach to predicting young adults' intention to invest in mHealth during the COVID-19 pandemic. *Telemat. Informatics* **2021**, *68*, 101765. [CrossRef] [PubMed]
40. Nahum-Shani, I.; Smith, S.N.; Spring, B.J.; Collins, L.M.; Witkiewitz, K.; Tewari, A.; Murphy, S.A. Just-in-Time Adaptive Interventions (JITAIs) in Mobile Health: Key Components and Design Principles for Ongoing Health Behavior Support. *Ann. Behav. Med.* **2018**, *52*, 446–462. [CrossRef]
41. Cechetti, N.P.; Bellei, E.A.; Biduski, D.; Rodriguez, J.P.M.; Roman, M.K.; de Marchi, A.C.B. Developing and implementing a gamification method to improve user engagement: A case study with an m-Health application for hypertension monitoring. *Telemat. Inform.* **2019**, *41*, 126–138. [CrossRef]
42. Wiström, V.; Johansson, C.; Svensson, A.; Högström, U. GAMIFICATION FOR LEARNING IN HEALTH PROMOTION WORK. In *INTED2022 Proceedings*; IATED: Valencia, Spain, 2022; pp. 3190–3195. [CrossRef]
43. Dias, L.P.S.; Barbosa, J.L.V.; Vianna, H.D. Gamification and serious games in depression care: A systematic mapping study. *Telemat. Inform.* **2018**, *35*, 213–224. [CrossRef]
44. Chiauzzi, E.; Newell, A. Mental Health Apps in Psychiatric Treatment: A Patient Perspective on Real World Technology Usage, JMIR Ment. *Health* **2019**, *6*, e12292. Available online: https://Mental.Jmir.Org/2019/4/e12292 (accessed on 22 December 2021). [CrossRef]
45. Firth, J.; Torous, J.; Nicholas, J.; Carney, R.; Pratap, A.; Rosenbaum, S.; Sarris, J. The efficacy of smartphone-based mental health interventions for depressive symptoms: A meta-analysis of randomized controlled trials. *World Psychiatry* **2017**, *16*, 287–298. [CrossRef]
46. Torous, J.; Andersson, G.; Bertagnoli, A.; Christensen, H.; Cuijpers, P.; Firth, J.; Haim, A.; Hsin, H.; Hollis, C.; Lewis, S.; et al. Towards a consensus around standards for smartphone apps and digital mental health. *World Psychiatry* **2019**, *18*, 97–98. [CrossRef] [PubMed]
47. Heffernan, K.J.; Chang, S.; MacLean, S.T.; Callegari, E.T.; Garland, S.M.; Reavley, N.J.; Varigos, G.A.; Wark, J.D. Guidelines and Recommendations for Developing Interactive eHealth Apps for Complex Messaging in Health Promotion. *JMIR Mhealth Uhealth* **2016**, *4*, e14. [CrossRef] [PubMed]
48. Eis, S.; Solà-Morales, O.; Duarte-Díaz, A.; Vidal-Alaball, J.; Perestelo-Pérez, L.; Robles, N.; Carrion, C. Mobile Applications in Mood Disorders and Mental Health: Systematic Search in Apple App Store and Google Play Store and Review of the Literature. *Int. J. Environ. Res. Public Health* **2022**, *19*, 2186. [CrossRef]
49. Clemente-Suárez, V.J.; Martínez-González, M.B.; Benitez-Agudelo, J.C.; Navarro-Jiménez, E.; Beltran-Velasco, A.I.; Ruisoto, P.; Arroyo, E.D.; Laborde-Cárdenas, C.C.; Tornero-Aguilera, J.F. The Impact of the COVID-19 Pandemic on Mental Disorders. *A Critical Review. Int. J. Environ. Res. Public Health* **2021**, *18*, 10041. [CrossRef] [PubMed]

life

MDPI

Systematic Review

The Impact of eHealth Interventions on the Improvement of Self-Care in Chronic Patients: An Overview of Systematic Reviews

Erika Renzi [1,*], Valentina Baccolini [1], Giuseppe Migliara [1], Corrado De Vito [1], Giulia Gasperini [2,3], Angelo Cianciulli [1], Carolina Marzuillo [1], Paolo Villari [1] and Azzurra Massimi [1]

1 Department of Public Health and Infectious Diseases, Sapienza University of Rome, 00185 Rome, Italy
2 Department of Translational and Precision Medicine, Umberto I Teaching Hospital, 00161 Rome, Italy
3 Department of Biomedicine and Prevention, University of Rome Tor Vergata, 00133 Rome, Italy
* Correspondence: erika.renzi@uniroma1.it; Tel.: +39-06-49914886; Fax: +39-06-49914449

Abstract: Promoting self-care is one of the most promising strategies for managing chronic conditions. This overview aimed to investigate the effectiveness of eHealth interventions at improving self-care in patients with type-2 diabetes mellitus, cardiovascular disease, and chronic obstructive pulmonary disease when compared to standard care. We carried out a review of systematic reviews on PubMed, Scopus, Cochrane, PsychInfo, and CINAHL. AMSTAR-2 was used for quality appraisal. Eight systematic reviews (six with meta-analysis) were included, involving a total of 41,579 participants. eHealth interventions were categorized into three subgroups: (i) reminders via messaging apps, emails, and apps; (ii) telemonitoring and online operator support; (iii) internet and web-based educational programs. Six systematic reviews showed an improvement in self-care measurements through eHealth interventions, which also led to a better quality of life and clinical outcomes (HbA1C, blood pressure, hospitalization, cholesterol, body weight). This overview provided some implications for practice and research: eHealth is effective in increasing self-care in chronic patients; however, it is required to designate the type of eHealth intervention based on the needed outcome (e.g., implementing telemonitoring to increase self-monitoring of blood pressure). In addition, there is a need to standardize self-care measures through increased use of validated assessment tools.

Keywords: eHealth; self-care; chronic diseases

Citation: Renzi, E.; Baccolini, V.; Migliara, G.; De Vito, C.; Gasperini, G.; Cianciulli, A.; Marzuillo, C.; Villari, P.; Massimi, A. The Impact of eHealth Interventions on the Improvement of Self-Care in Chronic Patients: An Overview of Systematic Reviews. *Life* **2022**, *12*, 1253. https://doi.org/10.3390/life12081253

Academic Editor: Daniele Giansanti

Received: 1 August 2022
Accepted: 16 August 2022
Published: 17 August 2022

1. Introduction

The sustainability of health systems depends to an extent on the ability of individuals to manage chronic diseases by taking responsibility for and participating in their care process [1,2]. In this context, continuity of care is a crucial element in the management of chronic diseases [3], particularly in the context of primary healthcare [4], where there is a need to promote a change in the care models adopted and to provide dedicated workforces. A transition is required toward a proactive approach that helps patients to achieve a higher level of autonomy in the management of their health conditions and that supports their self-care [5]. This approach can be effective at maintaining and improving health and quality of life and reduces complications, hospital admissions, and mortality [6,7].

eHealth, which is defined by WHO as the "use of information and communication technologies for health" [8], could be a fundamental tool for improving patient-centered care in health systems and self-care of non-communicable diseases (NCDs). Technologies that allow tracking and interventions and can be used by both healthcare professionals and patients/assistants include laptops, smartphones, Fitbit units, tablets, wearable sensors, videoconferencing, and GIS [9,10]. eHealth interventions for NCDs represent an opportunity to facilitate communication, stimulate the demand for services, and increase access to health information for disease management [11,12]. The use of digital technologies in the health context is a priority issue, especially considering the recent COVID-19 pandemic [13], which highlighted the need to implement sustainability of healthcare systems through digitization to enhance continuity of care. Indeed, the COVID-19 pandemic has led to a deterioration in self-care for patients suffering from NCDs [14], following a widespread disruption in chronic disease management services worldwide. In a WHO survey of the ability of different countries to address and respond to the growing burden of NCDs, 122 countries reported that the pandemic had caused delays and/or disruptions in their health services for chronic patients suffering from NCDs in 31% to 65% of cases. Services responsible for the management of patients with diabetes mellitus (T2DM), cardiovascular disease (CVD), and chronic obstructive pulmonary disease (COPD) showed a significant deterioration compared to baseline [15].

The implementation of eHealth and digital health measures appears to offer a viable means of improving the resilience of national health systems [16–18]. An understanding of the evidence on the effectiveness of eHealth interventions for the most common NCDs, with a focus on enhancing self-care, is crucial when planning and implementing person-centered care and interventions that involve patients in the management of their disease. However, most published systematic reviews have focused only on specific chronic diseases or single clinical outcomes (e.g., patients with heart failure, blood pressure levels). Here, our overview of systematic reviews aims to investigate the effectiveness of eHealth interventions at improving self-care in patients with chronic conditions, specifically those with type-2 diabetes mellitus, chronic obstructive pulmonary disease, and cardiovascular disease, when compared to standard care.

2. Materials and Methods

2.1. Selection Criteria and Search Strategy

We carried out a review of systematic reviews using the methodology of the Joanna Briggs Institute [19] to evaluate the efficacy of eHealth interventions in primary care, compared to standard care, at improving self-care in adult patients (>18 years old) with a diagnosis of type-2 diabetes mellitus, cardiovascular disease, or chronic obstructive pulmonary disease.

The primary outcome was the improvement of self-care levels in terms of self-maint enance, self-monitoring, and self-management, based on the definition provided by the middle-range theory of self-care of chronic illness [6] when associated with eHealth interventions that were evaluated through validated measurement tools. Secondary outcomes concerned the association between eHealth interventions and the improvement of observer-related outcomes (OROs) and patient-reported outcomes (PROs) [20].

Due to the recent implementation of technologies in healthcare, we limited our search to the last ten years (2010 to July 2020). We included only three groups of NCDs: T2DM, CVD, and COPD, which are the most common NCDs and are responsible for the majority of global deaths. In addition, due to the characteristics of these diseases (long duration and need for continuity of care), they were most affected during the pandemic by interruption or delay in the delivery of routine health services [15].

The search covered five electronic databases: PubMed, Cumulative Index to Nursing and Allied Health Literature (CINAHL), Scopus, PsycINFO, and the Cochrane Library. A manual search was performed through reference lists and relevant journals (JMIR). The search strategy keywords were based on the middle-range theory of self-care for chronic illness [6] (the search strategy is fully reported in Supplementary File S1).

The main inclusion criteria and their definitions [21–25] are detailed in Table 1. In particular, we included systematic reviews with or without meta-analysis of randomized controlled trials (RCTs), quasi-experimental studies, and cohort studies published in English or Italian and showing studies evaluating self-care using validated measurement tools. We excluded systematic reviews focused only on specific populations (e.g., pregnant women with diabetes, minorities). The reasons for the exclusion of specific populations are related to the difficulties of applying results to the general population with a chronic condition because specific populations are also characterized by peculiar features that distinguish them from other patients with NCDs (e.g., socioeconomic, geographic, and clinical characteristics).

Table 1. Main inclusion criteria and their definitions.

Inclusion Criteria	Definition	
Population	*People with T2DM, CVD, COPD*	The following are WHO definitions of the NCDs covered in this overview: (i) T2DM is a chronic, metabolic disease characterized by elevated levels of blood glucose (or blood sugar), which leads over time to serious damage to the heart, blood vessels, eyes, kidneys and nerves [21]. (ii) CVD are a group of disorders of the heart and blood vessels and include coronary heart disease, cerebrovascular disease, rheumatic heart disease and other conditions [22]. (iii) COPD is a disease characterized by chronic airflow limitation and a range of pathological changes in the lung [23].
Intervention	*eHealth interventions*	eHealth is an emerging field at the intersection of medical informatics, public health and business; it refers to health services and information delivered or enhanced through the internet and related technologies [24].
Comparison	*Standard care*	All interventions carried out without the use of the above digital technologies, particularly involving controlled visits (hospitals, outpatient clinics, general practitioners), paper-based information, and face-to-face interventions.
Outcome	*Self-care*	A process of maintaining health through health-promoting practices and managing illness. The middle range theory defines three key concepts: Self-care maintenance is defined as those behaviors used to maintain physical and emotional stability (daily physical activity). Self-care monitoring refers to the process of observing oneself for changes in signs and symptoms (for example, being able to monitor vital signs). Self-care management is defined as the response to signs and symptoms when they occur (for example, insulin administration in case of hyperglycemia) [6,25].
Setting	*Community*	Community setting includes patients' home, outpatient clinics and pharmacies, primary care clinics and community hospitals.
Type of study	*Systematic Review*	We included systematic reviews with or without meta-analysis of RCTs, quasi-experimental studies and cohort studies.

2.2. Data Extraction and Quality Assessment

Two reviewers independently screened the records. In case of disagreement that was not solved via consensus, a third reviewer arbitrated the decision process.

Articles were also selected via manual search from the reference list. For data extraction, we used a form that included the following features: population demographics, patient diseases (T2DM, COPD, CVD), eHealth providers, measurement tool, setting, primary outcome, in terms of self-care maintenance, self-care monitoring, and self-care management (Table 1 for definition), secondary outcomes and type of eHealth intervention. The latter, defined as the activities included in "telemedicine" according to the WHO classification of digital health interventions [26], were classified on the basis of the main component of eHealth technologies used to achieve the goal as reported in the included systematic review (e.g., goal: monitor vital signs, eHealth: telemonitoring; goal: improve therapeutic adherence, eHealth: reminders). Generally, eHealth activities include remote monitoring and data transmission, consultancy with remote health workers, and monitoring or training activities through online educational programs. When starting from this classification, three categories were identified by the end of the process: (i) reminders via SMS, MMS, messaging apps, emails, and/or mobile apps; (ii) telemonitoring and online operator support; (iii) internet and web-based educational programs for smartphones, PCs, apps.

Two reviewers independently assessed the methodological quality of the systematic reviews included in our overview using the updated version of A Measurement Tool to Assess Systematic Reviews (AMSTAR-2) [27], a 16-point tool designed for this purpose. Any disagreements were resolved by discussion among reviewers.

3. Results

3.1. Main Characteristics of the Included Studies

The selection process (title, abstract and full text) and the main reasons for full-text exclusion are shown in the flowchart in Figure 1 and were performed according to PRISMA (Preferred Reporting Items for Systematic Reviews and Meta-Analysis) [28]. Briefly, 637 articles were initially identified, of which, after the removal of duplicates, 452 papers went to the screening phase. Screening by title and abstract yielded 77 articles that were assessed for eligibility. A total of eight articles [29–36], comprising six systematic reviews with meta-analyses [29–33,36] and two systematic reviews [34,35], were finally included.

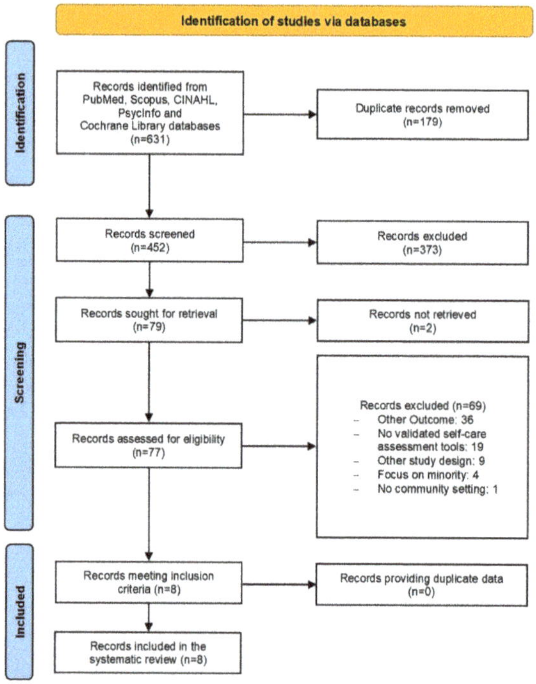

Figure 1. Flowchart diagram of study selection method.

The reviews encompassed 282 RCTs, one quasi-experimental study, and three cohort studies with a total of 41,579 participants aged 18 to 75 years. Three reviews included patients with CVD [30,32,36], one review concerned patients with T2DM [34], and one review included patients with COPD [33]. The other three systematic reviews were not related to a specific disease but included studies with patients who had at least one of the major chronic diseases (CVD, T2DM, or COPD) [29,31,35].

The eHealth interventions were mostly led by multi-professional teams. Healthcare professionals involved in the delivery of eHealth interventions were largely nurses, physicians, pharmacists, and research staff.

The self-care assessment questionnaires used in the studies covered by the systematic reviews were heterogeneous; the principal questionnaires used were as listed below: COPD-Self-Care Self-Efficacy Scale (SCES), Summary of Diabetes Self-Care Activities Measure (SDSCA), European Heart Failure Self-care Behaviour Scale (EHFScBS) and Self-Care Heart Failure Index (SCHFI).

The results of the AMSTAR-2 quality assessment show that, of the eight systematic reviews included, six were of high quality [29–33,36], and two were of critically low quality [34,35]. The main reason for the "critically low quality" classification of these two systematic reviews was the absence or incomplete implementation of methodological quality assessment of the single studies (item nine of the AMSTAR 2 tool). Detailed results of the quality assessment are reported in Supplementary File S2. A summary of the main characteristics of the included systematic reviews is reported in Table 2.

3.2. *Types of eHealth Intervention*

As already described, the eHealth interventions, even when multicomponent, have been categorized on the basis of the leading technological component characterizing the intervention itself and its main purpose.

1. Reminders via SMS, MMS, messaging apps, emails, and/or mobile apps (abbr. reminders): these interventions consisted of short-message reminders sent by healthcare providers through messaging apps, SMS, MMS, and/or emails with the aim of improving disease awareness and self-care of the chronic illness, and to remind individuals of therapy and daily activities (e.g., physical activity, daily glycemic control). Messaging apps allowed the person to communicate and give real-time feedback to the support operators and also facilitated emergency management by physicians and nurses. Two systematic reviews with meta-analysis [29,30] evaluated the effectiveness of reminder interventions in improving self-care. The eHealth interventions were compared with a traditional care approach that included routine home visits and face-to-face delivery of information only.
2. Telemonitoring and online operator support (abbr. telemonitoring): this method involves the patient transmitting clinical and physiological data via a phone or web-based automated electronic devices to healthcare professionals. Two systematic reviews with meta-analysis [31,32] examined telemonitoring alone or in association with videoconference educational sessions and online real-time operator support for symptom control. A comparison was made with standard care, which included face-to-face care, phone consultation, and routine visits.
3. Internet and web-based educational programs for smartphones, PCs, and apps (abbr. web-based education): these interventions consisted of structured online or offline programs designed to promote self-care using a set of resources that the patient must consult to achieve certain objectives. Four systematic reviews with [33,36] or without meta-analysis [34,35] evaluated this type of eHealth intervention. Standard care included no intervention, face-to-face interventions, education group sessions, and paper-based education materials.

Table 2. Main characteristics of the systematic reviews.

Author Year	Studies Included	Participants N (Mean Age)	NCDs	Intervention	Control	Assessment Tools	eHealth Providers	Setting	Primary Outcome	AMSTAR
De Jongh, 2012 [29]	Systematic review and meta-analysis of 4 RCTs	182 (44.7)	T2DM, CVD, COPD	Reminders	Standard care	(i) Summary of Diabetes Self-Care Activities Measure (SDSCA) (ii) Self-Efficacy for Diabetes test	(i) Research staff nurse and physician (2 RCTs) N.A. (2 RCTs)	Home	1. Health outcomes. 2. Capacity to self-manage long-term conditions	High Quality
Ma, 2019 [30]	Systematic review (15 RCTs) and meta-analysis of 14 RCTs	3889 (58)	Hypertension	Reminders	Standard care Attention control	Hill-Bone Compliance to High Blood Pressure Therapy Scale	(i) Nurse (5 RCTs). (ii) Physician (4 RCTs) (iii) Pharmacist (3 RCTs) (iv) Multi-professional team (3 RCTs)	(i) Primary care clinics (ii) Community health centers (iii) Clinics	1. Delivery mode and strategies of current eHealth interventions 2. Effectiveness of eHealth interventions on blood pressure control, self-care, and behavioral outcomes 3. Psychosocial well-being	High Quality
Flodgren, 2015 [31]	Systematic review of 93 RCTs and meta-analysis (66 RCTs on self-care)	22,047 (N.A.)	T2DM, COPD, Heart failure (27 studies on other chronic diseases)	Telemonitoring	Standard care Face-to-face Phone consultation	(i) European Heart Failure Self-care Behaviour Scale (EHFScBS) (ii) Self-Care Heart Failure Index (SCHFI)	N.A.	(i) Primary care clinics (ii) Community health centers (iii) Clinics	1. Mortality. 2. Disease-specific and general measures of health status 3.Healthcare resource use 4. Costs	High Quality
Inglis, 2015 [32]	Systematic review of 41 RCTs and meta-analysis (7 RCTs on self-care)	1062 (57.78)	Heart failure	Telemonitoring	Standard care	(i) Self-Care Heart Failure Index (SCHFI) (ii) European Heart Failure Self-care Behaviour Scale (EHFScBS)	(i) Multi-professional team (41 RCTs)	Home setting	1. All-cause mortality 2. All-cause hospitalizations 3. Heart failure-related hospitalizations	High Quality

Table 2. *Cont.*

Author Year	Studies Included	Participants N (Mean Age)	NCDs	Intervention	Control	Assessment Tools	eHealth Providers	Setting	Primary Outcome	AMSTAR
McCabe, 2017 [33]	Systematic review of 3 RCTs and meta-analysis	557 (64)	COPD	Web-based education	Standard care Face-to-face and/or hard copy/digital documentary educational/self-management support	(i) St. George's Respiratory Questionnaire (SGRQ) (ii) COPD-Self-Care Self-Efficacy Scale (SCES)	(i) Research staff (2 RCTs). (ii) Activity coach, researchers (1 RCT)	(i) Home setting (ii) Primary care clinics (iii) Community health centers (iv) Clinics	1. Hospital admissions 2. Acute exacerbations 3. Health-related quality of life	High Quality
Chrvala, 2016 [34]	Systematic review of 120 RCTs	11,093 (65.18)	T2DM	Web-based education	Standard care Waiting list	Summary of Diabetes Self-Care Activities Measure (SDSCA)	(i) Multi-professional team (53 RCTs) (ii) Physician (13 RCTs) (iii) Nurse (2 RCTs) (iv) Pharmacist (3 RCTs) (v) Nurses and physician (49 RCTs)	(i) Primary care clinics (ii) Community health centers. (iii) Clinics	Diabetes self-management and effect on glycemic control	Critically-Low Quality
Rush, 2018 [35]	Systematic review of 16 studies (12 RCTs, 3 cohort studies, 1 quasi-experimental study)	2870 (54)	T2DM, COPD, Heart failure	Web-based education	Standard care Routine visits. face-to-face education Paper copies of materials	(i) Summary of Diabetes Self-Care Activities Measure (SDSCA) (ii) Chronic Respiratory Questionnaire dyspnea (CRQ-D) subscale (iii) European Heart Failure Self-care Behaviour Scale (EHFScBS)	(i) Nurse (4 RCTs) (ii) Physician (3 RCTs) (iii) Research staff (1 RCT) (iv) Multi-professional team (8 RCTs, quasi-experimental, observational, cohort study)	(i) Home setting (ii) Primary care clinics	The efficacy of telehealth-delivered Educational approaches for patients with chronic diseases	Critically-Low Quality
Allida, 2020[36]	Systematic review and meta-analysis of 5 RCTs	921 (67.5)	Heart failure	Web-based education	Standard care	(i) European Heart Failure Self-care Behaviour Scale (EHFScBS) (ii) Self-Care Heart Failure Index (SCHFI)	(i) Multi-professional team (3 RCTs) (ii) Research nurse (1 RCT) N.A. (1 RCT)	Home settings	1. Heart-failure knowledge 2. Self-efficacy 3. Self-care 4. Adverse events	High Quality

3.3. Self-Care Improvements

Improvements in self-care measurements were associated with an eHealth intervention in six of the eight systematic reviews [29,31–35]. Interventions led by multi-professional teams reported more effective results in improving self-care than eHealth interventions led by single professionals and/or research staff [32,34,35]. The overview of the effectiveness of eHealth interventions in improving self-care is described in the following paragraph based on the classification given in Table 1 (self-maintenance, monitoring, and management). Table 3 summarizes the effectiveness of eHealth interventions in improving self-care.

Table 3. Summary of the effectiveness of eHealth interventions at self-care improvement.

Primary Outcome	Type of Intervention	Reference	Chronic Disease	Result	AMSTAR 2
Self-care improvement					
Self-maintenance	Web-based education	McCabe, 2017 [33]	COPD	+	High Quality
	Web-based education	Chrvala, 2016 [34] *	T2DM	+	Critically Low
	Web-based education	Rush, 2018 [35] *	T2DM-COPD	ns	Critically Low
	Web-based education	Allida, 2020 [36]	CVD	ns	High Quality
Self-monitoring	Reminders	De Jongh, 2012 [29]	T2DM	+	High Quality
	Telemonitoring	Flodgren, 2016 [31]	T2DM-COPD	ns	High Quality
	Telemonitoring	Inglis, 2015 [32]	CVD	+	High Quality
Self-management	Reminders	De Jongh, 2012 [29]	T2DM	+	High Quality
	Reminders	De Jongh, 2012 [29]	COPD	ns	High Quality
	Reminders	Ma, 2019 [30]	CVD	ns	High Quality
	Telemonitoring	Flodgren, 2016 [31]	T2DM-COPD	ns	High Quality
	Telemonitoring	Inglis, 2015 [32]	CVD	+	High Quality
	Web-based education	McCabe, 2017 [33]	COPD	+	High Quality
	Web-based education	Chrvala, 2016 [34] *	T2DM	+	Critically Low
	Web-based education	Rush, 2018 [35] *	COPD	+	Critically Low
	Web-based education	Rush, 2018 [35] *	T2DM	ns	Critically Low
	Web-based education	Allida, 2020 [36]	CVD	ns	High Quality

* No meta-analysis. +: Statistically significant results in favor of the intervention. ns: results not statistically significant. Reminders: Reminders via SMS, MMS, messaging apps, email, and/or mobile apps. Telemonitoring: Telemonitoring and online operator support. Web-based education: Internet and web-based educational programs for smartphones, PCs, and apps.

3.3.1. Self-Care Maintenance

Self-care maintenance was investigated in four systematic reviews (two with meta-analysis) [33–36] and consisted of interventions delivered via web-based education to a total of 15,441 patients. COPD patients registered a significant improvement in terms of self- maintenance, especially in terms of adherence to physical activity and stability of mental health [33]. These results were not confirmed in another systematic review, but this was one of the reviews assigned a "critically low quality" score in the AMSTAR assessment [35]. Web-based education also provided a statistically significant improvement in T2DM patients [34], especially in health education, on topics such as diet and how to monitor blood sugar. Improved self-care maintenance as a result of these eHealth interventions was especially marked in the elderly and those requiring home care [33].

3.3.2. Self-Care Monitoring

Self-care monitoring was evaluated in three systematic reviews [29,31,32] with a total of 23,291 patients. One systematic review included reminder interventions and showed a positive improvement in self-care monitoring in patients with T2DM [29], especially for monitoring blood sugar and weight. For telemonitoring, we included two systematic reviews [31,32], where we recorded an improvement in self-care monitoring only in patients with CVD for the daily assessment of blood pressure values, especially when interventions were provided by a multi-professional team (physician, nurse, pharmacist) [32].

3.3.3. Self-Care Management

Self-care management was evaluated in all the systematic reviews included here. Two systematic reviews with meta-analysis [29,30], with a total of 4071 patients, evaluated self-care management improvements with the use of reminders. No difference was found between the control and intervention groups in patients with COPD and CVD [29,30], whereas a statistically significant improvement was found in patients with T2DM [29] (increased self-management capacity in "Self-Efficacy for Diabetes—SED"—Mean Difference 6.10, 95% CI 0.45 to 11.75), particularly in patients of younger age and those requiring home care [29].

The effect of telemonitoring on self-care management of chronic diseases was studied in two systematic reviews with a total of 23,109 patients [31,32]. Only one systematic review showed significant improvements in self-care management in patients with CVD (heart failure) [32]. No difference was found between standard care and the experimental group for patients with T2DM and COPD [31].

The impact of web-based education on self-management was assessed by four systematic reviews (two with meta-analysis) [33–36] with a total of 15,441 patients. Only COPD patients recorded positive results for all dedicated programs, with statistically significant improvements in self-management of consulting behaviors, such as speaking to a healthcare provider if coughing/breathlessness increases [33,35].

3.4. *Secondary Outcomes*

Table 4 summarizes the effectiveness of eHealth interventions on secondary outcomes.

Table 4. Summary of the effectiveness of eHealth interventions at achievement of secondary outcomes.

Outcome Category	Type of Intervention	Reference	Chronic Disease	Result	AMSTAR 2
Observer-Reported Outcomes					
Systolic blood pressure	Reminders	De Jongh, 2012 [29]	CVD	ns	High Quality
	Reminders	Ma, 2019 [30]	CVD	+	High Quality
	Telemonitoring	Flodgren, 2015 [31]	CVD	+	High Quality
Diastolic blood pressure	Reminders	De Jongh, 2012 [29]	CVD	ns	High Quality
	Reminders	Ma, 2019 [30]	CVD	+	High Quality
	Telemonitoring	Flodgren, 2015 [31]	CVD	+	High Quality
HbA1c	Reminders	De Jongh, 2012 [29]	T2DM	ns	High Quality
	Telemonitoring	Flodgren, 2015 [31]	T2DM	+	High Quality
	Web-based education	Chrvala, 2016 [34] *	T2DM	+	Critically Low
	Web-based education	Rush, 2018 [35] *	T2DM	+	Critically Low
Total cholesterol	Reminders	Ma, 2019 [30]	CVD	ns	High Quality
	Web-based education	Rush, 2018 [35] *	CVD, T2DM	+	Critically Low
LDL cholesterol	Reminders	Ma, 2019 [30]	CVD	ns	High Quality
	Telemonitoring	Flodgren, 2015 [31]	CVD	+	High Quality
	Web-based education	Rush, 2018 [35] *	CVD, T2DM	+	Critically Low
HDL cholesterol	Reminders	Ma, 2019 [30]	CVD	ns	High Quality
Peak oxygen	Reminders	De Jongh, 2012 [29]	COPD	+	High Quality
Body weight	Reminders	Ma, 2019 [30]	CVD	+	High Quality
Hospitalizations	Reminders	De Jongh, 2012 [29]	CVD, T2DM, COPD	+	High Quality
	Telemonitoring	Inglis, 2015 [32]	CVD	+	High Quality
	Web-based education	Allida, 2020 [35]	CVD	ns	High Quality

Table 4. *Cont.*

Outcome Category	Type of Intervention	Reference	Chronic Disease	Result	AMSTAR 2
All-cause mortality	Telemonitoring	Flodgren, 2016 [31]	CVD, T2DM, COPD	ns	High Quality
	Telemonitoring	Inglis, 2015 [32]	CVD	+	High Quality
Patient-Reported Outcomes					
Quality of life	Telemonitoring	Flodgren, 2016 [31]	CVD, T2DM, COPD	+	High Quality
	Telemonitoring	Inglis, 2015 [32]	CVD	+	High Quality
	Web-based education	McCabe, 2017 [33]	COPD	+	Critically Low
	Web-based education	Rush, 2018 [35] *	T2DM, COPD	ns	Critically Low
	Web-based education	Allida, 2020 [36]	CVD	ns	High Quality
Medication adherence	Reminders	De Jongh, 2012 [29]	CVD, T2DM	ns	High Quality
	Web-based education	Rush, 2018 [35] *	T2DM, COPD	ns	Critically Low

* No meta-analysis. +: Statistically significant results in favor of the intervention. ns: result not statistically significant. Reminders: Reminders via SMS, MMS, messaging apps, email and/or mobile apps. Telemonitoring: Telemonitoring and online operator support. Web-based education: Internet and web-based educational programs for smartphones, PCs, apps.

3.4.1. Observer-Reported Outcomes

- Blood pressure levels: three systematic reviews with meta-analysis [29–31] assessed blood pressure levels in a total of 26,118 patients. Reminder interventions yielded statistically significantly lower systolic and diastolic blood pressure values in the experimental group compared to the control group in patients with hypertension. In particular, eHealth interventions significantly decreased the proportion of patients with inadequate blood pressure control (RR: 0.69, 95% CI: 0.57–0.84) [30]; however, no statistically significant changes were recorded in systolic (Mean Difference 1.10, 95% CI −4.37 to 6.57) and diastolic blood pressure (Mean Difference 1.84, 95% CI −2.14 to 5.82) in patients diagnosed with hypertension [29]. Telemonitoring interventions also showed a reduction in systolic and diastolic blood pressure values (Mean Difference—4.33, 95% CI −5.3 to −3.35: Mean Difference—2.75 95%, CI −3.28 to −2.22) in patients with CVD [31].
- HbA1c: this outcome was evaluated in four systematic reviews [29,31,34,35] in a total of 36,192 patients with T2DM. eHealth interventions, including reminders, showed no significant changes in glycemic values between the intervention and control groups (Mean Difference −0.15, 95% CI −0.77 to 0.47) [29]. In contrast, telemonitoring interventions did provide statistically significant improvements in the experimental group (Mean Difference −0.31, 95% CI −0.37 to −0.24) [31]. Two systematic reviews, rated as "critically low" quality according to AMSTAR−2, which analyzed web-based education, yielded a statistically significant improvement in glycemic control in patients with T2DM with or without other chronic conditions [34,35].
- Total cholesterol, LDL, HDL: three systematic reviews evaluated serum cholesterol levels [30,31,35] in 28,806 patients with chronic conditions. No improvement was reported with reminders in terms of total cholesterol (Mean Difference—0.20, 95% CI −0.49 to 0.08, $p = 0.16$), LDL (Mean Difference −0.14, 95% CI −0.39 to 0.11, $p = 0.27$) and HDL (Mean Difference −0.01, 95% CI −0.11 to 0.10, $p = 0.92$) [30]. However, significantly lower LDL cholesterol values were reported in patients with CVD following supervision via telemonitoring (LDL, Mean Difference 12.45, 95% CI −14.23 to −10.68; $p < 0.00001$) [31]. Another positive effect on LDL values was recorded with

interventions using the internet and web-based education, although this review was rated as of "critically low quality" [35].

- Peak oxygen: One systematic review was included with a total of 182 patients, in which the chosen eHealth intervention was the use of reminders. Peak oxygen levels were significantly higher in the intervention group of COPD patients [29].
- Body weight: A single review investigated this outcome in patients with CVD [30]. Reminders were associated with a statistically significant reduction in body mass index (Mean Difference −1.08, 95% CI −2.04 to −0.13).
- Hospitalizations: This outcome was investigated in three of the eight reviews [29,32,36] that included 2165 patients with CVD (heart failure) and T2DM. One review of the use of reminders showed a reduction in emergency hotline use for re-hospitalizations in T2DM patients (RR 0.32, 95% CI 0.09 to 1.08) [29]. One review of telemonitoring interventions showed a statistically significant reduction in heart failure-related hospitalizations (RR 0.85, 95% CI 0.77 to 0.93) [32], but another review showed no such difference (OR 0.74, 95% CI 0.52 to 1.06) [36].
- All-cause mortality: Two systematic reviews with meta-analysis investigated this outcome, both concerning telemonitoring interventions [31,32]. One of the reviews showed positive results in patients with heart failure (RR 0.80, 95% CI 0.68 to 0.94) [32]. However, no statistical significance in all-cause mortality was found in a meta-analysis (RR 0.89, 95% CI 0.76 to 1.03, p = 0.12) of patients with COPD, T2DM, and heart failure [31].

3.4.2. Patient-Reported Outcomes

- Quality of life (QoL): Five of the eight systematic reviews evaluated improvement in QoL by means of the SF-36 and SF-12 Health Status Questionnaires and the Kansas City Cardiomyopathy Questionnaire in a total of 27,457 patients [31–33,35,36]. Telemonitoring interventions were effective for CVD patients [33], particularly in the case of heart failure [32]. No difference was recorded between experimental and control groups when web-based educational programs were used [35,36].
- Adherence to medication regime: The effectiveness of reminders and web-based education interventions at ensuring adherence to a prescribed medication regime was evaluated by two systematic reviews [29,35]. No significant changes in chronic patients with T2DM, COPD, and CVD were uncovered.

4. Discussion

Healthcare systems worldwide face new health and organizational challenges as a result of two distinct phenomena: an aging population with an increased prevalence of chronic diseases and the need for healthcare systems to migrate outside of hospitals to promote proactive medicine and community support [37–39]. Chronic patients are, in fact, challenged with both an increase in their overall health needs and the necessity to guarantee continuity of care [40,41]. Primary care settings can help achieve these objectives by granting patients access to healthcare services and facilitating continuity of care [42].

According to the results of our overview of systematic reviews, community-wide eHealth interventions can indeed have a positive impact on self-care in patients with chronic diseases [29,31–35]. The eHealth approach also allows a higher degree of continuity of care than traditional methods delivered in community settings and/or at home and makes it possible to provide interventions founded on personalized care [31,32,35,43]. This is especially true in light of the recent COVID-19 pandemic [16], which highlighted, even more, the need to maintain close contact with chronic patients [44,45] to offer as much continuity of care as possible, despite a widespread reduction in the availability of access to healthcare services [16]. In fact, the eHealth interventions included in this overview appeared to be effective at improving self-care in chronic patients in six of the eight systematic reviews retrieved [29,31–35]. Self-care interventions in chronic patients were found to be effective when consistently monitored and maintained with the support of health services [46].

Thus, eHealth helps chronic patients in self-care by:

- Improving behavior that maintains physical and emotional stability (self-maintenance). This is particularly effective in COPD patients who use web-based education to ensure continuity in educational programs that maintain their physical and emotional status and allow them to control respiratory exacerbations [33,47,48];
- Providing early recognition of those signs and symptoms that suggest a deterioration in the patient's own health status (self-monitoring). In fact, this review has shown that the use of telemonitoring with the support of the operator, or reminders in patients with T2DM and CVD, allows early recognition of a deterioration in health status and prevents acute episodes, especially in patients with the decompensated disease [29,32,49,50];
- Allowing prompt action by means of lifestyle changes (self-management) [30–32,34,51] in all patients, particularly when using web-based education programs.

Our overview also showed that eHealth effectively enhances OROs and PROs in chronic patients. For example, telemonitoring interventions improve the quality of life for all chronic diseases and reduce hospitalization and mortality in patients with CVD [29,31–33,52,53]. Another interesting observation that emerged from this overview was that, although reminders are widely used to improve adherence to their medication regime in patients with chronic diseases, this improvement declined in the long term. Thus, patients using eHealth interventions for more than six months tended to return to "bad habits" once the novelty of telemedicine had worn off [29,35,54]. In fact, the duration of the intervention and engagement with it are also important factors that influence its effectiveness [55,56]. This evidence, in line with the literature, highlights the importance of the role of healthcare workers in encouraging patient adherence to eHealth programs [57,58].

Regarding the role of healthcare workers, this summary emphasizes the fact that eHealth interventions are most effective at improving self-care when they are led by multidisciplinary teams, especially when such teams work in primary care [30,32,34–36,59–62]. This is probably because specialized multidisciplinary teams can address both health and social health issues, ensuring that care is personalized and based on the perceived needs of the patient [63].

In conclusion, this overview carries some implications for practice, proving that eHealth is effective in increasing self-care in chronic patients with T2DM, CVD, and COPD; however, one must first be able to designate the most appropriate type of eHealth intervention based on the outcome to be achieved (e.g., implementing telemonitoring to increase self-monitoring of blood pressure). The results of this synthesis could help health care providers choose the most effective, outcomes-based eHealth interventions. In addition, this overview that included most of the major chronic diseases provided an overview of the effectiveness of eHealth on improving self-care, considering two aspects: (i) most population with chronic disease lives with multimorbidity, and designing an eHealth intervention on the basis of pathology could be a limitation (ii) eHealth interventions in increasing self-care should not be limited to disease, as self-care is a fundamental ability of patients with NCDs to live with their new life condition.

Finally, this overview of evidence brings to light two implications for the research: first, we observed that few systematic reviews in the literature use validated tools to assess the effectiveness of eHealth interventions in improving self-care. This might be because many of the self-care tools currently available have been developed for specific diseases and thus have limited applicability to other conditions, and also because transferring the data to the appropriate electronic platform can be a complex process [64]. Unfortunately, these limitations make it impossible to systematically evaluate those results that are not supported by standardized, validated instruments. Therefore, the quality of evidence would be markedly improved by the use of such standardized instruments across the scientific community to systematically evaluate self-care in all populations.

Secondly, none of the systematic reviews in our survey assessed the eHealth literacy of the patients involved, despite its importance for effective use of the interventions. The literature shows that people with high levels of eHealth literacy are empowered and enabled to fully participate in health decisions informed by eHealth resources and technologies [65]. Where eHealth literacy is at a low level, e.g., in elderly or rural populations [66], the ability to participate in eHealth interventions that aim to improve self-care is known to be reduced. If eHealth literacy levels are not assessed or if only technologically competent participants are selected, the results of any eHealth study or program are likely to be affected.

Finally, it should be noted as a limitation that this systematic review included only articles published in the last decade and up to July 2020. However, this choice allowed us to synthesize the most recent evidence by including systematic reviews with RCTs in chronic patient populations that were not affected by organizational changes resulting from the COVID-19 pandemic and are therefore more representative of care delivered in non-emergency settings.

5. Conclusions

eHealth interventions represent a means by which self-care and disease management in chronic patients can be increased. These interventions could also be applicable to the problems encountered during the COVID-19 pandemic. They might allow for much greater continuity of care during an emergency and non-emergency situation, supporting the sustainability of health care systems by reducing avoidable hospitalizations and re-hospitalizations and managing patients in primary care. However, it will be necessary to implement studies that investigate the effect of health inequality on the use of such eHealth interventions, considering the cost and availability of the electronic tools that this type of care requires. Furthermore, systematic reviews of higher methodological quality and with larger patient populations, particularly COPD patients, are urgently needed to assess the efficacy of eHealth in self-care programs. Wider adoption of standardized, validated tools for self-care assessment is also needed to achieve greater homogeneity of self-care measures, given that current evidence is based on a limited number of large studies of mixed methodological quality that can lack reliable self-care assessment tools.

Supplementary Materials: The following supporting information can be downloaded at: https://www.mdpi.com/article/10.3390/life12081253/s1, Supplementary File S1_search strategy; Supplementary File S2_AMSTAR2.0_Checklist.

Author Contributions: Conceptualization, E.R. and A.M.; methodology, E.R., V.B. and A.M.; data extraction E.R., G.G. and A.C.; data curation G.M. and C.D.V., writing—original draft preparation E.R. and C.M.; writing—review and editing, P.V. and A.M. All authors have read and agreed to the published version of the manuscript.

Funding: This research received no external funding.

Institutional Review Board Statement: Not applicable.

Informed Consent Statement: Not applicable.

Data Availability Statement: The data presented in this study are available on request from the corresponding author.

Conflicts of Interest: The authors declare no conflict of interest.

References

1. Nolte, E.; Knai, C.; Saltman, R.B. *Assessing Chronic Disease Management in European Health Systems: Concepts and Approaches*; European Observatory on Health Systems and Policies: Copenhagen, Denmark, 2014.
2. Grady, P.A.; Gough, L.L. Self-management: A comprehensive approach to management of chronic conditions. *Am. J. Public Health* **2014**, *104*, e25–e31. [CrossRef]
3. Facchinetti, G.; D'Angelo, D.; Piredda, M.; Petitti, T.; Matarese, M.; Oliveti, A.; De Marinis, M.G. Continuity of care interventions for preventing hospital readmission of older people with chronic diseases: A meta-analysis. *Int. J. Nurs. Stud.* **2020**, *101*, 103396. [CrossRef] [PubMed]
4. Health Quality Ontario. In-home care for optimizing chronic disease management in the community: An evidence-based analysis. *Ont. Health Technol. Assess. Ser.* **2013**, *13*, 1–65.
5. Lawn, S.; School, A. Supporting self-management of chronic health conditions: Common approaches. *Patient Educ. Couns.* **2010**, *80*, 205–211. [CrossRef] [PubMed]
6. Riegel, B.; Jaarsma, T.; Strömberg, A. A middle-range theory of self-care of chronic illness. *ANS Adv. Nurs. Sci.* **2012**, *35*, 194–204. [CrossRef]
7. Weingarten, S.R.; Henning, J.M.; Badamgarav, E.; Knight, K.; Hasselblad, V.; Gano, A., Jr.; Ofman, J.J. Interventions used in disease management programmes for patients with chronic illness-which ones work? Meta-analysis of published reports. *BMJ* **2002**, *325*, 925. [CrossRef]
8. World Health Organization. Using e-Health and Information Technology to Improve Health. Available online: https://www.who.int/westernpacific/activities/using-e-health-and-information-technology-to-improve-health (accessed on 28 July 2021).
9. Rooij, T.; Marsh, S. eHealth: Past and future perspectives. *Per. Med.* **2016**, *13*, 57–70. [CrossRef]
10. World Health Organization Regional Office for Europe. From Innovation to Implementation Ehealth in The WHO European Region. Available online: http://www.Euro.Who.Int/En/Ehealth (accessed on 27 July 2021).
11. Dinesen, B.; Nonnecke, B.; Lindeman, D.; Toft, E.; Kidholm, K.; Jethwani, K.; Young, H.M.; Spindler, H.; Oestergaard, C.U.; Southard, J.A.; et al. Personalized Telehealth in the Future: A Global Research Agenda. *J. Med. Internet Res.* **2016**, *18*, e53. [CrossRef]
12. Barbabella, F.; Melchiorre, M.G.; Quattrini, S.; Papa, R.; Lamura, G. How Can eHealth Improve Care for People with Multimorbidity in Europe? European Observatory on Health Systems and Policies: Copenhagen, Denmark, 2017.
13. Bitar, H.; Alismail, S. The role of eHealth, telehealth, and telemedicine for chronic disease patients during COVID-19 pandemic: A rapid systematic review. *Digit. Health* **2021**, *7*, 20552076211009396. [CrossRef]
14. Mirsky, J.B.; Horn, D.M. Chronic disease management in the COVID-19 era. *Am. J. Manag. Care* **2020**, *26*, 329–330.
15. World Health Organization. The Impact of the COVID-19 Pandemic on Noncommunicable Disease Resources and Services: Results of a Rapid Assessment. Available online: https://www.who.int/publications/i/item/9789240010291 (accessed on 28 July 2022).
16. The Lancet Respiratory Medicine. COVID-19 heralds a new era for chronic diseases in primary care. *Lancet Respir. Med.* **2020**, *8*, 647. [CrossRef]
17. Kretchy, I.A.; Asiedu-Danso, M.; Kretchy, J.P. Medication management and adherence during the COVID-19 pandemic: Perspectives and experiences from low-and middle-income countries. *Res. Soc. Adm. Pharm.* **2021**, *17*, 2023–2026. [CrossRef] [PubMed]
18. Bashshur, R.; Doarn, C.R.; Frenk, J.M.; Kvedar, J.C.; Woolliscroft, J.O. Telemedicine and the COVID-19 Pandemic, Lessons for the Future. *Telemed. J. E Health* **2020**, *26*, 571–573. [CrossRef] [PubMed]
19. Aromataris, E.; Fernandez, R.; Godfrey, C.; Holly, C.; Khalil, H.; Tungpunkom, P. Chapter 10: Umbrella Reviews. In *JBI Manual for Evidence Synthesis*; Aromataris, E., Munn, Z., Eds.; JBI: Adelaide, Australia, 2020.
20. U.S. Department of Health and Human Services FDA Center for Drug Evaluation and Research; U.S. Department of Health and Human Services FDA Center for Biologics Evaluation and Research; U.S. Department of Health and Human Services FDA Center for Devices and Radiological Health. Guidance for industry: Patient-reported outcome measures: Use in medical product development to support labeling claims: Draft guidance. *Health Qual. Life Outcomes* **2006**, *4*, 79.
21. World Health Organization. Diabetes. Available online: https://www.who.int/health-topics/diabetes#tab=tab_1 (accessed on 17 May 2022).
22. World Health Organization. Cardiovascular Diseases (CVD). Available online: https://www.who.int/news-room/fact-sheets/detail/cardiovascular-diseases-(CVD) (accessed on 17 May 2022).
23. World Health Organization. Chronic Obstructive Pulmonary Disease (COPD). Available online: https://www.who.int/news-room/fact-sheets/detail/chronic-obstructive-pulmonary-disease-(copd) (accessed on 17 May 2022).
24. Eysenbach, G. What is e-health? *J. Med. Internet Res.* **2001**, *3*, e20. [CrossRef]
25. Riegel, B.; Jaarsma, T.; Lee, C.S.; Strömberg, A. Integrating symptoms into the middle-range theory of self-care of chronic illness. *ANS Adv. Nurs. Sci.* **2019**, *42*, 206–215. [CrossRef]
26. World Health Organization. Classification of Digital Health Interventions, 2018. WHO/RHR/18.06. Licence: CC BY-NC-SA 3.0 IGO. Available online: https://apps.who.int/iris/handle/10665/260480 (accessed on 17 May 2022).

27. Shea, B.J.; Reeves, B.C.; Wells, G.; Hamel, C.; Moran, J.; Moher, D.; Tugwell, P.; Welch, V.; Kristjansson, E.; Henry, D.A. AMSTAR 2: A critical appraisal tool for systematic reviews that include randomised or non-randomised studies of healthcare interventions, or both. *BMJ* **2017**, *358*, j4008. [CrossRef]
28. Moher, D.; Liberati, A.; Tetzlaff, J.; Altman, D.G.; PRISMA Group. Preferred reporting items for systematic reviews and meta-analyses: The PRISMA statement. *Ann. Intern. Med.* **2009**, *151*, 264–269. [CrossRef]
29. de Jongh, T.; Gurol-Urganci, I.; Vodopivec-Jamsek, V.; Car, J.; Atun, R. Mobile phone messaging for facilitating self-management of long-term illnesses. *Cochrane Database Syst. Rev.* **2012**, *2017*, CD007459. [CrossRef]
30. Ma, Y.; Cheng, H.Y.; Cheng, L.; Sit, J.W.H. The effectiveness of electronic health interventions on blood pressure control, self-care behavioural outcomes and psychosocial well-being in patients with hypertension: A systematic review and meta-analysis. *Int. J. Nurs. Stud.* **2019**, *92*, 27–46. [CrossRef]
31. Flodgren, G.; Rachas, A.; Farmer, A.J.; Inzitari, M.; Shepperd, S. Interactive telemedicine: Effects on professional practice and health care outcomes. *Cochrane Database Syst. Rev.* **2015**, *2016*, CD002098. [CrossRef] [PubMed]
32. Inglis, S.C.; Clark, R.A.; Dierckx, R.; Prieto-Merino, D.; Cleland, J.G.F. Structured telephone support or non-invasive telemonitoring for patients with heart failure. *Cochrane Database Syst. Rev.* **2015**, *2015*, CD007228. [CrossRef]
33. McCabe, C.; McCann, M.; Brady, A.M. Computer and mobile technology interventions for self-management in chronic obstructive pulmonary disease. *Cochrane Database Syst. Rev.* **2017**, *2020*, CD011425. [CrossRef]
34. Chrvala, C.A.; Sherr, D.; Lipman, R.D. Diabetes self-management education for adults with type 2 diabetes mellitus: A systematic review of the effect on glycemic control. *Patient Educ. Couns.* **2016**, *99*, 926–943. [CrossRef] [PubMed]
35. Rush, K.L.; Hatt, L.; Janke, R.; Burton, L.; Ferrier, M.; Tetrault, M. The efficacy of telehealth delivered educational approaches for patients with chronic diseases: A systematic review. *Patient Educ. Couns.* **2018**, *101*, 1310–1321. [CrossRef] [PubMed]
36. Allida, S.; Du, H.; Xu, X.; Prichard, R.; Chang, S.; Hickman, L.D.; Davidson, P.M.; Inglis, S.C. mHealth education interventions in heart failure. *Cochrane Database Syst. Rev.* **2020**, *2020*, CD011845.
37. Hofer, A.N.; Abraham, J.M.; Moscovice, I. Expansion of coverage under the Patient Protection and Affordable Care Act and primary care utilization. *Milbank Q.* **2011**, *89*, 69–89. [CrossRef]
38. Maresova, P.; Javanmardi, E.; Barakovic, S.; Barakovic Husic, J.; Tomsone, S.; Krejcar, O.; Kuca, K. Consequences of chronic diseases and other limitations associated with old age—A scoping review. *BMC Public Health* **2019**, *19*, 1431. [CrossRef]
39. Wagner, E.H.; Groves, T. Care for chronic diseases. *BMJ* **2002**, *325*, 913–914. [CrossRef]
40. Yang, F.; Xiong, Z.F.; Yang, C.; Qiao, G.; Wang, Y.; Zheng, T.; He, H.; Hu, H. Continuity of Care to Prevent Readmissions for Patients with Chronic Obstructive Pulmonary Disease: A Systematic Review and Meta-Analysis. *COPD J. Chronic Obstr. Pulm. Dis.* **2017**, *14*, 251–261. [CrossRef]
41. Russell, D.; Rosati, R.J.; Rosenfeld, P.; Marren, J.M. Continuity in home health care: Is consistency in nursing personnel associated with better patient outcomes? *J. Healthc. Qual.* **2011**, *33*, 33–39. [CrossRef] [PubMed]
42. Bodenheimer, T.; Wagner, E.H.; Grumbach, K. Improving primary care for patients with chronic illness. *J. Am. Med. Asso.* **2002**, *288*, 1775–1779. [CrossRef] [PubMed]
43. Tebeje, T.H.; Klein, J. Applications of e-Health to support person-centered health care at the time of COVID-19 pandemic. *Telemed. J. E Health* **2021**, *27*, 150–158. [CrossRef] [PubMed]
44. Chudasama, Y.V.; Gillies, C.L.; Zaccardi, F.; Coles, B.; Davies, M.J.; Seidu, S.; Khunti, K. Impact of COVID-19 on routine care for chronic diseases: A global survey of views from healthcare professionals. *Diabetes Metab. Syndr. Clin. Res. Rev.* **2020**, *14*, 965–967. [CrossRef]
45. Kendzerska, T.; Zhu, D.T.; Gershon, A.S.; Edwards, J.D.; Peixoto, C.; Robillard, R.; Kendall, C.E. The Effects of the health system response to the COVID-19 pandemic on chronic disease management: A narrative review. *Risk Manag. Healthc. Policy* **2021**, *14*, 575–584. [CrossRef]
46. Lancaster, K.; Abuzour, A.; Khaira, M.; Mathers, A.; Chan, A.; Bui, V.; Lok, A.; Thabane, L.; Dolovich, L. The use and effects of electronic health tools for patient self-monitoring and reporting of outcomes following medication use: Systematic Review. *J. Med. Internet Res.* **2018**, *20*, e294. [CrossRef]
47. Clari, M.; Matarese, M.; Ivziku, D.; De Marinis, M.G. Self-Care of People with Chronic Obstructive Pulmonary Disease: A Meta-Synthesis. *Patient* **2017**, *10*, 407–427. [CrossRef]
48. Tossaint-Schoenmakers, R.; Versluis, A.; Chavannes, N.; Talboom-Kamp, E.; Kasteleyn, M. The Challenge of Integrating eHealth Into Health Care: Systematic Literature Review of the Donabedian Model of Structure, Process, and Outcome. *J. Med. Internet Res.* **2021**, *23*, e27180. [CrossRef]
49. Piotrowicz, E. The management of patients with chronic heart failure: The growing role of e-Health. *Expert Rev. Med. Devices* **2017**, *14*, 271–277. [CrossRef]
50. Seo, H.J.; Kim, S.Y.; Sheen, S.S.; Cha, Y. e-Health Interventions for Community-Dwelling Type 2 Diabetes: A Scoping Review. *Telemed. J. E Health* **2021**, *27*, 276–285. [CrossRef]
51. Kennedy, A.; Rogers, A.; Bower, P. Support for self-care for patients with chronic disease. *BMJ* **2007**, *335*, 968–970. [CrossRef] [PubMed]
52. Lin, M.H.; Yuan, W.L.; Huang, T.C.; Zhang, H.F.; Mai, J.T.; Wang, J.F. Clinical effectiveness of telemedicine for chronic heart failure: A systematic review and meta-analysis. *J. Investig. Med.* **2017**, *65*, 899–911. [CrossRef] [PubMed]

53. Jha, R.; Karnes, A.; Oates, P.; Wojciechowski, K.F.; Dvorak, A.; Grindle, A.; Feitell, S. Innovative ehealth at home program reduces hospitalization and readmission rates of patients with heart failure. *J. Card. Fail.* **2019**, *25*, S127.
54. Timpel, P.; Oswald, S.; Schwarz, P.E.H.; Harst, L. Mapping the evidence on the effectiveness of telemedicine interventions in diabetes, dyslipidemia, and hypertension: An umbrella review of systematic reviews and meta-analyses. *J. Med. Internet Res.* **2020**, *22*, e16791. [CrossRef]
55. Ro Hamine, S.; Gerth-Guyette, E.; Faulx, D.; Green, B.B.; Ginsburg, A.S. Impact of mHealth chronic disease management on treatment adherence and patient outcomes: A systematic review. *J. Med. Internet Res.* **2015**, *17*, e52. [CrossRef]
56. Vandelanotte, C.; Spathonis, K.M.; Eakin, E.G.; Owen, N. Website-delivered physical activity interventions a review of the literature. *Am. J. Prev. Med.* **2007**, *33*, 54–64. [CrossRef]
57. Linn, A.J.; Vervloet, M.; van Dijk, L.; Smit, E.G.; Van Weert, J.C. Effects of eHealth interventions on medication adherence: A systematic review of the literature. *J. Med. Internet Res.* **2011**, *13*, e103. [CrossRef]
58. Pouls, B.P.H.; Vriezekolk, J.E.; Bekker, C.L.; Linn, A.J.; van Onzenoort, H.A.W.; Vervloet, M.; van Dulmen, S.; van den Bemt, B.J.F. Effect of interactive eHealth interventions on improving medication adherence in adults with long-term medication: Systematic Review. *J. Med. Internet Res.* **2021**, *23*, e18901. [CrossRef]
59. Gorina, M.; Limonero, J.T.; Álvarez, M. Effectiveness of primary healthcare educational interventions undertaken by nurses to improve chronic disease management in patients with diabetes mellitus, hypertension and hypercholesterolemia: A systematic review. *Int. J. Nurs. Stud.* **2018**, *86*, 139–150. [CrossRef]
60. Silva-Cardoso, J.; Juanatey, J.R.G.; Comin-Colet, J.; Sousa, J.M.; Cavalheiro, A.; Moreira, E. The Future of Telemedicine in the Management of Heart Failure Patients. *Card. Fail. Rev.* **2021**, *7*, e11. [CrossRef]
61. Jiménez-Marrero, S.; Yun, S.; Cainzos-Achirica, M.; Enjuanes, C.; Garay, A.; Farre, N.; Verdú, J.M.; Linas, A.; Ruiz, P.; Hidalgo, E.; et al. Impact of telemedicine on the clinical outcomes and healthcare costs of patients with chronic heart failure and mid-range or preserved ejection fraction managed in a multidisciplinary chronic heart failure programme: A sub-analysis of the iCOR randomized trial. *J. Telemed. Telecare* **2020**, *26*, 64–72. [PubMed]
62. Massimi, A.; De Vito, C.; Brufola, I.; Marzuillo, C.; Migliara, G.; Rega, M.L.; Ricciardi, W.; Villari, P.; Damiani, G. Are community-based nurse-led self-management support interventions effective in chronic patients? Results of a systematic review and meta-analysis. *PLoS ONE* **2017**, *12*, e0173617. [CrossRef] [PubMed]
63. Fildes, K.; Stefoska-Needham, A.; Atkinson, J.; Lambert, K.; Lee, A.; Pugh, D.; Smyth, M.; Turner, R.; Wallace, S.; Nealon, J. Optimising health care for people living with chronic kidney disease: Health-professional perspectives. *J. Ren. Care* **2022**, *48*, 168–176. [CrossRef]
64. Caro-Bautista, J.; Martín-Santos, F.J.; Morales-Asencio, J.M. Systematic review of the psychometric properties and theoretical grounding of instruments evaluating self-care in people with type 2 diabetes mellitus. *J. Adv. Nurs.* **2014**, *70*, 1209–1227. [CrossRef] [PubMed]
65. Smith, B.; Magnani, J.W. New technologies, new disparities: The intersection of electronic health and digital health literacy. *Int. J. Cardiol.* **2019**, *292*, 280–282. [CrossRef]
66. Baccolini, V.; Rosso, A.; Di Paolo, C.; Isonne, C.; Salerno, C.; Migliara, G.; Prencipe, G.P.; Massimi, A.; Marzuillo, C.; De Vito, C.; et al. What is the Prevalence of Low Health Literacy in European Union Member States? A Systematic Review and Meta-analysis. *J. Gen. Intern. Med.* **2021**, *36*, 753–761. [CrossRef]

life

MDPI

Article

Physical Activity Monitoring and Classification Using Machine Learning Techniques

Saeed Ali Alsareii [1,*], Muhammad Awais [2,*], Abdulrahman Manaa Alamri [1], Mansour Yousef AlAsmari [1], Muhammad Irfan [3], Nauman Aslam [4] and Mohsin Raza [2]

1 Department of Surgery, College of Medicine, Najran University Saudi Arabia, Najran 61441, Saudi Arabia; manaa_880@hotmail.com (A.M.A.); dr.aboyousef@hotmail.com (M.Y.A.)
2 Department of Computer Science, Edge Hill University, St Helens Rd, Ormskirk L39 4QP, UK; razam@edgehill.ac.uk
3 Electrical Engineering Department, College of Engineering, Najran University Saudi Arabia, Najran 61441, Saudi Arabia; miditta@nu.edu.sa
4 Department of Computer and Information Sciences, Northumbria University, Newcastle upon Tyne NE1 8ST, UK; nauman.aslam@northumbria.ac.uk
* Correspondence: alsareii@nu.edu.sa (S.A.A.); awaism@edgehill.ac.uk (M.A.)

Abstract: Physical activity plays an important role in controlling obesity and maintaining healthy living. It becomes increasingly important during a pandemic due to restrictions on outdoor activities. Tracking physical activities using miniature wearable sensors and state-of-the-art machine learning techniques can encourage healthy living and control obesity. This work focuses on introducing novel techniques to identify and log physical activities using machine learning techniques and wearable sensors. Physical activities performed in daily life are often unstructured and unplanned, and one activity or set of activities (sitting, standing) might be more frequent than others (walking, stairs up, stairs down). None of the existing activities classification systems have explored the impact of such class imbalance on the performance of machine learning classifiers. Therefore, the main aim of the study is to investigate the impact of class imbalance on the performance of machine learning classifiers and also to observe which classifier or set of classifiers is more sensitive to class imbalance than others. The study utilizes motion sensors' data of 30 participants, recorded while performing a variety of daily life activities. Different training splits are used to introduce class imbalance which reveals the performance of the selected state-of-the-art algorithms with various degrees of imbalance. The findings suggest that the class imbalance plays a significant role in the performance of the system, and the underrepresentation of physical activity during the training stage significantly impacts the performance of machine learning classifiers.

Keywords: digital health; e-health; pandemic; physical activity; machine learning; performance evaluation

Citation: Alsareii, S.A.; Awais, M.; Alamri, A.M.; AlAsmari, M.Y.; Irfan, M.; Aslam, N.; Raza, M. Physical Activity Monitoring and Classification Using Machine Learning Techniques. *Life* **2022**, *12*, 1103. https://doi.org/10.3390/life12081103

Academic Editor: Yudong Cai

Received: 16 June 2022
Accepted: 18 July 2022
Published: 22 July 2022

1. Introduction

Regular physical activity plays a vital role in improving the health of individuals, whether it is a child under 5 or an elderly above 65. Physical activity has well-documented health benefits and can extensively improve the health and well-being of individuals and reduce the risks from noncommunicable diseases. Both moderate- and vigorous-intensity physical activity improve health. Physical inactivity increases the risk of noncommunicable disease mortality and puts inactive people at a 20–30% higher risk of death in comparison to physically active people [1]. Physical inactivity is among the leading factors which cause mortality and is estimated to contribute to 6% of worldwide deaths [2]. Therefore, World Health Organization (WHO) also recommends people of all ages indulge in physical activity and recommends the duration and intensity of physical activity for different age groups [1]. It has been noted that physical activity improves muscular and cardiorespiratory fitness, bone health and mental fitness while reducing the risk of heart diseases, diabetes, hypertension, obesity and fractures [1].

Physical activity and the promotion of healthy living can significantly lower the risks of non-communicable diseases. It also serves as the best remedy for obesity [3]. Obesity is one of the major chronic illnesses and increases the risk of developing many serious comorbidities, such as hypertension, sleep apnea, type 2 diabetes, depression, etc. [3]. Furthermore, obesity is becoming an increasingly prevalent issue. Obesity has become a global epidemic, with global stats suggesting nearly one-third of the world population is obese or overweight. Obesity has also added a significant burden to healthcare services, with nearly 10% of the medical costs in the US being spent on obesity-related issues. It also has been among the major causes of death in the US. Similarly, in Saudi Arabia, with 36% of the population being obese and 69% being categorized as overweight, nearly 20,000 lives are claimed to obesity each year. Therefore, under such circumstances, the provision of physical activity (PA) as a measure to control obesity has become increasingly important [4].

Obesity is one of the prevailing problems responsible for several health issues and medical conditions. Weight loss surgery, also referred to as bariatric or metabolic surgery, is one of the possible solutions for extremely overweight people. While the surgery can result in significant weight losses, it is still not termed a cure for obesity. Obesity is not a matter of concern only for the younger and older adults as it has become very common in children as well [5,6]. Therefore, suitable lifestyle changes should be introduced to avoid regaining weight. Patients who have undergone weight loss surgery need a balanced diet along with regular exercise once they have recovered from surgery. They also need to maintain a regular appointment schedule to keep everything in check. It is therefore important that a technology-driven framework for long-term support is developed to assist these patients in prolonging their healthy living choices and balancing exercise and diet accordingly. With the emergence of digital technologies, information and communications technology (ICT) solutions, machine intelligence and system analytics, post-surgery and long-term support can be efficiently managed with technology-driven solutions. This work primarily focuses on devising effective solutions for monitoring the physical activity levels of the patients in the post-surgery phase to maintain healthy living and discourage weight gain.

The increasing stress on the healthcare systems and the need to promote healthy living urge new measures to promote physical activities. The initial step in encouraging the physical activity is the ability to be able to quantify the physical activity into individual components of tangible impact. As such, physical activity classification can serve as a foundation by recording and transforming physical activities of an individual to give accurate quantification of a daily routine, thus encouraging active and healthy living. This highlights a clear need to develop feasible solutions to monitor the activities of daily living (ADLs) as a measure to avoid/overcome obesity.

Physical activities and exercise both serve as necessary measures for healthy living and maintaining healthy weights. Exercise is the subbranch of physical activity, and it is more structured, repetitive and planned with an intention to maintain or improve body fitness [7]. The promotion of physical activities is towards establishing and maintaining healthy living habits, such as walking to work, using stairs instead of lifts, use of muscles instead of motorized tools, etc. While promoting physical activities offer a more sustainable solution for staying active, it still needs to be quantized to give a better estimation of the efforts put in by the individuals and how these have impacted their healthy living. Quantifying the physical activities performed offers a means to relay the impact to the individuals as well as the medical staff to better evaluate the active status and suggest/intervene accordingly.

The physical activities are logged in several ways where questionnaires and direct observations are conventionally used. The logging of activities requires information on the type of activity performed, the duration for which it was performed and the intensity of the activity. An example could be walking, where the information about how much time is spent walking in a day/week, walking pace, etc. However, these are not as accurate and add additional time commitments from the observee and observer. Therefore, novel techniques are needed to use technology-driven solutions to log the type of activity performed, its duration and intensity.

The recent developments in the miniaturization of inertial sensors equipped with state-of-the-art processing and communication capabilities lay the foundations for the smart health and activity monitoring using machine learning techniques [8,9]. Wearable inertial measurement units (IMUs) use accelerometers and gyroscopes to measure acceleration and angular velocities to offer unobtrusive, reliable, and low-cost measurement of sensory data for physical activity classification. Single or multiple wearable IMUs can be placed on various body locations to classify daily life activities [10].

These small battery-operated wearable IMUs not only offer ease of use but are also equipped with transceivers to accumulate the vitals and activity data of patients to fog/cloud. The data accumulated at cloud or fog devices can be further processed using machine learning techniques [11,12] to identify the activity performed, its duration and intensity. While some existing works offer activity classification, however, there is still much room for improvement.

In [13], the authors proposed a solution for activity classification to identify strange behavior using support vector machines (SVM); however, it used surveillance videos instead of wearable sensors, and the focus of the work was security. Another similar study was carried out in [14], where abnormal behavior of a person was identified using pose estimation. Both these techniques, while detecting physical attributes, are still much further from the objectives of this work and use visual sensors/cameras instead of wearable devices.

In [15], the authors use the asymmetric 3D Convolutional Neural Networks for action recognition. The work was tested on the UCF-101 dataset, which combines actions from YouTube videos. While the claimed results were promising, the work was more tilted towards the general-purpose activity classification and use of visual sensing. Another work presented in [16] provides a unified framework for exploring multidimensional features in conjunction with body part models for pose estimation. A maximum entropy Markov model was used as a recognition engine which was claimed to have accurately detected body parts and recognized physical activity performed.

In [17], the authors used multimodal feature-level fusion for activity recognition. K-nearest neighbor and SVM were used for the classification of activities. As an input to the classification system, RGB camera, depth and inertial sensors data were used. While diversity was exploited, the camera usually conflicts with personal and security preferences and offers a limited field of view. Similarly, the depth sensor can also work only in a constrained field of view, which limits the scope of the work. In addition, the study was not focused on activities inspiring healthy living and controlling obesity.

In [18], the authors examined the relationship between physical activity and weight status. The performance of several machine learning techniques was evaluated on a largescale dataset. The objective of the study was to link physical activity with obesity. However, no sensory data were used to classify or log physical activities.

The existing literature and research studies use diverse techniques for activity classification with a wide scope of applications [18–21]. These applications range from security, autonomous transportation, expression evaluation, healthcare, etc. A relatively wide variety of sensors are also used, with some less suitable for the proposed work. While there are a variety of studies focusing on activity classification in healthcare using wearable IMUs [11,12,22–25], these focus on well-balanced data where all the physical activities performed are of equal samples. However, it is important to mention that in real life setting, physical activities (e.g., sitting, standing, walking, lying, stairs up, stairs down, etc.) are unstructured. Therefore, the natural occurrence and frequency of each activity cannot be controlled. This can lead to an imbalanced set of activities where certain classes of activities have more samples, data instances and sensory data than others [10,26]. Joana et al. [27] also found that underrepresented physical activities can affect the performance of the machine learning classifiers due to the availability of limited data at the training stage of the classifier. Therefore, it is important to not only study the impact of class imbalance on the performance of machine learning classifier when classifying physical activities but also investigate how such machine learning classifiers behave when more than one class of physical activities are imbalanced at their training stage. To the best of our knowledge, none of the existing studies have investigated the effect of multi class imbalance induced at the classifier training stage and its impact on the performance of physical activity classification. Moreover, the study also investigated a variety of machine learning classifiers to observe, which are more sensitive to class imbalance than others considering the overall performance of the physical activity classification system. Therefore, the work presented in this paper offers a unique contribution to evaluating physical activity to support healthcare professionals and medical staff in making correct interventions, maintaining diet and mandatory active living style for overweight and obese patients.

The main contributions of the paper are:

1. The paper compares several machine learning techniques to identify the best-suited activity classification techniques on a balanced dataset.
2. The physical activity dataset is intentionally skewed to introduce class imbalance and to evaluate the abilities of six well-known machine learning classifiers.
3. The proposed work compares the performance of the selected state-of-the-art machine learning algorithms with different training splits and various degrees of imbalance and identifies the best-suited machine learning techniques.

The rest of the paper is organized as follows: Section 2 presents the proposed system model including a data communication and activity classification framework, dataset used, feature computation, experimentation and implementation of machine learning algorithms. Research and discussion are covered in Sections 3 and 4, respectively, whereas the concluding remarks and future directives are presented in Section 5.

2. System Model

In this work, a data communications and activity classification framework is presented, as shown in Figure 1. The proposed data communication and machine-learning-based activity classification framework lays out a communication infrastructure capable of sampling and relaying patients' sensory data (vitals + activity-related data such as accelerometer readings etc.) over the internet to accumulate data from virtually infinite number of patients. It also proposes a machine-learning-based activity classification framework to process the accumulated patients' data on the cloud and translate the sensory data into physical activities, thus maintaining the exercise/activity logs for each patient.

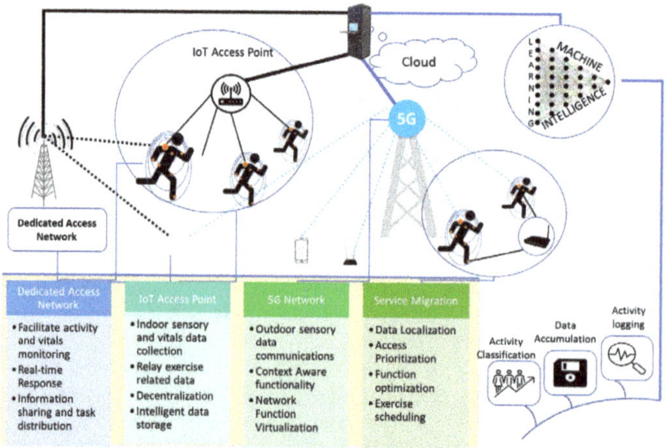

Figure 1. Post-surgery patient's sensory data communications and activity classification framework.

2.1. *Wireless Communications Framework*

The wireless communications framework intends to collect the data from different sensory elements mounted on the patient's body. In addition, these data need to be relayed to the access points to be transferred over longer distances. Considering the applicability of the proposed framework in medical facilities, private indoor environments as well as public outdoor places, a multi-layer hybrid network is recommended. A modular approach must be followed to define a hybrid network to enable flexibility and scalability in network sizes. The two main operational blocks in the hybrid network framework consist of the body network and the Internet of Things (IoT) framework.

The sensory data collected from the potential multisensory agents on the body are communicated to the body communications hub (BCH) using the body network. A graphical representation of the body network is presented in Figure 2.

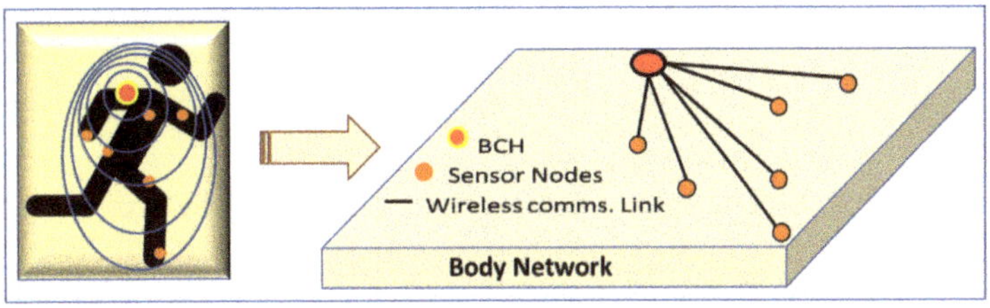

Figure 2. A representation of body network to collect sensory data from body-mounted sensors.

2.2. *Machine Learning Paradigm for Physical Activity Classification*

2.2.1. Dataset

The study utilizes a publicly available dataset [28] to detect the activities of daily living (ADLs). The dataset is collected using a Galaxy S II smartphone by using its triaxial (3D) gyroscope and triaxial (3D) accelerometer sensors. The smartphone phone was mounted at a waist level to carry on data collection. Thirty subjects participated in the data collection experiment, aged from 19 to 48 years, and performed a variety of ADLs. The ADLs performed were lying, sitting, standing, walking upstairs, walking downstairs and normal walking. Data collection was conducted in a laboratory environment, and the participants were instructed to perform ADLs as naturally as possible. The sampling frequency of the accelerometer and gyroscope sensors was set to 50 Hz. The ground truth of the ADLs performed was maintained through visual observation.

2.2.2. Feature Computation

Raw 3D accelerometer and gyroscope sensory signals obtained from the smartphone underwent several preprocessing and feature extraction steps to derive features that are fed to a machine learning algorithm later on to profile and classify ADLs.

The preprocessing steps involved are (i) low pass filtering using butter worth 3rd order filter at cutoff 20 Hz and (ii) median filtering. The acceleration signal is then divided into body acceleration and gravitational acceleration signals. To achieve this, a frequency of 0.3 Hz is used to separate the gravitational signals ($<$0.3 Hz) from the body acceleration signal ($>$0.3 Hz). Furthermore, jerk signals are derived from the acceleration signal and gyroscope signal by taking their derivates. The cadence is also derived from these signals. To analyze the frequency components, fast Fourier transform (FFT) is also computed to detect the trends and variations occurring in the frequency domain when different ADLs are performed. These derivations resulted in a total of 17 signals, including the original 3D gyroscope and 3D accelerometer signals. Further details about the feature extraction process can be found in [28].

The originals signals (3D accelerometer and 3D gyroscope) and the aforementioned derived signals are further processed using the windowing method to extract more features. The window length is set to 2.56 sec (128 samples of data) with 50% overlap (64 samples).

Several statistical, time and frequency domains features are obtained from these derived signals, and each time, window of 128 samples is as follows: mean, standard deviation, median, signal magnitude area, maximum value, minimum value, angle between two signals, frequency domain band energy, skewness, kurtosis, average frequency component, maximum frequency component, correlation coefficient between signals, autoregressive correlation coefficients, interquartile range, sum of squares (energy) and band energy [28].

2.2.3. Experiments

Several experiments have been conducted using a different split of the training and testing dataset to train the machine learning classifiers and to observe the performance of different machine learning classifiers in overbalanced and imbalanced datasets. Class imbalance is a critical issue in machine learning, and this often occurs when one or few classes are underrepresented (having fewer samples) than other classes. This often creates biases during the training stage of the machine learning algorithm, and the performance of underrepresented class or classes is highly affected by these imbalanced distributions.

Therefore, in this study, we also investigated the impact of imbalanced classes on the performance of the different machine learning classifiers by conducting different experiments. In addition, we also investigated which machine learning algorithms are relatively less sensitive to class imbalance or performed better than others.

The class distribution used in the first experiment or experiment 1 (E1) is presented in Table 1. The class distribution shown in Table 1 is the original class distribution obtained after the actual data collection. Each instance (number or sample) in Table 1 represents the number of time windows obtained for that particular class. Column 1 represents the activity type (walk, sit, stand, etc.), column 2 represents the total number of data instances obtained originally, column 3 represents the proportion of each activity class with respect to the total dataset, column 5 represents the train split or the number of time instances used to train the machine learning model and the last column represents the test split or the number instance used to test the performance of the machine learning algorithms. This original distribution or the balanced distribution of the ADLs in the train/test split is named experiment 1 (E1).

Table 1. Class distribution of different ADLs in experiment 1 (E1).

Activity Type	Total Dataset	Percentage (Total Dataset)	Train Split	Test Split
Walk	1722	16.72%	1226	496
Upstairs	1544	14.99%	1073	471
Downstairs	1406	13.65%	986	420
Sit	1777	17.25%	1286	491
Stand	1906	18.51%	1374	532
Lie	1944	18.88%	1407	537

After designing experiment 1, six further experiments are conducted by inducing class imbalance in each class to observe the performance of the machine learning classifier in classifying different imbalanced ADLs. Table 2 represents the further six experiments conducted (from E2 to E7) in addition to E1 (please see Table 1).

Table 2. Class distribution of different ADLs in training samples during experiments 1–7 (E1, E2, E3, E4, E5, E6 and E7).

Activity Type	E1	E2	E3	E4	E5	E6	E7
	Train Split	Train Split	Train Split	Train Split	Train Split	Train Split	Train Split
Walk	1226	100	100	100	100	100	100
Upstairs	1073	1073	100	100	100	100	100
Downstairs	986	986	986	100	100	100	100
Sit	1286	1286	1286	1286	100	100	100
Stand	1374	1374	1374	1374	1374	100	100
Lie	1407	1407	1407	1407	1407	1407	100

It is important to note that training samples of the underrepresented classes are different in each of these experiments (from E1 to E7), while the test samples remain the same as per the original distribution presented in Table 1. This is due to the fact that the class imbalance added in the training samples may affect the performance of the machine learning classifier and the testing samples have no influence on the trained machine learning model.

2.2.4. Machine Learning Algorithms Used

We implemented several machine learning algorithms on the dataset generated from 7 experiments and observed the performance of the different machine learning in classifying the ADLs in a balanced class distribution scenario (E1) and imbalanced class distribution scenarios (from E2 to E7).

The classifiers used in this study are support vector machine (SVM), Gradient boosting (GB) classifier, Extreme Gradient boosting (XGB) classifier, catboost (CB) classifier, AdaBoost classifier using decision tree (ADA-DT) and AdaBoost classifier using random forest (ADA-RF) [10,29–33]. The choice of classifiers is influenced by the fact that some of these are preferred due to their ensemble properties of combining the weak learners and improving the performance by collective or majority learning, while others, such as SVM, are widely used due to their hyper plane properties to create significant margin, thus achieving high performance [34,35].

All the simulations are performed in Python using its associated libraries. The parameters used to train these classifiers are as follows. The XGB parameters are maximum depth = 50, minimum child weight = 2, number of estimators = 100 and learning rate = 0.16. The GB parameters are objective function = multiclass, maximum depth= 50, learning rate = 0.1 and number of estimators = 100. The CB parameters are learning rate = 0.15, depth = 10 and loss function = Multi Class. The SVM parameters are kernel = linear, class weight = balanced and complexity = 1. The ADA(DT) parameters are Tree = Decision Tree Classifier with maximum depth = 10 and number of estimators = 100. The ADA(RF) parameters are Tree = Random Forest Classifier with number of estimators = 100, maximum features = auto and number of estimators = 100. Macro averaged F-score is used as a performance metric to compute the performance of the different classifiers in classifying the ADLs of daily living. The expression to calculate the F-score is expressed in Equation (1).

$$F - score = \frac{2 * TP_C}{2 * TP_C + FP_C + FN_C} * 100 \quad (1)$$

where TP_C represents true positive, FP_C represents false positive, FN_C represents false negative and the subscript c represents the class it is computed for, such as sit, stand, waling, lie, etc.

3. Results

The performance achieved using various machine learning classifiers for seven experiments (E1–E7) are presented in Figures 3–9, and the respective performance by classes are presented in Tables A1–A7 in Appendix A.

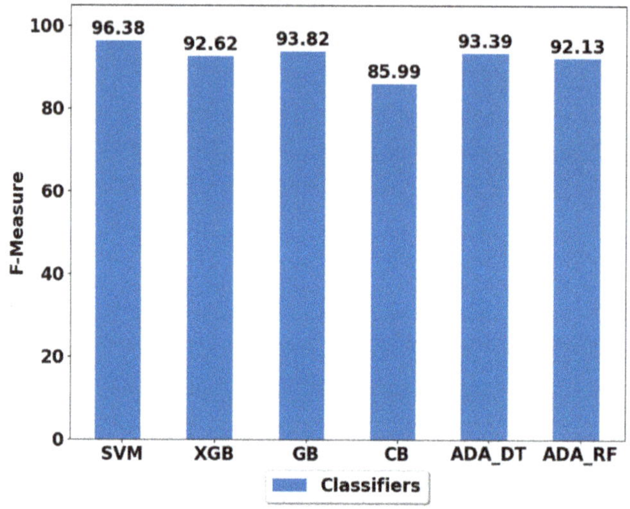

Figure 3. Performance analysis of classifiers using the train/test split in experiment 1 (E1).

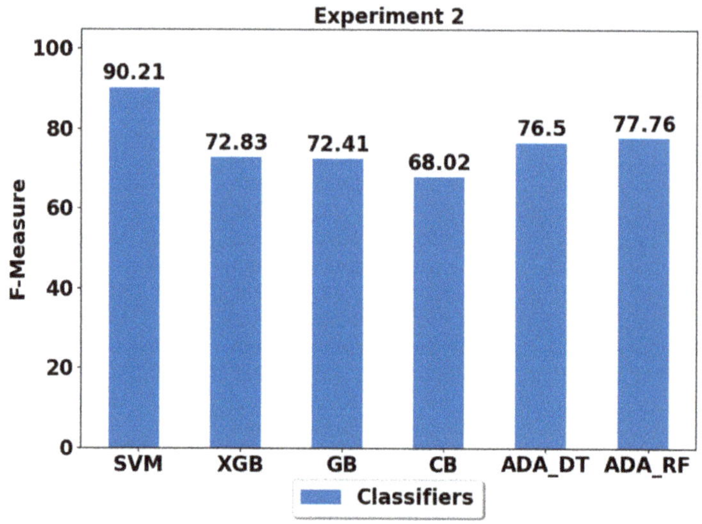

Figure 4. Performance analysis of classifiers using the train/test split in experiment 2 (E2).

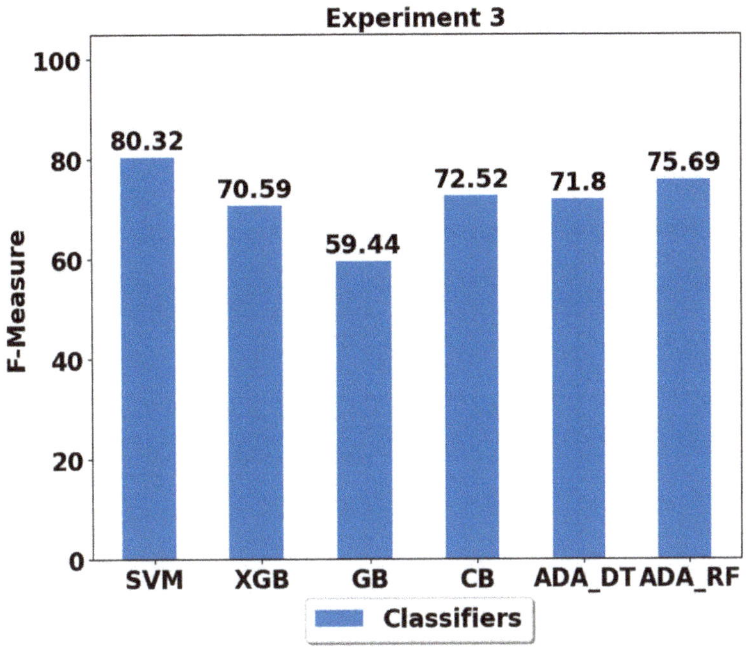

Figure 5. Performance analysis of classifiers using the train/test split in experiment 3 (E3).

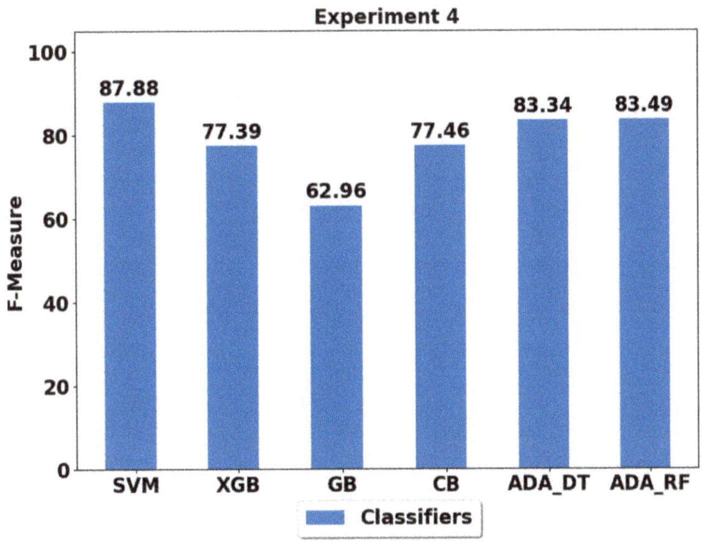

Figure 6. Performance analysis of classifiers using the train/test split in experiment 4 (E4).

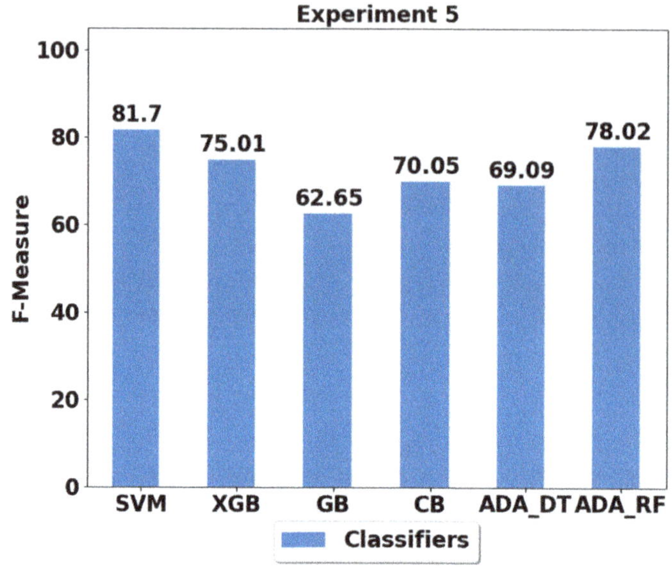

Figure 7. Performance analysis of classifiers using the train/test split in experiment 5 (E5).

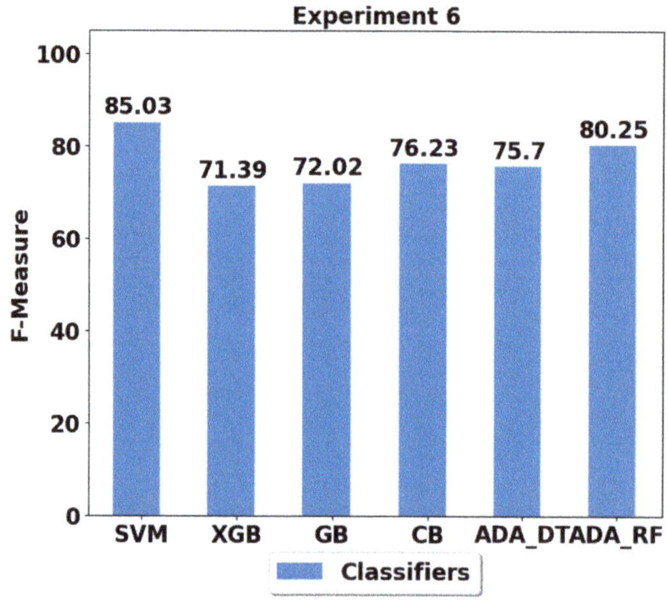

Figure 8. Performance analysis of classifiers using the train/test split in experiment 6 (E6).

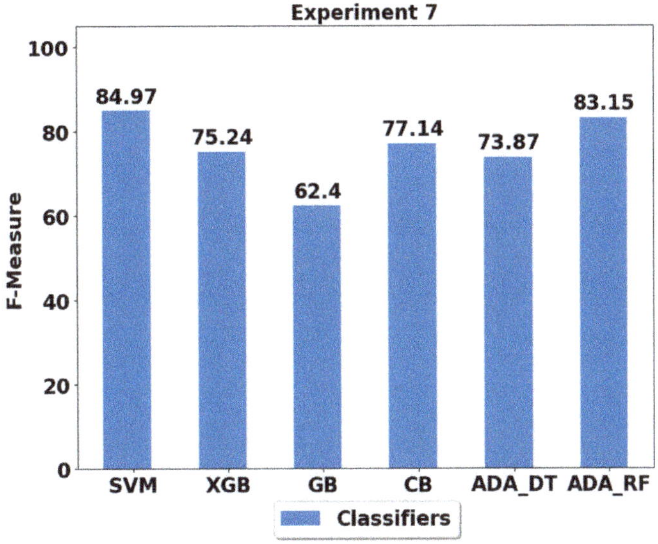

Figure 9. Performance analysis of classifiers using the train/test split in experiment 7 (E7).

It is fairly evident from Figure 3 that all the classifiers have achieved the performance above 85% in classifying the ADLs (sitting, standing, walking, lying, walking upstairs and walking downstairs). These findings show the strength of the proposed machine-learning-based activity classification methods to classify ADLs. The best performer among all classifiers appeared to be SVM which achieved a performance of 96.38% (please see Figure 3). The SVM also outperformed the activity classification method proposed by Anguita et al. [28], thus confirming performance improvement when compared to the existing works. The second-best performer is GB, with the performance of 93.82%. All other classifiers also performed considerably well except the CB, whose performance is worst among all (85.99%). The detailed performance by class is visualized in Table A1. It is evident from Table A1 that most low-performing classifiers largely struggled in distinguishing between sitting and standing activities and struggled to distinguish between upstairs and downstairs walking. This could be due to the fact the smartphone is waist mounted during the data collection and the standing and sitting postures with respect to the accelerometer and gyroscope signals are relatively similar considering the smartphone orientation. The same is the case during upstairs and downstairs activities, which could make it hard for the classifiers to distinguish between different postures and locomotive activities. However, the SVM performed well in this scenario, and this could be due to the fact that SVM use high margin and hyperplanes to distinguish between different classes during the training stage, which assisted in better distinguishing these ADLs (sit vs. stand, upstairs vs. downstairs).

In experiment 2 (E2), only the walking class is imbalanced during the training stage with a total of 100 samples, while number of samples of all other classes remained the same as per the original or balanced distribution (please see Table 2). As expected and evident from Figure 4 and Table A2, most classifiers struggled in classifying the walking activity due to its low representation in the training stage. This suggests that class imbalance has serious consequences on the overall performance of the activity classification system and on the performance of the underrepresented/imbalanced class (es). Our findings suggest that the best performance of 90.21% is obtained using the SVM classifier, while none of the other classifiers are able to achieve the performance above 80%. This is an interesting finding and suggests the strength of SVM in handling class imbalance. The SVM inherently possess the properties of adaptive weighting, which provides more weight to the imbalanced classes and fewer weights to the over represented or balanced classes [10], thus improving classification performance.

In experiment 3 (E3), walking and walking upstairs are imbalanced classes at the training stage with a total of 100 samples each. The results in Figure 5 and Table A3 suggest that the best performer is still SVM in classifying the different ADLs with performance of around 80%, and the second best is ADA (RF) with performance of 75.69%.

In experiment 4 (E4), the underrepresented classes or imbalanced classes are walking, walking upstairs and walking downstairs. The best performance of 87.88% is achieved by the SVM classifier and the second best candidates are ADA (DT) and ADA (RF), with performance of above 83%, as shown in Figure 6 and Table A4. The worst performer is GB, with an F-score of 62.96%. It is important to note that the performance of all the classifiers is generally improved in E4 as compared to in E3. This could be due to the fact that only walking and walking upstairs are imbalanced in E3, while in E4, walking, walking upstairs and walking downstairs are imbalanced. This suggests that in E3, class sample mismatch between walking upstairs and walking downstairs could have more biased induced in training classifiers due to class imbalance in only walking upstairs class and not in the walking downstairs class. However, this has been reduced in E4 since both walking upstairs and walking downstairs are imbalanced with equal proportion when compared to other classes. Thus, giving equal opportunities to most of the classifiers to train properly.

In experiment 4 (E4), walking, walking upstairs, walking downstairs and sitting are underrepresented and imbalanced as compared to other majority classes. The results shown in Figure 7 and Table A5 suggest that the SVM again outperformed all the other classifiers with an F-score of 81.7%, and ADA (RF) is the second-best classifier with an F-score of 78.02%.

During experiment 6 (E6), all classes are underrepresented except the majority class represented class, which is lying. The performance analysis of the classifiers using the E6 train/test split is shown in Figure 8 and Table A6. All the classifiers are able to achieve the performance of above 70%. Similar to previous experiments' results, SVM outperformed all the classifiers with performance of 85.03%, and the second-best performance was obtained by the ADA (RF) classifier.

In experiment 7 (E7), all the classes are balanced with equal samples; however, the samples are very low (100, please see Table 2) when compared to the original samples (around 500 for each class, please see Table 2) in E1. Lower number of training samples can influence the performance of the machine learning classifiers since supervised machine learning is all about feeding sufficient data to the classifiers. Therefore, fewer samples mean fewer training opportunities for the classifier to estimate and quantify the underlying trends from the data. The performances of the different classifiers using the E7 dataset are depicted in Figure 9 and Table A7. The SVM and ADA (RF) classifiers performed well with the performance of 84.97% and 83.15%, respectively, while the lowest performance of 62.4% was achieved by the GB classifier.

4. Discussion

The findings of the study are rather interesting and suggest the effect of class imbalance on system performance and how different classifiers behave when training classes are highly imbalanced. The SVM proved itself to be the best performance among all classifiers, and the second-best classifier is the ADA (RF) classifier. The possible rationale behind the high performance of the SVM in all the experiments could be due to the fact that it uses an adaptive weighting approach at the training stage [10]. This adaptive weighting reduced the bias induced at the training stage due to the class imbalance and underrepresentation and penalized the majority of classes with weighted samples. Moreover, the ADA (RF) uses a more sophisticated random-forest-based method to train, which could have been able to handle class imbalance to some extent.

The analysis of the class imbalanced datasets also suggested that most of the machine learning classifiers investigated in this work are sensitive to class balance except SVM, which is less sensitive to class imbalance due to its inherited property of adaptive weighting at the training stage to compensate for class imbalance up to some extent. The direction that can be opted in future works is to investigate the methods that can deal with class imbalance by performing a variety of data-handling techniques. These methods include synthetic minority over-sampling technique (SMOTE) [36], adaptive synthetic sampling technique (ADASYN) [37], under sampling and over sampling [38]. Therefore, such methods should be implemented on physical activity classification dataset collected in real life conditions with more natural settings. It is also worth mentioning that treating class imbalance can be harmful in some scenarios, as reported by Goorbergh et al. [39]. This is because treating class imbalance also depends on the type of classifier implemented, application domain and type of class imbalance dataset, as highlighted in [40].

Nevertheless, it is worth mentioning that the findings of the study are very encouraging and suggest that the proposed methods can obtain very high performance of above 96% in classifying the activities of daily living (sitting, standing, walking, lying, walking upstairs and waking downstairs). This provides the strength of the proposed physical activity classification system and its applicability in real life conditions. Promoting quality of life and tracking daily life activities are strongly correlated with obesity since active life patterns discourage sedentary behaviors and reduce the onset of several diseases (hypertension, diabetes, cardiovascular diseases), including obesity. Profiling such ADLs for a relatively longer duration (weeks, months, years, etc.) not only provides a detailed insight to individuals but also provides a detailed overview of the activity behaviors to the healthcare care practitioners, who can then tailor and customize the treatment to those suffering from obesity and other severe conditions.

The proposed physical activity classification system is applicable to a variety of different application scenarios in daily life conditions. Since the dataset used in this study utilizes the in-built motion sensors (accelerometer and gyroscope) of smartphones, there is no need for a separate sensing unit or equipment to acquire the activity patterns and retrieve sensory data. This sensory dataset acquired through smartphone can benefit from the on-device processing unit to compute the task requiring low computational power. Further processing can benefit from the scenario presented in Figure 1, where IoT assessment points can transmit the data to the cloud and storage units, where more sophisticated machine learning models can be implemented to classify the activity patterns. These activity patterns can then be profiled (e.g., 2% running, 10% walking, 20% sitting, 25% lying, 10% standing, 33% other sedentary or active activity over the day) and provide the distribution of activities performed by any individual over the course of a day, week, months and even years. This will not only benefit the general population to adopt a healthier lifestyle and well-being but also tracks the individuals with health issues such as obesity. The profiling of obese individuals with health disorders can then be linked with the healthcare services via IoT to track the activity patterns of individuals and to develop be-spoke exercise and therapy plans to effectively reduce obesity and to become healthy and active members of society. As the proposed system only used the smartphone for data gathering, its applicability in large-scale studies would not require resource-intensive equipment to track activity patterns. Moreover, such large-scale studies should be practiced in the future to develop big datasets in real life conditions and to train data-intensive deep learning classifiers for the efficient classification of daily life activities.

While the proposed research offers great to possibly deal with real life situations, there are certain limitations. One of such limitations is that it uses the dataset of only healthy individuals due to the unavailability of the sensory datasets collected from overweight individuals. Therefore, future works should focus on collecting and analyzing the dataset of only obese or overweight individuals to classify the activity patterns. It is important to mention that conducting longitudinal studies for overweight cohorts to record sensory data requires significant resources. This is one of the reasons why the publicly available dataset is used for the analysis and classification of physical activities in the present work. In future work, it would also be interesting to investigate how the deep-learning-based machine learning classifiers' (such as convolutional neural network (CNN) [41], long-short term memory (LSTM) [42] or other deep learning classifiers') behaves on the imbalanced dataset. In future research, a broad range of deep learning techniques will be evaluated for imbalanced dataset to investigate their performance. Moreover, cloud-based computing paradigms can be explored in the future to enable scalability and remote accessibility. The future work should also focus on reducing the impact of class imbalance on the classifier's performance by implementing data-handling techniques such as over-sampling, under sampling, SMOTE, ADASYN, etc.

5. Conclusions

The study developed a novel physical activity classification system and investigated the impact of class imbalance on the performance of machine learning classifiers. The findings concluded that the proposed system is capable of classifying daily life activities such as sitting, standing, walking, lying, walking upstairs and walking downstairs with very high accuracy (above 96%). In addition, a thorough analysis of the impact of class imbalance on the performance of classifiers' is also investigated. A number of experiments are conducted with class imbalance. The findings also suggested that the weighted SVM with penalized approach offered the best classification performance, followed by the ADA(RF) in most of the experiments. Out of the six classifiers evaluated, the SVM, with an overall performance of above 80% in all the class imbalance experiments, depicts its ability to deal with real life situations with certain types of activities being underrepresented.

Author Contributions: Conceptualization, S.A.A., M.R., M.I. and M.A.; methodology, M.R, M.I. and M.A.; software, M.R. and M.A.; validation, M.R. and N.A.; formal analysis, M.R. and M.A.; investigation, M.I. and N.A.; resources, S.A.A.; data curation, M.R. and M.A.; writing—original draft preparation, M.R. and M.A.; writing—review and editing, S.A.A., M.R., M.I., N.A. and M.A.; visualization, M.R. and M.A.; supervision, S.A.A., N.A. and M.A.; project administration, S.A.A., A.M.A. and M.Y.A.; funding acquisition, S.A.A., A.M.A. and M.Y.A. All authors have read and agreed to the published version of the manuscript.

Funding: The authors would like to acknowledge the support of the Deputy for Research and Innovation-Ministry of Education, Kingdom of Saudi Arabia, for this research through a grant (NU/IFC/ENT/01/020) under the institutional Funding Committee at Najran University, Kingdom of Saudi Arabia.

Acknowledgments: The authors would like to acknowledge the support of the Deputy for Research and Innovation-Ministry of Education, Kingdom of Saudi Arabia, for this research through a grant (NU/IFC/ENT/01/020) under the institutional Funding Committee at Najran University, Kingdom of Saudi Arabia. The authors would like to acknowledge Saeed Saad Alahamri from Najran University for their valuable feedback on the draft to improve the flow and quality of the work.

Conflicts of Interest: The authors declare no conflict of interest.

Appendix A

Table A1. Performance by class of different classifiers using the train/test split in Experiment 1 (E1).

Activity Type	SVM (%)	XGB (%)	GB (%)	CB (%)	ADA (DT) (%)	ADA (RF) (%)
Walk	97.62	94.09	95.19	86.74	95.65	93.42
Upstairs	96.78	90.66	92.44	86.37	91.91	89.10
Downstairs	98.08	93.97	94.49	81.03	92.37	89.82
Sit	92.26	87.45	89.55	80.36	89.36	89.78
Stand	93.55	89.57	91.26	81.42	91.04	90.64
Lie	100.00	100.00	100.00	100.00	100.00	100.00
Overall	96.38	92.62	93.82	85.99	93.39	92.13

Table A2. Performance by class of different classifiers using the train/test split in Experiment 2 (E2).

Activity Type	SVM (%)	XGB (%)	GB (%)	CB (%)	ADA (DT) (%)	ADA (RF) (%)
Walk	76.92	15.96	9.23	6.25	25.35	34.11
Upstairs	90.80	76.23	78.80	72.26	80.81	82.22
Downstairs	87.73	70.76	74.00	65.22	70.03	69.82
Sit	92.26	85.93	85.53	81.76	90.74	89.73
Stand	93.55	88.13	86.90	82.60	92.05	90.67
Lie	100.00	100.00	100.00	100.00	100.00	100.00
Overall	90.21	72.83	72.41	68.02	76.50	77.76

Table A3. Performance by class of different classifiers using the train/test split in Experiment 3 (E3).

Activity Type	SVM (%)	XGB (%)	GB (%)	CB (%)	ADA (DT) (%)	ADA (RF) (%)
Walk	81.92	51.04	13.83	44.41	47.15	60.22
Upstairs	49.68	39.11	27.09	65.26	47.36	53.82
Downstairs	64.55	59.53	61.23	62.82	56.04	59.89
Sit	92.28	85.99	84.05	80.81	89.36	89.66
Stand	93.47	87.87	70.42	81.84	90.88	90.55
Lie	100.00	100.00	100.00	100.00	100.00	100.00
Overall	80.32	70.59	59.44	72.52	71.80	75.69

Table A4. Performance by class of different classifiers using the train/test split in Experiment 4 (E4).

Activity Type	SVM (%)	XGB (%)	GB (%)	CB (%)	ADA (DT) (%)	ADA (RF) (%)
Walk	84.30	62.19	23.49	69.53	77.59	75.24
Upstairs	79.21	67.58	40.00	71.26	74.44	73.49
Downstairs	77.99	76.32	77.01	73.67	78.60	72.94
Sit	92.28	72.36	78.74	73.70	82.92	89.05
Stand	93.47	85.91	58.54	76.87	86.48	90.19
Lie	100.00	100.00	100.00	99.72	100.00	100.00
Overall	87.88	77.39	62.96	77.46	83.34	83.49

Table A5. Performance by class of different classifiers using the train/test split in Experiment 5 (E5).

Activity Type	SVM (%)	XGB (%)	GB (%)	CB (%)	ADA (DT) (%)	ADA (RF) (%)
Walk	84.30	71.29	35.79	60.89	65.68	78.91
Upstairs	79.35	65.12	45.05	72.67	42.90	72.75
Downstairs	77.99	75.60	78.09	68.01	67.74	73.81
Sit	67.29	61.67	61.06	46.13	61.02	63.20
Stand	81.23	77.69	55.93	77.30	77.17	79.48
Lie	100.00	98.71	100.00	95.30	100.00	100.00
Overall	81.70	75.01	62.65	70.05	69.09	78.02

Table A6. Performance by class of different classifiers using the train/test split in Experiment 6 (E6).

Activity Type	SVM (%)	XGB (%)	GB (%)	CB (%)	ADA (DT) (%)	ADA (RF) (%)
Walk	84.30	58.55	65.23	61.04	F-score	69.80
Upstairs	79.35	55.60	48.78	67.14	64.38	71.53
Downstairs	77.99	76.61	79.96	67.58	57.96	72.94
Sit	83.12	65.34	66.55	79.92	70.77	84.11
Stand	85.43	74.15	80.73	81.68	80.08	83.47
Lie	100.00	98.08	90.86	100.00	80.99	99.63
Overall	85.03	71.39	72.02	76.23	100.00	80.25

Table A7. Performance by class of different classifiers using the train/test split in Experiment 7 (E7).

Activity Type	SVM (%)	XGB (%)	GB (%)	CB (%)	ADA (DT) (%)	ADA (RF) (%)
Walk	84.30	63.61	21.72	71.23	59.02	83.08
Upstairs	79.35	56.58	43.16	62.70	53.74	76.52
Downstairs	77.99	77.26	75.66	72.36	70.94	76.15
Sit	82.88	73.37	75.18	81.34	79.68	81.57
Stand	85.27	80.70	58.96	78.95	80.30	81.60
Lie	100.00	99.91	99.72	96.23	99.53	100.00
Overall	84.97	75.24	62.40	77.14	73.87	83.15

References

1. World Health Organization. *Global Status Report on Noncommunicable Diseases 2014 (No. WHO/NMH/NVI/15.1)*; World Health Organization: Geneva, Switzerland, 2014.
2. Awais, M.; Chiari, L.; Ihlen, E.; Helbostad, J.; Palmerini, L. Classical Machine Learning versus Deep Learning for the Older Adults Free-Living Activity Classification. *Sensors* **2021**, *21*, 4669. [CrossRef]
3. Wolfenden, L.; Barnes, C.; Jones, J.; Finch, M.; Wyse, R.J.; Kingsland, M.; Tzelepis, F.; Grady, A.; Hodder, R.K.; Booth, D. Strategies to improve the implementation of healthy eating, physical activity and obesity prevention policies, practices or programmes within childcare services. *Cochrane Database Syst. Rev.* **2016**, *10*, CD011779. [CrossRef]
4. Ding, D.; Gebel, K. Built environment, physical activity, and obesity: What have we learned from reviewing the literature? *Health Place* **2012**, *18*, 100–105. [CrossRef]
5. Sacchetti, R.; Dallolio, L.; Musti, M.A.; Guberti, E.; Garulli, A.; Beltrami, P.; Castellazzi, F.; Centis, E.; Zenesini, C.; Coppini, C. Effects of a school based intervention to promote healthy habits in children 8–11 years old, living in the lowland area of Bologna Local Health Unit. *Ann. Ig.* **2015**, *27*, 432–446.
6. La Torre, G.; Mannocci, A.; Saulle, R.; Sinopoli, A.; D'Egidio, V.; Sestili, C.; Manfuso, R.; Masala, D. Improving knowledge and behaviors on diet and physical activity in children: Results of a pilot randomized field trial. *Ann. Ig. Med. Prev. Comunita* **2017**, *29*, 584–594.
7. Caspersen, C.J.; Powell, K.E.; Christenson, G.M. Physical activity, exercise, and physical fitness: Definitions and distinctions for health-related research. *Public Health Rep.* **1985**, *100*, 126–131.
8. Novaes, M.T.; de Carvalho, O.L.; Ferreira, P.H.; Tiraboschi, T.L.; Silva, C.S.; Zambrano, J.C.; Gomes, C.M.; de Paula Miranda, E.; de Carvalho Júnior, O.A.; de Bessa Júnior, J. Prediction of secondary testosterone deficiency using machine learning: A comparative analysis of ensemble and base classifiers, probability calibration, and sampling strategies in a slightly imbalanced dataset. *Inform. Med. Unlocked* **2021**, *23*, 100538. [CrossRef]
9. Singh, L.K.; Garg, H.; Khanna, M.; Bhadoria, R.S. An Analytical Study on Machine Learning Techniques. In *Multidisciplinary Functions of Blockchain Technology in AI and IoT Applications*; IGI Global: Hershey, PA, USA, 2021; pp. 137–157.
10. Awais, M.; Chiari, L.; Ihlen, E.A.F.; Helbostad, J.L.; Palmerini, L. Physical Activity Classification for Elderly People in Free-Living Conditions. *IEEE J. Biomed. Health Inform.* **2018**, *23*, 197–207. [CrossRef]
11. Kerdjidj, O.; Ramzan, N.; Ghanem, K.; Amira, A.; Chouireb, F. Fall detection and human activity classification using wearable sensors and compressed sensing. *J. Ambient Intell. Humaniz. Comput.* **2019**, *11*, 349–361. [CrossRef]
12. Qi, J.; Yang, P.; Newcombe, L.; Peng, X.; Yang, Y.; Zhao, Z. An overview of data fusion techniques for Internet of Things enabled physical activity recognition and measure. *Inf. Fusion* **2019**, *55*, 269–280. [CrossRef]
13. Roy, P.K.; Om, H. Suspicious and Violent Activity Detection of Humans Using HOG Features and SVM Classifier in Surveillance Videos. In *Advances in Soft Computing and Machine Learning in Image Processing*; Springer: Berlin/Heidelberg, Germany, 2018; pp. 277–294.

14. Thyagarajmurthy, A.; Ninad, M.G.; Rakesh, B.G.; Niranjan, S.; Manvi, B. Anomaly Detection in Surveillance Video Using Pose Estimation. In *Emerging Research in Electronics, Computer Science and Technology*; Springer: Berlin/Heidelberg, Germany, 2019; pp. 753–766.
15. Yang, H.; Yuan, C.; Li, B.; Du, Y.; Xing, J.; Hu, W.; Maybank, S.J. Asymmetric 3D Convolutional Neural Networks for action recognition. *Pattern Recognit.* **2018**, *85*, 1–12. [CrossRef]
16. Nadeem, A.; Jalal, A.; Kim, K. Accurate Physical Activity Recognition using Multidimensional Features and Markov Model for Smart Health Fitness. *Symmetry* **2020**, *12*, 1766. [CrossRef]
17. Ehatisham-Ul-Haq, M.; Javed, A.; Azam, M.A.; Malik, H.M.A.; Irtaza, A.; Lee, I.H.; Mahmood, M.T. Robust Human Activity Recognition Using Multimodal Feature-Level Fusion. *IEEE Access* **2019**, *7*, 60736–60751. [CrossRef]
18. Cheng, X.; Lin, S.-Y.; Liu, J.; Liu, S.; Zhang, J.; Nie, P.; Fuemmeler, B.; Wang, Y.; Xue, H. Does Physical Activity Predict Obesity—A Machine Learning and Statistical Method-Based Analysis. *Int. J. Environ. Res. Public Health* **2021**, *18*, 3966. [CrossRef] [PubMed]
19. Li, J.; Siegrist, J. Physical Activity and Risk of Cardiovascular Disease—A Meta-Analysis of Prospective Cohort Studies. *Int. J. Environ. Res. Public Health* **2012**, *9*, 391–407. [CrossRef] [PubMed]
20. Awais, M.; Palmerini, L.; Bourke, A.K.; Ihlen, E.A.F.; Helbostad, J.L.; Chiari, L. Performance Evaluation of State of the Art Systems for Physical Activity Classification of Older Subjects Using Inertial Sensors in a Real Life Scenario: A Benchmark Study. *Sensors* **2016**, *16*, 2105. [CrossRef]
21. Pereira, L.M.C.; Aidar, F.J.; de Matos, D.G.; Neto, J.P.D.F.; de Souza, R.F.; Sousa, A.C.S.; de Almeida, R.R.; Nunes, M.A.P.; Nunes-Silva, A.; Júnior, W.M.D.S. Assessment of Cardiometabolic Risk Factors, Physical Activity Levels, and Quality of Life in Stratified Groups up to 10 Years after Bariatric Surgery. *Int. J. Environ. Res. Public Health* **2019**, *16*, 1975. [CrossRef]
22. Hernando, C.; Hernando, C.; Collado, E.J.; Panizo, N.; Martinez-Navarro, I.; Hernando, B. Establishing cut-points for physical activity classification using triaxial accelerometer in middle-aged recreational marathoners. *PLoS ONE* **2018**, *13*, e0202815. [CrossRef]
23. Qi, J.; Yang, P.; Hanneghan, M.; Tang, S.; Zhou, B. A Hybrid Hierarchical Framework for Gym Physical Activity Recognition and Measurement Using Wearable Sensors. *IEEE Internet Things J.* **2018**, *6*, 1384–1393. [CrossRef]
24. Voicu, R.-A.; Dobre, C.; Bajenaru, L.; Ciobanu, R.-I. Human Physical Activity Recognition Using Smartphone Sensors. *Sensors* **2019**, *19*, 458. [CrossRef]
25. Sanhudo, L.; Calvetti, D.; Martins, J.P.; Ramos, N.M.; Mêda, P.; Gonçalves, M.C.; Sousa, H. Activity classification using accelerometers and machine learning for complex construction worker activities. *J. Build. Eng.* **2020**, *35*, 102001. [CrossRef]
26. Pizot, C.; Boniol, M.; Mullie, P.; Koechlin, A.; Boniol, M.; Boyle, P.; Autier, P. Physical activity, hormone replacement therapy and breast cancer risk: A meta-analysis of prospective studies. *Eur. J. Cancer* **2015**, *52*, 138–154. [CrossRef] [PubMed]
27. Chong, J.; Tjurin, P.; Niemelä, M.; Jämsä, T.; Farrahi, V. Machine-learning models for activity class prediction: A comparative study of feature selection and classification algorithms. *Gait Posture* **2021**, *89*, 45–53. [CrossRef] [PubMed]
28. Anguita, D.; Ghio, A.; Oneto, L.; Parra-Llanas, X.; Reyes-Ortiz, J. A public domain dataset for human activity recognition using smartphones. In Proceedings of the ESANN 2013 Proceedings, European Symposium on Artificial Neural Networks, Computational Intelligence and Machine Learning, Bruges, Belgium, 24–26 April 2013; Volume 3, p. 3.
29. Peter, S.; Diego, F.; Hamprecht, F.A.; Nadler, B. Cost efficient gradient boosting. In Proceedings of the 31st Conference on Neural Information Processing Systems (NIPS 2017), Long Beach, CA, USA, 4–9 December 2017; Advances in Neural Information Processing Systems; Volume 30.
30. Hancock, J.T.; Khoshgoftaar, T.M. CatBoost for big data: An interdisciplinary review. *J. Big Data* **2020**, *7*, 94. [CrossRef]
31. Wu, Y.; Ke, Y.; Chen, Z.; Liang, S.; Zhao, H.; Hong, H. Application of alternating decision tree with AdaBoost and bagging ensembles for landslide susceptibility mapping. *Catena* **2019**, *187*, 104396. [CrossRef]
32. Suthaharan, S. Support vector machine. In *Machine Learning Models and Algorithms for Big Data Classification*; Springer: Berlin/Heidelberg, Germany, 2016; pp. 207–235.
33. Chen, T.; He, T.; Benesty, M.; Khotilovich, V.; Tang, Y.; Cho, H.; Chen, K. *Xgboost: Extreme Gradient Boosting*; R Package Version 0.4-2; 2015; Volume 1, pp. 1–4. Available online: https://cran.microsoft.com/snapshot/2017-12-11/web/packages/xgboost/vignettes/xgboost.pdf (accessed on 1 February 2022).
34. Ghori, K.M.; Ayaz, A.R.; Awais, M.; Imran, M.; Ullah, A.; Szathmary, L. Impact of feature selection on non-technical loss detection. In Proceedings of the 2020 6th Conference on Data Science and Machine Learning Applications (CDMA), 4–5 March 2020; IEEE: Piscataway, NJ, USA, 2020; pp. 19–24.
35. Ghori, K.M.; Awais, M.; Khattak, A.S.; Imran, M.; Amin, F.E.; Szathmary, L. Treating Class Imbalance in Non-Technical Loss Detection: An Exploratory Analysis of a Real Dataset. *IEEE Access* **2021**, *9*, 98928–98938. [CrossRef]
36. Feng, W.; Dauphin, G.; Huang, W.; Quan, Y.; Bao, W.; Wu, M.; Li, Q. Dynamic Synthetic Minority Over-Sampling Technique-Based Rotation Forest for the Classification of Imbalanced Hyperspectral Data. *IEEE J. Sel. Top. Appl. Earth Obs. Remote Sens.* **2019**, *12*, 2159–2169. [CrossRef]
37. He, H.; Bai, Y.; Garcia, E.A.; Li, S. ADASYN: Adaptive synthetic sampling approach for imbalanced learning. In Proceedings of the 2008 IEEE International Joint Conference on Neural Networks (IEEE World Congress on Computational Intelligence), Hong Kong, China, 1–8 June 2008; IEEE: Piscataway, NJ, USA, 2008; pp. 1322–1328.
38. Ramentol, E.; Caballero, Y.; Bello, R.; Herrera, F. SMOTE-RSB: A hybrid preprocessing approach based on oversampling and undersampling for high imbalanced data-sets using SMOTE and rough sets theory. *Knowl. Inf. Syst.* **2012**, *33*, 245–265. [CrossRef]

39. Goorbergh, R.V.D.; van Smeden, M.; Timmerman, D.; Van Calster, B. The harm of class imbalance corrections for risk prediction models: Illustration and simulation using logistic regression. *J. Am. Med. Inform. Assoc.* **2022**, ocac093. [CrossRef]
40. Japkowicz, N. The class imbalance problem: Significance and strategies. In Proceedings of the 2000 International Conference on Artificial Intelligence, Acapulco, Mexico, 11–14 April 2000; Volume 56, pp. 111–117.
41. Arya, K.V.; Bhadoria, R.S. *The Biometric Computing: Recognition and Registration*; CRC Press: Boca Raton, FL, USA, 2019.
42. Awais, M.; Raza, M.; Singh, N.; Bashir, K.; Manzoor, U.; Islam, S.U.; Rodrigues, J.J.P.C. LSTM-Based Emotion Detection Using Physiological Signals: IoT Framework for Healthcare and Distance Learning in COVID-19. *IEEE Internet Things J.* **2020**, *8*, 16863–16871. [CrossRef]

life

MDPI

Article

Robust Gaussian and Nonlinear Hybrid Invariant Clustered Features Aided Approach for Speeded Brain Tumor Diagnosis

Yassir Edrees Almalki [1,†], Muhammad Umair Ali [2,†], Waqas Ahmed [3], Karam Dad Kallu [4], Amad Zafar [5,*], Sharifa Khalid Alduraibi [6], Muhammad Irfan [7], Mohammad Abd Alkhalik Basha [8], Hassan A. Alshamrani [9] and Alaa Khalid Alduraibi [6]

1 Division of Radiology, Department of Internal Medicine, Medical College, Najran University, Najran 61441, Saudi Arabia; yealmalki@nu.edu.sa
2 Department of Unmanned Vehicle Engineering, Sejong University, Seoul 05006, Korea; umair@sejong.ac.kr
3 Secret Minds, Entrepreneurial Organization, Islamabad 44000, Pakistan; engnr.waqasahmed@gmail.com
4 Department of Robotics and Intelligent Machine Engineering (RIME), School of Mechanical and Manufacturing Engineering (SMME), National University of Sciences and Technology (NUST), H-12, Islamabad 44000, Pakistan; karamdad.kallu@smme.nust.edu.pk
5 Department of Electrical Engineering, The Ibadat International University, Islamabad 54590, Pakistan
6 Department of Radiology, College of Medicine, Qassim University, Buraidah 52571, Saudi Arabia; salduraibi@qu.edu.sa (S.K.A.); al.alderaibi@qu.edu.sa (A.K.A.)
7 Electrical Engineering Department, College of Engineering, Najran University, Najran 61441, Saudi Arabia; miditta@nu.edu.sa
8 Radiology Department, Faculty of Human Medicine, Zagazig University, Zagazig 44631, Egypt; maatya@zu.edu.eg
9 Radiological Sciences Department, College of Applied Medical Sciences, Najran University, Najran 61441, Saudi Arabia; hamalshamrani@nu.edu.sa
* Correspondence: amad.zafar@iiui.edu.pk
† These authors contributed equally as a first author to this work.

Citation: Almalki, Y.E.; Ali, M.U.; Ahmed, W.; Kallu, K.D.; Zafar, A.; Alduraibi, S.K.; Irfan, M.; Basha, M.A.A.; Alshamrani, H.A.; Alduraibi, A.K. Robust Gaussian and Nonlinear Hybrid Invariant Clustered Features Aided Approach for Speeded Brain Tumor Diagnosis. *Life* **2022**, *12*, 1084. https://doi.org/10.3390/life12071084

Academic Editor: Yudong Cai

Received: 20 June 2022
Accepted: 17 July 2022
Published: 20 July 2022

Abstract: Brain tumors reduce life expectancy due to the lack of a cure. Moreover, their diagnosis involves complex and costly procedures such as magnetic resonance imaging (MRI) and lengthy, careful examination to determine their severity. However, the timely diagnosis of brain tumors in their early stages may save a patient's life. Therefore, this work utilizes MRI with a machine learning approach to diagnose brain tumor severity (glioma, meningioma, no tumor, and pituitary) in a timely manner. MRI Gaussian and nonlinear scale features are extracted due to their robustness over rotation, scaling, and noise issues, which are common in image processing features such as texture, local binary patterns, histograms of oriented gradient, etc. For the features, each MRI is broken down into multiple small 8 × 8-pixel MR images to capture small details. To counter memory issues, the strongest features based on variance are selected and segmented into 400 Gaussian and 400 nonlinear scale features, and these features are hybridized against each MRI. Finally, classical machine learning classifiers are utilized to check the performance of the proposed hybrid feature vector. An available online brain MRI image dataset is utilized to validate the proposed approach. The results show that the support vector machine-trained model has the highest classification accuracy of 95.33%, with a low computational time. The results are also compared with the recent literature, which shows that the proposed model can be helpful for clinicians/doctors for the early diagnosis of brain tumors.

Keywords: magnetic resonance imaging (MRI); brain tumor; machine learning

1. Introduction

The brain is the most complex organ in the human body. It has over 100 billion nerve cells with trillions of synapses [1]. In other words, the human brain is the primary command and control center of the neurological system. Therefore, an injury in the brain has a catastrophic influence on human health. For example, in a brain tumor, the development of abnormal brain cells may damage the brain and may even threaten a patient's life. Because brain tumors have long-term and life-altering physical and psychological implications, they can significantly influence a patient's living quality and affect their entire life [2]. According to a World Health Organization (WHO) report [3], cancer is the second greatest cause of mortality globally. It is responsible for around 10 million fatalities. Therefore, early cancer identification improves the patient's survival chances. According to a National Brain Tumor Foundation (NBTF) report [4], around 29,000 persons in the USA have primary malignant tumors, and 13,000 people die due to this type of brain tumor.

The location, progression stage, type, and rate of growth of brain tumors determine whether they are benign or malignant [5,6]. The affected cells rarely attack nearly healthy cells in benign brain tumors. They also progress slowly and have clear limits, such as in meningioma and pituitary tumors. In contrast, neighboring healthy cells are influenced by affected cells in malignant brain tumors. These tumors also have a fast advancement rate with broad limitations, such as gliomas. Furthermore, brain tumors may be divided into two types based on their origin: primary and secondary brain tumors [7]. The brain tumors that start in the brain tissues are known as primary tumors. In contrast, secondary brain tumors develop in many areas of the central nervous system (CNS) and move to the brain via the blood vessels. Therefore, early cancer type detection (meningioma, pituitary, and glioma) is crucial for cancer treatment to save the patient's life.

For brain tumor detection, several diagnostic methods, both invasive and non-invasive, are utilized [8]. A biopsy is an invasive approach: a sample is retrieved by an incision and is inspected under a microscope to assess malignancy. Unlike other tumors in other areas of the body, the biopsy is usually delayed until the final brain surgery. Due to this, computer-aided diagnostics (CAD) (non-invasive) such as computed tomography (CT), positron emission tomography (PET), and magnetic resonance imaging (MRI) are thought to be faster and safer than a biopsy for diagnosing brain tumors. Brain MRI is considered to be the most recommended method owing to its ability to provide extensive information regarding the position, extension, nature, and size of the brain tumor [9]. Meanwhile, manual MRI scan interpretation takes a long time and has a significant risk of mistakes. Therefore, an automatic computer-aided diagnostic approach is required for injury detection in the brain.

The evolution of machine learning methods has increased CAD systems' efficiency in assisting doctors in identifying brain tumors [7,10,11]. Numerous learning methods have been presented in the literature to diagnose brain tumors; they can be further categorized as deep learning and classical learning methods based on the literature [12]. In deep learning approaches, convolution neural networks (CNNs) are generally utilized to identify brain tumors using MRI [13]. Various researchers have used pre-trained and developed learning models to classify MRI images. In one work [14], the authors developed a CNN model to classify brain MRI images into two classes (tumor and no tumor). The main shortcoming of their model was the detection of the subclasses of the tumor. Abiwinanda et al. [15] designed a CNN model to detect brain tumor subclasses (glioma, meningioma, and pituitary). However, their model had a low accuracy of only 84.19%. Recently, a new CNN model was developed to classify brain MRI images into three subclasses [8]. The authors also performed data augmentation to enhance the classification accuracy of brain MRI images. A classification accuracy of 96.56% was achieved using a 10-fold cross-validation approach. Irmak [16] developed a 25-layer CNN model to classify brain images into five classes, with an accuracy of 92.66%. Pre-trained networks such as GoogLeNet and ResNet-50 are also used to classify brain images [17–19]. However, the deep networks require long training times, have a complex architecture, high memory requirements, a strong processing unit (GPU), etc.

In contrast to deep learning models, classical models require the most basic features of brain MRI images to diagnose a brain tumor. Therefore, they require less time to train the models; methods include support vector machine (SVM), tree, Naïve Bayes, etc. Kumari et al. [20] computed the gray-level co-occurrence matrix of brain MRI images to classify them into two classes. The model's accuracy was high; however, the authors only detected the tumors on the brain MRI images. The accuracy of these global-level features is not high due to the high similarity in the brain MRI images. Therefore, local-level features such as the bag of words [21], Fisher vector [22], and scale-invariant feature transformation [23] are also used to classify brain MRI images. In one study [24], the authors hybridized the gray-level co-occurrence matrix, histogram intensity, and bag of words to classify brain MRI images. They achieved a classification accuracy of 91.28% for the three-class classification MRI dataset. In a recent study [25], the authors calculated the deep features of brain MRI image datasets using pre-trained CNN models. The results showed that the hybrid features of the pre-trained model had the best accuracy of 93.72% when using an SVM classifier. However, the size of their dataset was large, and it required a long training time. Moreover, in machine learning images/MRI feature extraction approaches, features such as texture (extracted through gray-level co-occurrence matrix), local binary pattern, histogram of oriented gradient, etc., are quite sensitive to noise, scaling, rotation, visibility, etc., which affect the performance, memory requirement, execution time, etc.

Considering the shortcomings of deep and machine learning approaches, the following are the main contributions of this work:

1. This study presents a fast automatic approach for brain tumor detection and differentiation using brain MRI images to increase the accuracy, grading, robustness to noise, rotation, and scaling with the least memory and processing system requirements.
2. The Gaussian scale-space features are extracted through speeded up robust features (SURF) and nonlinear scale-space features are extracted through KAZE of brain MRI images.
3. Each MRI is divided into sub-MRIs of 8 × 8-pixel images to capture the small details/tumor information.
4. Afterwards, to reduce the memory requirements, the strongest features are selected based on variance and subjected to segmentation into 400 Gaussian features and 400 nonlinear features against each brain MRI scan (a total of 800 features).
5. Various classical machine learning models are trained to check their performance.
6. Finally, two available online datasets are used to validate the proposed approach.
7. The findings of the work are also compared with the approaches present in the literature.

The paper's organization is as follows: Section 2 presents the feature extraction and the workings of the proposed approach. Then, the dataset and results are presented in the third section. Finally, the results are discussed and concluded in Sections 4 and 5.

2. Materials and Methods

2.1. Feature Extraction

In computer image processing, feature detection and description are hot topics. In image classification applications, computing features that are repeatable and distinct in the face of various image transformations are of high importance. The classification of brain tumors also mainly relies on retrieving the relevant and relatable features from brain MRI images. Therefore, many global [20] and local features [22,23] are used to classify brain MRI images. The global-level features have accuracy problems in a multiclass environment, as discussed in Section 1. Various local features such as scale-invariant feature transform (SIFT) [26], speeded up robust features (SURF) [27], and KAZE [28] compute distinctive features at various interest point locations. These distinctive features primarily relate to the local maxima/minima/mean in regard to the computed feature. A descriptor vector represents the intensity patterns surrounding these interest points. Lowe [26] introduced the SIFT feature descriptor. It gained much attention owing to its translation invariance, robustness to image noise, invariance to scale, and rotation invariance properties. However, the computational cost of SIFT feature extraction is very high, so it is not recommended for real-time applications [29].

2.1.1. Speeded up Robust Feature (SURF)

To overcome the issues related to SIFT, Bay et al. [27] introduced the SURF method to tackle the robustness issues of the SIFT approach. The SURF approach is based on Gaussian scale-space image analysis, similar to the SIFT method. Unlike the SIFT detector, the SURF approach depends on the Hessian Matrix determinant. It employs integrated images to enhance the speed of feature detection. SURF's 64-bin descriptor characterizes each detected feature using a dispersion of Haar wavelet responses within a specific area. Unlike SIFT, the SURF features show limited affine invariance. However, to deal with more considerable viewpoint shifts, the descriptor can be expanded to 128-bin values. The Hessian Matrix is generated at the point "$m = (m,n)$" at scale "σ".

$$H(m,\sigma) = \begin{bmatrix} L_{mm}(m,\sigma) & L_{mn}(m,\sigma) \\ L_{mn}(m,\sigma) & L_{nn}(m,\sigma) \end{bmatrix} \quad (1)$$

where $L_{mm}(m,\sigma)$ is the Gaussian second-order derivate convolution $\frac{\partial^2}{\partial x^2} g(\sigma)$ with the image I at a point m, similar to $L_{mn}(m,\sigma)$ and $L_{nn}(m,\sigma)$.

2.1.2. KAZE

KAZE is a revolutionary 2D feature identification and description approach that works entirely in nonlinear scale-space using nonlinear diffusion and the additive operator splitting method [28]. Thus, blurring in images becomes locally adaptable to feature points, resulting in noise reduction without affecting the image region boundaries. The KAZE is derived by the Hessian Matrix determinant with a normalized scale and is calculated at different scale levels. A moving window identifies the maxima/minima/mean of detector response as feature points (mean is used in this work). In the feature description, the rotation invariance property is introduced by determining the prevalent orientation in a rounded region surrounding each detected feature. It has the properties of scale and rotation invariance, little invariance to affine, and has greater distinctness at different scales, with a slight increase in computational cost. The nonlinear diffusion equation is presented below.

$$\frac{\partial L}{\partial t} = div(c(m,n,t).\nabla L) \quad (2)$$

where c, div, ∇, and L are the conductivity function, divergence, gradient operator, and luminance of the image, respectively.

2.2. Support Vector Machine (SVM)

Cortes and Vapnik [30] proposed the SVM model in 1995, and it is a very popular and powerful classifier used in various fields [31–33]. The SVM algorithm uses kernel functions $K(x, x_a)$ to transfer the nonlinear low-dimensional input data space into a high-dimensional linear data space. The hyperplane function used to separate the transferred data (high-dimensional linear data) is presented in Equation (3).

$$y(x) = \sum_{a=1}^{n} \beta_a K(x, x_a) + b_1 \quad (3)$$

Meanwhile, various kernel functions, such as linear kernel, sigmoid kernel, and RBF kernel, can be used to classify the data. Further details about SVM can be found in [30,32].

2.3. Proposed Framework

This section discusses the overall framework of the proposed approach in detail. The proposed approach consists of 4 main components, namely brain MRI image acquisition, pre-processing, feature extraction, and model training, as shown in Figure 1.

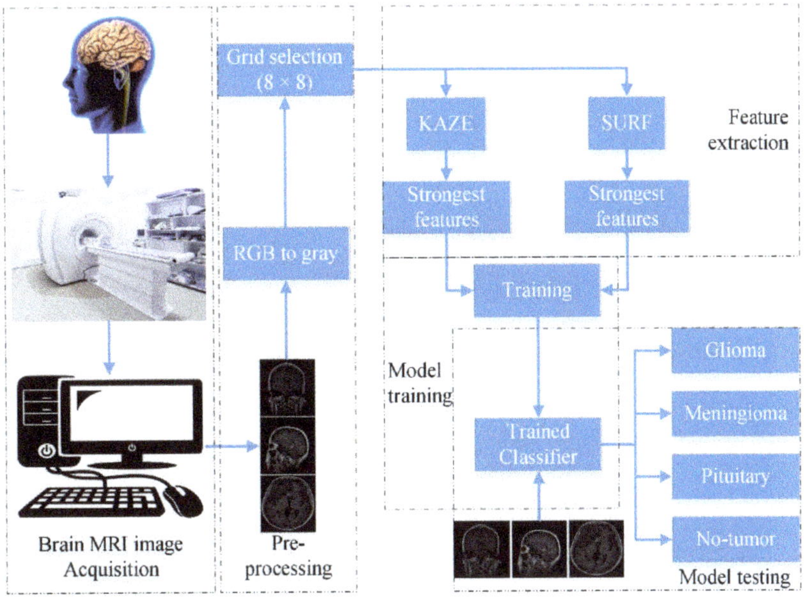

Figure 1. Framework of the proposed hybrid brain MRI image classification model.

The brain images were acquired using the brain MRI machine in the first step. Next, the acquired brain MRI images were pre-processed from the RGB images into grayscale images. Then, an 8 × 8-pixel grid was defined as a selection point for the feature extraction of the brain MRI images. Variations in pixel size affect the computational cost and feature vector size. Furthermore, the four-element vectors ([16, 32, 48, 64] and [17, 34, 51, 68]) were used to extract the KAZE and SURF features, respectively. The details of KAZE and SURF extraction were already provided in Section 2.1. After this, 20% of the redundant features were discarded to reduce the feature vector size. Finally, based on the simplicity and robustness, the *k*-means clustering algorithm was utilized for feature segmentation. Furthermore, it kept observations inside each cluster as close to each other and as far away from objects in other clusters as possible Therefore, 400-feature histograms were created using the *k*-means clustering approach. Further details about the *k*-means clustering approach can be found in [34,35]. After this, various machine learning classifiers, such as SVM, tree [36], Naïve Bayes [37], k-nearest neighbors (K-NN) [38], ensemble, and neural network (NN), were used to train the models. The results of the proposed method are presented in the subsequent section.

3. Brain MRI Dataset and Results

This study validates the suggested paradigm using an online collection of brain MRI images [39]. The dataset for this study was obtained from the Kaggle website [39]. It contains three tumor classes (glioma, pituitary, and meningioma) and one class of no tumor. It has 2870 brain MRI images in total. Additionally, 80% of the data of each class were utilized for the training of the models. The remaining 20% of the data were used to test the trained models. The brain MRI images and percentage distribution of images per class are shown in Figure 2.

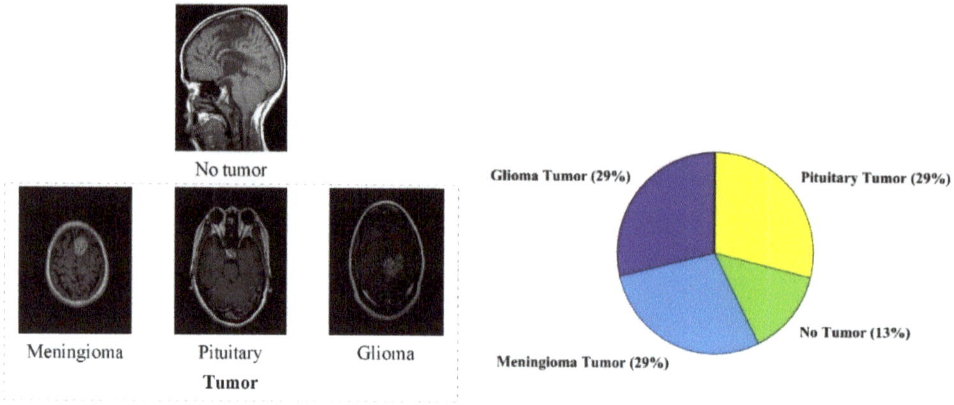

(**a**) (**b**)

Figure 2. (**a**) The brain MRI images of each class; (**b**) the percentage distribution of MRI images per class.

In this work, MATLAB 2021 was utilized for training the models in the 64-bit Windows 11 operating system (core i7, 11th generation, 32 GB RAM, NVIDIA GeForce GTX 1060, and 1 TB SSD). In addition, the classification accuracy was used as a comparison metric for the various trained models (SVM, tree, Naïve Bayes, K-NN, ensemble, and NN). The results of the KAZE- and SURF-trained models are presented in Figure 3.

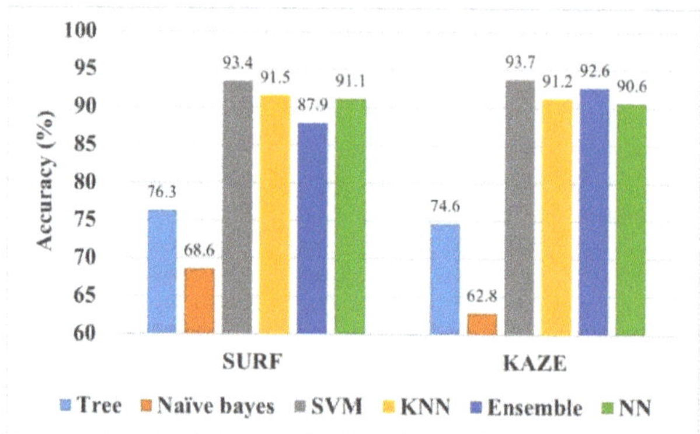

Figure 3. The comparison of various machine learning models for SURF and KAZE features.

It is evident from Figure 3 that the SVM model trained with SURF and KAZE features shows accuracies of 93.4% and 93.7%, respectively, which are the highest among all methods. Therefore, it may be fruitful to concatenate the features of SURF and KAZE to determine the model's performance in classifying brain MRI images. Furthermore, the confusion matrixes of the SURF-, KAZE-, and SURF + KAZE- (hybrid) trained SVM models are shown in Figure 4.

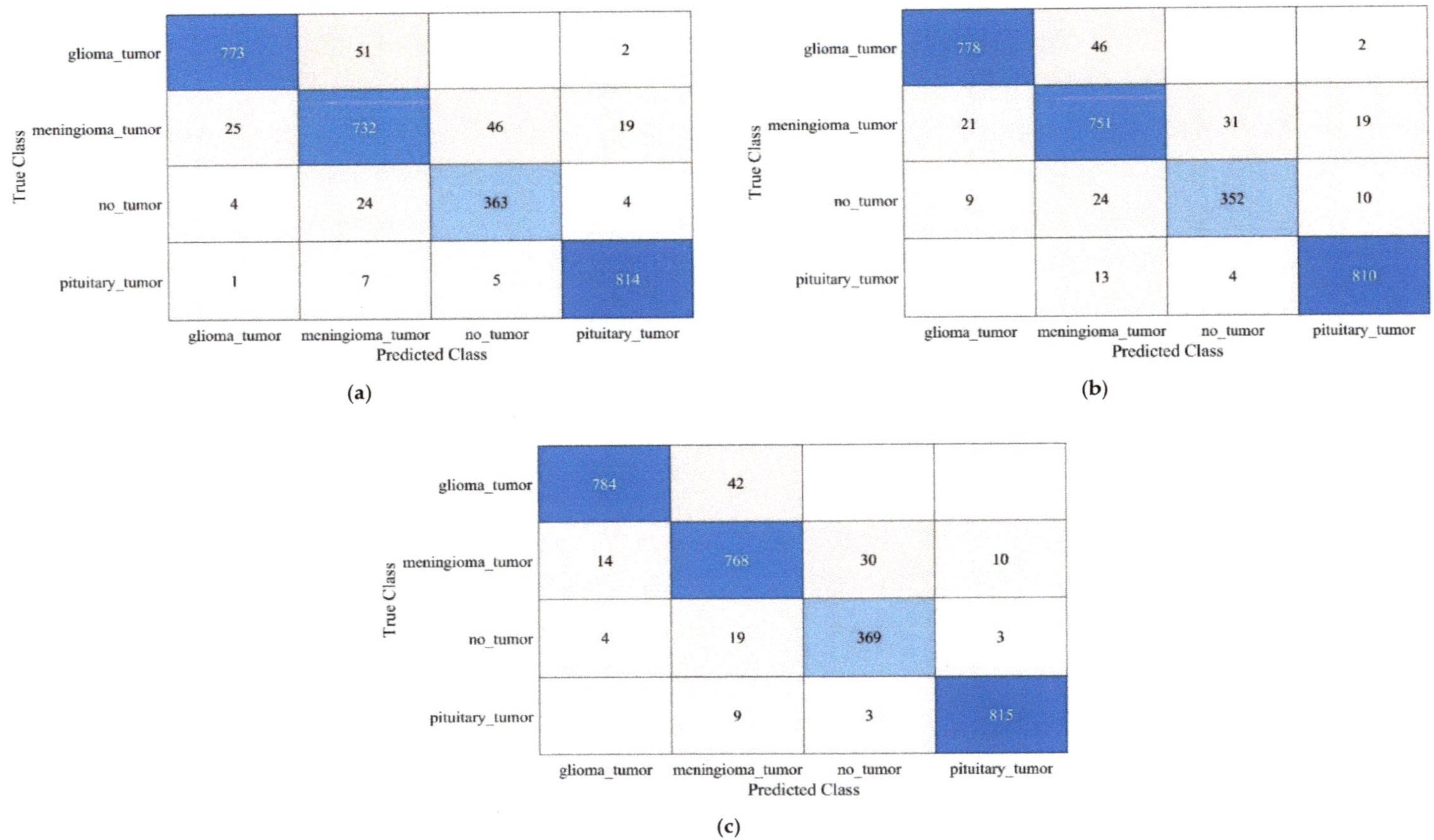

Figure 4. Confusion matrixes of various models: (**a**) SURF-trained SVM; (**b**) KAZE-trained SVM; (**c**) SURF + KAZE (hybrid)-trained SVM (proposed model).

The SVM model trained with concatenation features shows the highest accuracy of 95.33%, almost 2% higher than the SVM model trained with SURF features. Therefore, the proposed SURF + KAZE-trained SVM model has true positive rates (TPRs) of 97.75% and 98.42% for the glioma and pituitary tumor classes. Furthermore, the proposed model correctly classifies 19 more MRI brain images for the no tumor class than the KAZE-trained model. Similarly, 36 more brain MRI images are correctly classified for the meningioma tumor class compared to the SURF-trained model. Finally, the proposed model is compared with the pre-trained deep-feature-trained SVM model presented by Kang et al. [25]. The comparison results of various SVM models are presented in Figure 5.

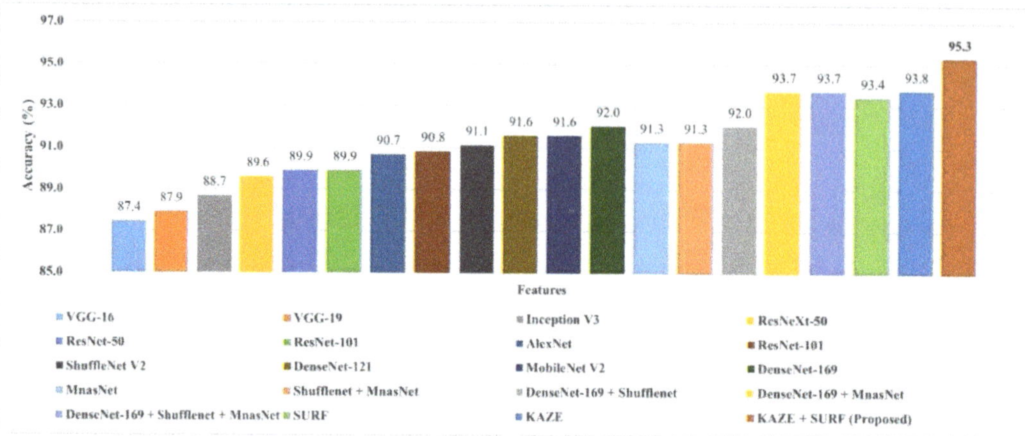

(a)

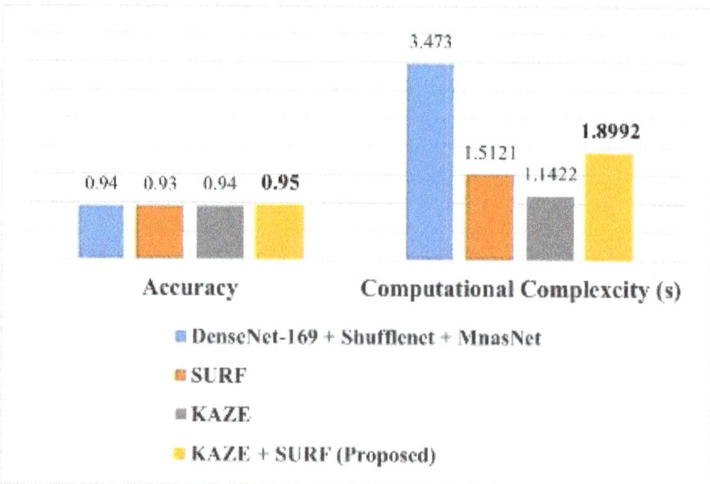

(b)

Figure 5. Comparison of SVM model trained with deep features with the proposed model: (**a**) accuracy comparison; (**b**) accuracy and computational complexity.

For further validation of the proposed approach, another public dataset is utilized [40]. The dataset contains a total of 3064 brain MRI scans. Further details about the dataset are shown in Figure 6. The classification result of the new dataset is presented in Figure 7.

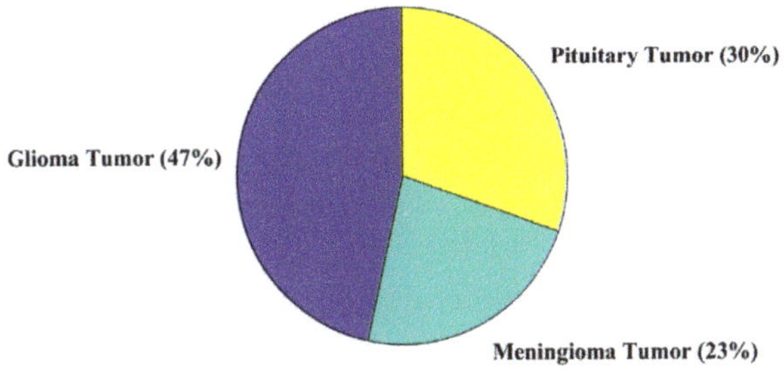

Figure 6. The percentage distribution per class of brain MRI dataset [40].

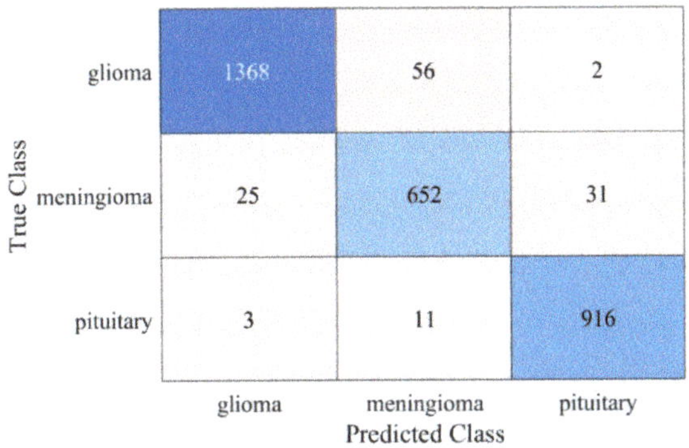

Figure 7. Confusion matrix of the proposed model for new dataset [40].

4. Discussion

Computer-aided detection/diagnosis involves a computer-based system that assists clinicians in making quick judgments in the field of medical imaging. Several studies have reported several training methods for categorizing brain MRI images [8,16,22,25,41].

In this work, an SVM model for brain MRI images trained with hybrid SURF and KAZE features is proposed for brain tumor classification. First, the acquired brain MRI images were processed using the 8×8-pixel uniform grid to extract the SURF and KAZE features, as discussed in Sections 2.1.1 and 2.1.2. As a result, 16,577,120 features were extracted for the whole dataset containing 2870 brain MRI images of various classes (see Section 3 for details). In addition, 80% of the strongest features were computed using the computer vision toolbox of MATLAB, which reduced the feature vector size to 7,300,864 for all of the brain MRI images. Finally, *k*-means clustering was utilized to form feature vectors with a size of 400 for each image. As a result, the SVM-trained model showed the best accuracies of 93.4% and 93.7% for SURF and KAZE, respectively (see Figure 3). Furthermore, the concatenation of both the SURF and KAZE features resulted in a better accuracy of 95.3% for brain MRI multiclass classification.

Kang et al. [25] trained the SVM model using pre-trained network deep features. The results suggested that the DenseNet-169 + Shufflenet + MnasNet-trained SVM model had the best classification accuracy of 93.72% for a similar dataset (see Figure 5). The proposed SURF + KAZE-trained SVM model showed an accuracy of 95.33%, almost 1.5% higher than the model proposed by Kang et al. (see Figure 5a). The computational cost of the proposed model was also almost two times lower than their proposed model (see Figure 5b). In a study [41], pre-trained CNN models (GoogleNet, VGGNet, and AlexNet) were utilized to classify brain MRI images. The model showed high classification accuracy with a high training time of around 1 h and 30 min for the fine-tuned VGGNet CNN model. The model presented in our study (SURF + KAZE) showed an accuracy of 95.33% and had a computational complexity of only 1.8992 s. For further validation, a new public dataset that had three classes was used to check the performance of the proposed framework (see Figure 6). The proposed approach showed similar accuracy (95.9%) for the classification in the new brain MRI dataset, as shown in Figure 7. The results validate the adeptness, robustness, and high classification accuracy of the proposed approach. This demonstrates that the presented model is relatively straightforward to implement for real-time applications. As a result, the suggested technique has the potential to play a critical role in assisting clinicians/doctors for early brain cancer detection.

5. Conclusions

This study presents an automatic brain tumor diagnostic approach using brain MRI images. First, the proposed approach computes the SURF and KAZE features using a grid of 8×8 pixels in size of brain MRI images. Then, 80% of the strongest features are considered for segmentation using k-means clustering. The final feature vector has a size of 400 per image for each feature (SURF and KAZE). Finally, the proposed hybrid feature vector is used to train the SVM model. The classification accuracies of the proposed model (SURF + KAZE) are 95.33% and 95.9%, almost 2% higher than the SURF-trained SVM model. The comparison of the proposed approach with the findings presented in the literature also shows its superiority due to its high accuracy and lower computational time. Thus, the proposed approach can be used for the automatic detection of brain tumors.

Author Contributions: Conceptualization, Y.E.A., M.U.A., W.A. and A.Z.; data curation, S.K.A.; formal analysis, Y.E.A., M.U.A., M.I. and H.A.A.; investigation, A.Z. and S.K.A.; methodology, Y.E.A., M.U.A. and W.A.; project administration, A.Z., S.K.A. and M.A.A.B.; resources, M.I.; software, K.D.K. and A.K.A.; supervision, H.A.A.; validation, M.U.A.; writing—original draft, Y.E.A., M.U.A. and A.Z.; writing—review and editing, K.D.K., A.Z., M.I., M.A.A.B. and A.K.A. All authors have read and agreed to the published version of the manuscript.

Funding: The authors are thankful to the Deanship of Scientific Research, Najran University, Kingdom of Saudi Arabia.

Institutional Review Board Statement: Not applicable.

Informed Consent Statement: Not applicable.

Data Availability Statement: Not applicable.

Acknowledgments: The authors world like to express their gratitude to the Deanship of Scientific Research, Najran University, Kingdom of Saudi Arabia, for their financial and technical support under code number (NU/-/MRC/10/388).

Conflicts of Interest: The authors declare no conflict of interest.

References

1. Louis, D.N.; Perry, A.; Reifenberger, G.; von Deimling, A.; Figarella-Branger, D.; Cavenee, W.K.; Ohgaki, H.; Wiestler, O.D.; Kleihues, P.; Ellison, D.W. The 2016 World Health Organization Classification of Tumors of the Central Nervous System: A summary. *Acta Neuropathol.* **2016**, *131*, 803–820. [CrossRef]
2. Arabahmadi, M.; Farahbakhsh, R.; Rezazadeh, J. Deep Learning for Smart Healthcare—A Survey on Brain Tumor Detection from Medical Imaging. *Sensors* **2022**, *22*, 1960. [CrossRef] [PubMed]
3. Cancer. Available online: https://www.who.int/news-room/fact-sheets/detail/cancer (accessed on 9 September 2021).
4. Singh, V.; Gourisaria, M.K.; GM, H.; Rautaray, S.S.; Pandey, M.; Sahni, M.; Leon-Castro, E.; Espinoza-Audelo, L.F. Diagnosis of Intracranial Tumors via the Selective CNN Data Modeling Technique. *Appl. Sci.* **2022**, *12*, 2900. [CrossRef]
5. Society, A.C. Available online: www.cancer.org/cancer.html (accessed on 9 September 2021).
6. Diagnosis, B.T. Available online: https://www.cancer.net/cancer-types/brain-tumor/diagnosis (accessed on 9 September 2021).
7. Tandel, G.S.; Biswas, M.; Kakde, O.G.; Tiwari, A.; Suri, H.S.; Turk, M.; Laird, J.R.; Asare, C.K.; Ankrah, A.A.; Khanna, N.N.; et al. A Review on a Deep Learning Perspective in Brain Cancer Classification. *Cancers* **2019**, *11*, 111. [CrossRef]
8. Badža, M.M.; Barjaktarović, M.Č. Classification of Brain Tumors from MRI Images Using a Convolutional Neural Network. *Appl. Sci.* **2020**, *10*, 1999. [CrossRef]
9. Pereira, S.; Pinto, A.; Alves, V.; Silva, C.A. Brain Tumor Segmentation Using Convolutional Neural Networks in MRI Images. *IEEE Trans. Med. Imaging* **2016**, *35*, 1240–1251. [CrossRef]

10. Doi, K. Computer-aided diagnosis in medical imaging: Historical review, current status and future potential. *Comput. Med. Imaging Graph.* **2007**, *31*, 198–211. [CrossRef]
11. Munir, K.; Elahi, H.; Ayub, A.; Frezza, F.; Rizzi, A. Cancer Diagnosis Using Deep Learning: A Bibliographic Review. *Cancers* **2019**, *11*, 1235. [CrossRef]
12. Wadhwa, A.; Bhardwaj, A.; Verma, V.S. A review on brain tumor segmentation of MRI images. *Magn. Reson. Imaging* **2019**, *61*, 247–259. [CrossRef]
13. Nazir, M.; Shakil, S.; Khurshid, K. Role of deep learning in brain tumor detection and classification (2015 to 2020): A review. *Comput. Med. Imaging Graph.* **2021**, *91*, 101940. [CrossRef]
14. Pereira, S.; Meier, R.; Alves, V.; Reyes, M.; Silva, C.A. Automatic Brain Tumor Grading from MRI Data Using Convolutional Neural Networks and Quality Assessment. *arXiv* arXiv:1809.09468, 2018.
15. Abiwinanda, N.; Hanif, M.; Hesaputra, S.T.; Handayani, A.; Mengko, T.R. *Brain Tumor Classification Using Convolutional Neural Network*; Springer: Singapore, 2019; pp. 183–189.
16. Irmak, E. Multi-Classification of Brain Tumor MRI Images Using Deep Convolutional Neural Network with Fully Optimized Framework. *Iran. J. Sci. Technol. Trans. Electr. Eng.* **2021**, *45*, 1015–1036. [CrossRef]
17. Deepak, S.; Ameer, P.M. Brain tumor classification using deep CNN features via transfer learning. *Comput. Biol. Med.* **2019**, *111*, 103345. [CrossRef] [PubMed]
18. Çinar, A.; Yildirim, M. Detection of tumors on brain MRI images using the hybrid convolutional neural network architecture. *Med. Hypotheses* **2020**, *139*, 109684. [CrossRef]
19. Alanazi, M.F.; Ali, M.U.; Hussain, S.J.; Zafar, A.; Mohatram, M.; Irfan, M.; AlRuwaili, R.; Alruwaili, M.; Ali, N.H.; Albarrak, A.M. Brain Tumor/Mass Classification Framework Using Magnetic-Resonance-Imaging-Based Isolated and Developed Transfer Deep-Learning Model. *Sensors* **2022**, *22*, 372. [CrossRef]
20. Kumari, R. SVM classification an approach on detecting abnormality in brain MRI images. *Int. J. Eng. Res. Appl.* **2013**, *3*, 1686–1690.
21. Ayadi, W.; Elhamzi, W.; Charfi, I.; Atri, M. A hybrid feature extraction approach for brain MRI classification based on Bag-of-words. *Biomed. Signal Processing Control.* **2019**, *48*, 144–152. [CrossRef]
22. Cheng, J.; Yang, W.; Huang, M.; Huang, W.; Jiang, J.; Zhou, Y.; Yang, R.; Zhao, J.; Feng, Y.; Feng, Q.; et al. Retrieval of Brain Tumors by Adaptive Spatial Pooling and Fisher Vector Representation. *PLoS ONE* **2016**, *11*, e0157112. [CrossRef]
23. Bosch, A.; Munoz, X.; Oliver, A.; Marti, J. Modeling and Classifying Breast Tissue Density in Mammograms. In Proceedings of the 2006 IEEE Computer Society Conference on Computer Vision and Pattern Recognition (CVPR'06), New York, NY, USA, 17–22 June 2006; pp. 1552–1558.
24. Cheng, J.; Huang, W.; Cao, S.; Yang, R.; Yang, W.; Yun, Z.; Wang, Z.; Feng, Q. Enhanced Performance of Brain Tumor Classification via Tumor Region Augmentation and Partition. *PLoS ONE* **2015**, *10*, e0140381. [CrossRef]
25. Kang, J.; Ullah, Z.; Gwak, J. MRI-Based Brain Tumor Classification Using Ensemble of Deep Features and Machine Learning Classifiers. *Sensors* **2021**, *21*, 2222. [CrossRef]
26. Lowe, D.G. Distinctive Image Features from Scale-Invariant Keypoints. *Int. J. Comput. Vis.* **2004**, *60*, 91–110. [CrossRef]
27. Bay, H.; Ess, A.; Tuytelaars, T.; Van Gool, L. Speeded-Up Robust Features (SURF). *Comput. Vis. Image Underst.* **2008**, *110*, 346–359. [CrossRef]
28. Alcantarilla, P.F.; Bartoli, A.; Davison, A.J. KAZE features. In Proceedings of the European Conference on Computer Vision, Florence, Italy, 7–13 October 2012; pp. 214–227.
29. Hongpeng, Y.; Chao, P.; Yi, C.; Qu, F. A robust object tracking algorithm based on surf and Kalman filter. *Intell. Autom. Soft Comput.* **2013**, *19*, 567–579. [CrossRef]
30. Cortes, C.; Vapnik, V. Support-vector networks. *Mach. Learn.* **1995**, *20*, 273–297. [CrossRef]
31. Ali, M.U.; Khan, H.F.; Masud, M.; Kallu, K.D.; Zafar, A. A machine learning framework to identify the hotspot in photovoltaic module using infrared thermography. *Sol. Energy* **2020**, *208*, 643–651. [CrossRef]
32. Ali, M.U.; Zafar, A.; Nengroo, S.H.; Hussain, S.; Park, G.-S.; Kim, H.-J. Online Remaining Useful Life Prediction for Lithium-Ion Batteries Using Partial Discharge Data Features. *Energies* **2019**, *12*, 4366. [CrossRef]
33. Ali, M.U.; Saleem, S.; Masood, H.; Kallu, K.D.; Masud, M.; Alvi, M.J.; Zafar, A. Early hotspot detection in photovoltaic modules using color image descriptors: An infrared thermography study. *Int. J. Energy Res.* **2022**, *46*, 774–785. [CrossRef]
34. Hartigan, J.A.; Wong, M.A. Algorithm AS 136: A k-means clustering algorithm. *J. R. Stat. Society Ser. C (Appl. Stat.)* **1979**, *28*, 100–108. [CrossRef]
35. k-Means Clustering. Available online: https://www.mathworks.com/help/stats/k-means-clustering.html (accessed on 17 March 2022).
36. Safavian, S.R.; Landgrebe, D. A survey of decision tree classifier methodology. *IEEE Trans. Syst. Man Cybern.* **1991**, *21*, 660–674. [CrossRef]
37. Niazi, K.A.K.; Akhtar, W.; Khan, H.A.; Yang, Y.; Athar, S. Hotspot diagnosis for solar photovoltaic modules using a Naive Bayes classifier. *Sol. Energy* **2019**, *190*, 34–43. [CrossRef]
38. Ali, N.; Neagu, D.; Trundle, P. Evaluation of k-nearest neighbour classifier performance for heterogeneous data sets. *SN Appl. Sci.* **2019**, *1*, 1559. [CrossRef]

39. Brain Tumor Classification (MRI). Available online: https://www.kaggle.com/datasets/sartajbhuvaji/brain-tumor-classification-mri?select=Training (accessed on 17 March 2022).
40. Jun, C. Brain Tumor Dataset. 2017. Available online: https://figshare.com/articles/dataset/brain_tumor_dataset/1512427 (accessed on 17 March 2022).
41. Rehman, A.; Naz, S.; Razzak, M.I.; Akram, F.; Imran, M. A Deep Learning-Based Framework for Automatic Brain Tumors Classification Using Transfer Learning. *Circuits Syst. Signal Processing* **2020**, *39*, 757–775. [CrossRef]

life

Review

Healthcare Professionals' Experience of Performing Digital Care Visits—A Scoping Review

Ieva Lampickienė and Nadia Davoody *

Health Informatics Centre, Department of Learning, Informatics, Management and Ethics, Karolinska Institute, Tomtebodavägen 18A, 171 77 Solna, Sweden; ieva.lampickiene@gmail.com
* Correspondence: nadia.davoody@ki.se

Abstract: The use of digital care visits has been increasing during the COVID-19 pandemic. Learning more about healthcare professionals' technology experiences provides valuable insight and a basis for improving digital visits. This study aimed to explore the existing literature on healthcare professionals' experience performing digital care visits. A scoping review was performed following Arksey & O'Malley's proposed framework using the Preferred Reporting Items for Systematic reviews and Meta-Analyses. The collected data were analyzed using thematic content analysis. Five main themes were identified in the literature: positive experiences/benefits, facilitators, negative experiences/challenges, barriers, and suggestions for improvement. Healthcare professionals mostly reported having an overall positive experience with digital visits and discovered benefits for themselves and the patients. However, opinions were mixed or negative regarding the complexity of decision making, workload and workflow, suitability of this type of care, and other challenges. The suggestions for improvement included training and education, improvements within the system and tools, along with support for professionals. Despite overall positive experiences and benefits for both professionals and patients, clinicians reported challenges such as physical barriers, technical issues, suitability concerns, and others. Digital care visits could not fully replace face-to-face visits.

Keywords: digital care visit; online consultation; medical staff; healthcare personnel; user experience

Citation: Lampickienė, I.; Davoody, N. Healthcare Professionals' Experience of Performing Digital Care Visits—A Scoping Review. *Life* **2022**, *12*, 913. https://doi.org/10.3390/life12060913

Academic Editor: Daniele Giansanti

Received: 6 May 2022
Accepted: 16 June 2022
Published: 17 June 2022

1. Introduction

Currently, Information and Communications Technology (ICT) plays a significant role in all industries and people's everyday lives. The healthcare field is no exception. Medical institutions have been using advanced ICT for health records, telemedicine, various forms of e-learning, as well as other tools. The increasing accessibility to the internet and smart devices has influenced the use of applications and implementation of telemedicine in healthcare [1].

One of the concepts used in today's health care is video-conferencing [2]. Videoconferencing is often described in different terms, such as video meetings [3], digital/virtual meetings, digital visits [4], or video teleconferencing [5]. The concept is rather broad. It includes consultations not only between patients and healthcare professionals [6] but also consultations between two or more healthcare professionals. In which a patient and a healthcare professional are present, on a clinical site or in the home, and together they are consulted by an included specialist from another clinical site [7,8]. So, the consultation may be referred to as video-conferencing [9,10], even though it is broader than just patient-to-healthcare professional consultations. In this review, a narrower concept of video-conferencing is considered central, which is video consultations initiated by patients consulted by health care professionals. This type of consultation is referred to differently in the literature; virtual visits [11], telehealth which can mean both telephone and video consultations [12,13], digital visits, or video consultations [13], to name a few.

It is important to note that the use of video visits has increased due to both its advantages, such as providing timely care to patients in rural areas or homebound chronically ill patients, and the occurrence of the COVID-19 pandemic. While the pandemic has not been the only factor driving the adoption of digital care visits, it still played a significant role in the process. Due to the widespread infection, most countries implemented public restrictions and recommendations such as minimizing or banning gatherings, countrywide lockdowns, social distancing, wearing protective masks, and paying special attention to hand hygiene to control the contagion [14]. The infection rates increased exponentially during the first and second waves of the pandemic, and there were large numbers of severely ill patients that needed immediate hospitalization and even intensive care [14]. This resulted in an unusually high workload for the healthcare sector; multiple hospital wards were transformed into COVID-19 wards due to the shortage of beds in intensive care units [15–17]. The situation was so severe that routine visits and other non-emergency procedures had to be postponed, prioritizing COVID-19 patients [14,17].

On account of the circumstances, health care institutions were required to rapidly adopt and implement digital care visits in their practice to be able to provide telemedicine services [18,19]. The urgency of the situation sped up the process of authorization and regulation regarding legal matters such as new payment models for remote health care services and health information privacy [19]. Digital care visits got implemented in various areas of health care—primary care, mental health [20,21], orthopedic care [22], neurology [18,23,24], palliative care [25], pharmacy [26], dentistry [27], and others. Even though digital care visits do not provide possibilities for physical examinations where a healthcare professional would need to examine a patient physically, video consultations allow specialists to evaluate and sometimes diagnose by inspecting the patient through video. The pandemic has brought massive challenges and burdens to this world. Still, it also stimulated people to adapt and seek quick and creative solutions, speeding up technology implementation in different areas, including the health care sector.

Some research has been done regarding the use of video conferencing, implementation issues, policies, etc. [11], along with patients' experiences and perceptions of using video-conferencing for healthcare visits [6,28–30]. However, the number of studies on healthcare professionals' experience with patient-initiated digital visits is limited.

A broader overview, including healthcare professionals from different specializations and the latest literature, could contribute to a better understanding of what is known on this topic, what the research gaps are, and what should be studied more. Finding out what the benefits and challenges of using digital care visits are from the healthcare professionals' perspective could help optimize the service for both health workers and patients. Thus making it safer and more usable, resulting in higher satisfaction with the service as well as more efficient use of limited healthcare staff resources.

Aim

The aim of this study is to explore the existing literature concerning the user experience of digital care visits from different healthcare professionals' points of view.

2. Materials and Methods

A scoping review design was chosen for this study. The review was conducted using the methodological framework by H. Arksey and L. O'Malley [31] and adapted PRISMA-ScR checklist by Tricco et al. [32]. Scoping reviews are "a type of knowledge synthesis, follow a systematic approach to map evidence on a topic and identify main concepts, theories, sources and knowledge gaps" according to A.C.Tricco et al. [32]. This type of review may vary in the breadth of the literature coverage and the depth of the information elicited from it [32].

2.1. Search Strategy and Timeframe

Specific search terms and their combinations for finding the literature were thoroughly researched and tested. The search terms were chosen based on the aim of this study and were adjusted to retrieve the most relevant studies that fall under the scope of the selected topic. MeSH term "telemedicine" was used in the test searches and retrieved a large number of results, out of which many were irrelevant as there were publications on phone consultations, remote monitoring, wearable tracking/monitoring devices, etc. Therefore, to narrow down the search and retrieve more relevant results, this term was not used and was replaced with more specific keywords. In addition, the queries were adapted to match each chosen database's syntax. Three databases were chosen for the search: PubMed, Web of Science, and IEEE Xplore. The search queries used for the selected databases are presented in Table 1. Special tags and MeSH terms were used for targeting the most relevant studies—tag TIAB was used in PubMed for searching in the title, abstract, and keywords, MH for MeSH terms, TS for searching in the title, abstract, and keywords in Web of Science, and "All metadata" for searching in IEEE Xplory. IEEE Xplory digital library does not use MeSH terms. Thus, additional synonyms to some keywords were added to expand the search. The filters applied for the searches were 10 years' time span, English language, and full text available.

Table 1. Search strategy and the number of papers retrieved from the databases. The asterisk (*) in the search quies in PubMed and Web of Science represents any group of characters. It also represents no character.

Database	Search Words	Number of Papers
PubMed	("digital visit*" [TIAB] OR "remote visit*" [TIAB] OR "remote consult*" [TIAB] OR teleconsultation [TIAB] OR "online consult*" [TIAB] OR "video consult*" [TIAB] OR videoconferencing [MH] OR videoconferencing [TIAB] OR "digital consult*" [TIAB] OR e-consultation* [TIAB] OR "electronic visit" [TIAB] OR "virtual visit" [TIAB]) AND ("medical professional*" [TIAB] OR "medical staff*" [MH] OR "medical staff*" [TIAB] OR "health personnel*" [TIAB] OR "health personnel*" [MH]) AND (experience* [TIAB] OR "user experience*" [TIAB] OR "user satisfaction" [TIAB])	n = 122
Web of Science	TS = ("digital visit*" OR "remote visit*" OR "remote consult*" OR teleconsultation OR "online consult*" OR "video consult*" OR "electronic visit*" OR "virtual visit*" OR "telemedicine*" OR "telehealth*" OR video conference* OR e-consult* OR e-health) AND TS = ("medical professional*" OR "medical staff*" OR "health* personnel" OR physician* OR nurs* OR therapist* OR midwi* OR "health* professional" OR "dentist*" OR "caregiver*" OR "pharmacist*") AND TS = (experience* OR "user experience*" OR "user satisfaction")	n = 1289
IEEE Xplore	(("All Metadata": "digital visit" OR "All Metadata": "remote visit" OR "All Metadata": "remote consult" OR "All Metadata": teleconsultation OR "All Metadata": "online consult*" OR "All Metadata": "video consult*" OR "All Metadata": "electronic visit" OR "All Metadata": "virtual visit" OR "All Metadata": telemedicine OR "All Metadata": telehealth OR "All Metadata": videoconferenc* OR "All Metadata": e-consult* OR "All Metadata": e-health) AND ("All Metadata": "medical professional" OR "All Metadata": "medical staff" OR "All Metadata": "health personnel" OR "All Metadata": "health professional" OR "All Metadata": physician OR "All Metadata": nurs OR "All Metadata": therapist OR "All Metadata": midwi* OR "All Metadata": dentist OR "All Metadata": caregiver OR "All Metadata": pharmacist) AND ("All Metadata": experience* OR "All Metadata": "user experience" OR "All Metadata": "user satisfaction"))	n = 59

Apart from the database search, grey literature ("includes a range of documents not controlled by commercial publishing organizations") [33] was searched using similar search terms through Google Scholar and reviewing the reference lists of included studies to identify literature that has not been formally published in scientific journals. Manuscripts that were not yet published, conference papers, dissertations, government documents, and other types of grey literature [34] were searched and screened for eligibility. The search was carried out from 1 March 2021–15 April 2021.

2.2. Study Selection

The literature was screened for eligibility based on inclusion and exclusion criteria. The inclusion criteria were original articles, conference proceedings, review articles, and reports published within the last 10 years in English, focused on healthcare professionals' experience using digital care visits for patient consultations. Papers that fell under the scope and were published within a specified time frame and were retrieved during the "grey literature" search were also included. The exclusion criteria were articles published in languages other than English and earlier than 2011. They focused on patients' experiences using digital care visits or covering healthcare professionals' willingness to use digital care visits rather than their experience using it.

For this review, 1440 studies were retrieved and 44 duplicates were removed—more detailed numbers can be found in the flowchart (Figure 1). Citations were handled using the referencing program Mendeley. The initial screening was performed by reading the titles and abstracts of the retrieved results. After the screening, 97 studies were read in full to decide which to include in the review. Out of those 97, 28 studies met the inclusion criteria and were deemed eligible for this study. Studies were excluded if they focused on remote consultations via phone, asynchronous telemedicine using store and forward technology, clinician attitudes towards telemedicine or willingness to use it, healthcare professionals' experience of using telemedicine for professional-to-professional consultations, or remote monitoring. Studies that explored healthcare professionals' and patients' or caregivers' experience with digital care visits were included if separating clinicians' experience from the results was possible. Papers in which healthcare professionals' experience using several methods for providing telemedicine were studied and deemed eligible for the review if it was possible to separate the experience from digital care visits.

2.3. Data Analysis

The 28 studies were read again and the information was charted in an MS Excel spreadsheet. Details such as title, publication date, study design, the type of healthcare professionals who participated in the study, country/region, main findings, and other relevant information were documented in the spreadsheet. The emergent themes are presented in the results.

The collected information was analyzed using thematic content analysis. Several themes had emerged, including positive experiences/advantages, facilitators, negative experiences/challenges, barriers, and possible improvements in using digital care visits from healthcare professionals' experiences. The themes were divided into categories and sub-categories.

2.4. Ethical Considerations

Ethical issues were considered for this study, although no human subjects were involved due to the nature of this study. The data analyzed in this review is from published articles and reports that are freely or institutionally accessible. No sensitive data such as real medical records were used, meaning that no one's integrity was compromised.

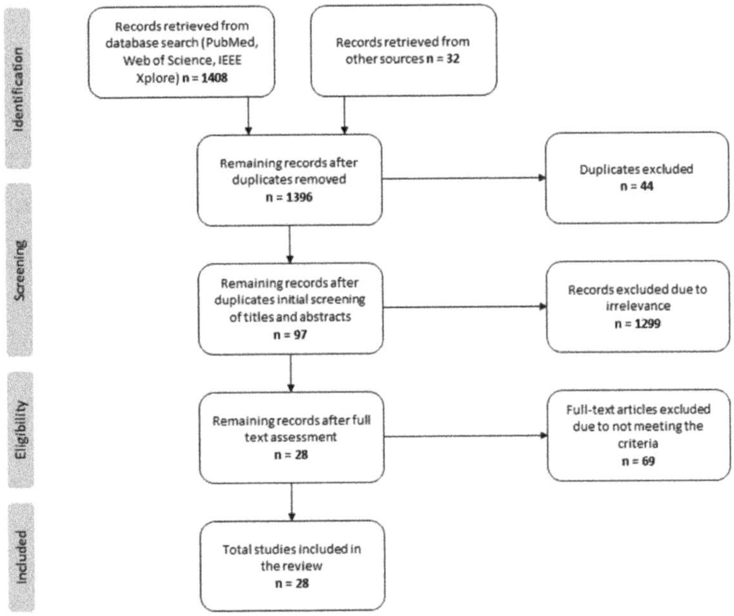

Figure 1. The process of study selection—Preferred Reporting Items for Systematic reviews and Meta-Analyses (PRISMA) flow chart.

3. Results

3.1. General Characteristics of the Reviewed Studies

Most of the selected studies were published in the last 5 years, while only one study was published earlier in 2015. More than two-thirds of the included papers were recent and published in 2020 or 2021. The studies were carried out in different countries, Australia (n = 4) and Europe (n = 10), while half of the studies originated from the United States of America (n = 14). More than a half of the studies explored the experiences of physicians, among which were medical oncology professionals [35,36], general practitioners [13,37–39], otolaryngologists [40], urologists [41], cardiologists [42], and sports medicine professionals (physiatrists) [43,44]. Another considerable group of professionals was mental health professionals—therapists and psychotherapists—who participated in eight studies. The experiences of other healthcare professionals such as nurses, advanced practice professionals, dieticians, and physical therapists were studied in nine papers. Non-medical professionals, patients, and caregivers were included in some studies; however, their experiences were separated in the results, and findings regarding their experience were not included in this review. Fifteen studies were related to the ongoing COVID-19 pandemic (Table 2).

Table 2. General characteristics of the included studies.

Characteristics	Number of Studies	Reference Number
Year of publication		
2021	n = 13	[38–40,43,45–53]
2020	n = 9	[13,35,37,41,42,44,54–56]
2019	n = 2	[36,57]
2018	n = 1	[58]

Table 2. *Cont.*

Characteristics	Number of Studies	Reference Number
2017	n = 2	[59,60]
2015	n = 1	[61]
Country		
Australia	n = 3	[36,49,59]
Belgium	n = 1	[42]
France	n = 1	[40]
Italy	n = 1	[54]
Norway	n = 2	[38,60]
Sweden	n = 3	[13,37,39]
The Netherlands	n = 1	[55]
USA	n = 14	[35,43–48,50–53,57,58,61]
Worldwide	n = 2	[41,56]
Methodology/type		
Qualitative study—semi-structured interviews	n = 5	[36,37,39,51,59]
Qualitative study—Focus groups	n = 1	[60]
Web-based survey	n = 12	[13,38,40,41,43,44,48,50,53,54,56,61]
Randomized controlled trial	n = 1	[58]
Descriptive study	n = 1	[52]
Observational survey	n = 1	[55]
Mixed methods study	n = 5	[35,45–47,57]
Design thinking—customer journey	n = 1	[42]
Review	n = 1	[49]
Study participants		
Mental health professionals	n = 8	[36,45,46,49,52,54,56,57]
Physicians	n = 15	[13,35,37–44,48,50,53,58,61]
Surgeons	n = 1	[55]
Nurses/nurse assistants/advanced practice professionals/residents/physical therapists /speech pathologists etc.	n = 8	[36,42,47,48,51,59–61]
Patients and/or caregivers	n = 7	[40,43,55,58,60,61]
Non-medical professionals (education staff, IT workers, social workers, care coordinators)	n = 4	[36,43,48,61]
Studies related to COVID-19 pandemic	n = 15	[35,36,40–50,52,54,56]

Five major themes emerged from the data—positive experiences/benefits, facilitators, negative experiences/challenges, barriers, and possible improvements in digital care visits. Each of these themes had multiple categories and sub-categories. The categories will be used as subheadings further on. The results of each will be presented and explained in more detail. (Tables 3–5).

3.2. Positive Experiences/Benefits of Digital Care Visits

Numerous benefits and aspects of a positive experience have been reported in the reviewed literature. Aspects such as benefits of remote work, efficiency, satisfaction with digital care visits, and benefits for the patient were identified (Table 3).

Table 3. Positive experiences/advantages and facilitators of the digital care visits use.

Theme	Category	Sub-Category	Reference
Positive experiences/ advantages	Benefits of remote work	Flexible working hours and/or place	[13,37,39,45,46,54]
		Saved travel time/costs	[39,41,46]
		Feeling more relaxed and at ease	[39,54]
		Convenience	[45,46]
		Reduced workload	[37,39]
	Efficiency	Shorter visits	[13,47,48,50]
		Increased productivity/efficiency	[35,37,39,41,47,49,54,59]
	Satisfaction	Overall positive experience	[13,35,40,41,43–46,48,50,55,58,59]
		Easy to learn how to use	[41,44]
		Easy to use	[13,39,41–43,46,48,50,55,59–61]
		Satisfaction with the system/ platform and/or its features	[13,46,47,58]
		Comfortable treating patients via digital care visits	[39,47]
		The interaction between healthcare professional and patient was satisfactory/effective	[39,43,48,55,58]
	Convenient, accessible care and saved resources for patients	Increased flexibility	[13,39,46,51,57,59]
		Greater accessibility	[13,39,41,46,51,52,58,59]
		Convenience	[45,58]
		Reduced costs and/or time for traveling	[13,46,51,52,57–59]
		Eliminated other costs	[51,57,58]
		Protection from communicable diseases	[39,51]
		Family inclusion and/or education	[51]
		Proper care for patients	[39,41,58]
	Patients' emotional state	Reduced stress, empowerment	[39,45,46,49,58,59]
		Confidence and increased cooperation	[45,46,59,60]
	Patient satisfaction	Satisfaction with digital care	[37,38,48,53]
Facilitators	New perspectives in remote care	Ability to get instant non-verbal feedback	[59]
		Ability to intervene in real-time	[57]
		Focusing on what is most important	[49,59]
		Less demanding	[39,59]
		Increased personal safety	[39,46]
		Observing themselves on video is helpful	[52]
		Insight into patient's home environment	[35,39,45,46,57]
		More frequent visits	[46,49,51]
		Continuity of care	[45,46]
		More personal visits	[59]
		Visits can be intimate	[52]
		Better than phone call consultations	[13,60]
	Technical qualities	Video and audio quality is acceptable/good	[13,37,38,43,48,55,59,61]
		No connectivity issues	[55]
	Possibilities of digital care visits	Possibility to consult/examine/diagnose/treat patients	[35,39,43,45,48,57,58,61]
		Possibility to work with patients' emotions	[52]
		Possibility to build rapport with patients	[45,46]
		The relationship with patients was authentic	[48,56]
	Suitability	Suitable for delivering sensitive/bad news	[47]
		Suitable for follow-up visits	[38,40,54,55]

Table 3. *Cont.*

Theme	Category	Sub-Category	Reference
		Suitable to treat mental health problems	[38,39]
		Suitable for treating some skin conditions	[39,51]
		Suitable for administrative purposes	[38]
		Physical contact was not necessary	[40,55]
		Suitable for chronic disease management	[40]
		Suitable for palliative care	[35]
		Suitable for pediatric care	[35]

Table 4. Negative experiences/challenges and barriers of the digital care visits use.

Theme	Category	Sub-Category	Reference
Negative experiences/ Challenges	Complicated decision making	Difficulties making decisions regarding patient's diagnosis, treatment, or referrals	[13,37–39,46,50,55]
		Difficulties in guiding the right patients to digital care visits	[13,37,39]
		The need to rely on patient's observations and descriptions	[35,39,51,52,59]
	Clinicians' professional competence development	Concerns regarding loss of competence	[37]
		Lack of medical skills practice	[37]
	Work environment	Loneliness and isolation working from home	[37]
	Workload and workflow	Requires higher concentration	[45,54,57]
		More tiring	[45,49,54,56,57]
		Difficulties structuring time	[37,54]
		More stressful	[50]
		Administration or preparation takes time	[36,49]
		Lack of administrative support	[35,36,41]
	Dissatisfaction	Overall dissatisfaction with digital care visits	[53]
		Felt that patients' needs were not adequately addressed	[53,60]
		Digital care visits are inferior to in-person visits	[35,45,50,52,53]
	Patient-professional relationship	Difficulty fostering rapport	[44,56]
		Difficulty in dealing with emotional situations	[45,49,52,56]
		Digital care visits are less personal	[52,60]
		Digital care visits were less intimate	[57]
		Difficulty in maintaining patient's attention/engagement	[45,46,56,57]
	Unmet patients' expectations	Patient's desire for physical consultation was unmet	[51,60]
		Unrealistic patient expectations and poor understanding	[13]
		Patients are reluctant to pay for digital care visits	[51]
	Technical challenges	Patients lack technical skills	[45,51]
		Patients lack comfort in using technology	[45,49,51,57]
		Restricted access to technology due to socioeconomic status	[51]

Table 4. *Cont.*

Theme	Category	Sub-Category	Reference
	Complications from the patient's side	More visit cancellations or rescheduling by patients	[57,59]
		Disruptions from patients' side	[45,57]
	Patient safety and privacy	Safety concerns	[37,38,45]
		Privacy concerns	[45,52,56]
Barriers	Physical barriers	Inability to apply certain treatment techniques	[36,43,54]
		Inability to provide written information	[51]
		Lack of physical examination is problematic	[36,38,39,43,44,46,48,50,51,55,58,59, 61]
		Inability to see non-verbal cues clearly	[45,48–50,52,57]
	Suitability	Inapplicable for some types of patients	[35,38,46,50,54]
		Inappropriate for sensitive conversations	[51]
	Technical issues	Connectivity issues	[36,42,45,46,50,52,53,58,60]
		Poor quality or lost audio and/or video	[42,46,52,53,58,60,61]
		Lack of technical support when working off office hours	[37]
		Lack of unified documentation system	[37]
		Difficult or uncomfortable to use	[35]
	Reimbursement issues	Ambiguity of insurance coverage status	[51]
		Training and administration time are not compensated	[36]
		Reimbursement model needs to be adapted	[13,41,44]

Table 5. Suggestions for improvement.

Theme	Category	Sub-Category	Reference
Suggestions for improvement	Training and education	Provide proper training in using the technology	[13,36,42,44,47,56]
		Tutorial materials on how to use the technology for professionals and/or patients	[55,59]
		Promotion and education on digital care visits	[41,44]
	System and tools	Standardized equipment for providers	[47]
		Incorporate video-conferencing tools into the EHR system	[37]
		Implement triage system	[13,37,50]
		Enhanced data security	[42]
		Use double web-cameras	[49]
	Clinician support	Promotion of self-care for healthcare professionals	[46]
		Incorporate administration/ coordination support	[42,44,47]
		Ensure access to a suitable work environment and tools	[45]

3.2.1. Benefits of Remote Work

Flexibility regarding working hours and the workplace has repeatedly been reported in the literature. Cioffi's study has found that nearly 60% of psychotherapists reported greater flexibility [54]. In Koch and Guhres's research, physicians expressed that digital care visits allow "flexibility to work from home" and "flexibility regarding working time" [13]. Hardy et al., in their mixed-methods study, found that therapists providing teletherapy for couples felt similarly "Flexibility in scheduling and location" [45]. Björndell & Premberg have found that "the flexibility of work and the regulated assignment online were positive for the physicians' work situation and well-being" [39]. Sugarman et al. have also discovered that therapists identify "scheduling/flexibility" as an advantage [46]. The same findings mentioned in Fernemark et al. study—flexibility with work hours and the ability to choose where to work from, were considered advantages by general practitioners in Sweden [37].

Less travel time and costs were also mentioned as benefits in several publications. Physicians from three studies think the use of digital care visits saves commuting time [39,41,46]. This plays a role in enhancing healthcare professionals' quality of life: "saving travel time, being present at home, and participating in family activities, etc., was considered beneficial" [39].

Digital care visits provide flexibility which contributes to healthcare professionals feeling less stressed and more at ease. A small percentage of psychotherapists in an Italian study agreed that they felt more relaxed during online sessions [54]. Björndell & Premberg wrote that "working from home was appreciated by the physicians because it let them work in peace, feel less stressed, and enjoy being at home" [39].

Sugarman et al. reported that digital care visit "supports personal safety concerns, including COVID-19 risk" [46], and Björndell & Premberg mentioned "reduced risk of infection transmission" in their paper [39].

3.2.2. Efficiency

It was indicated in several studies that working with digital care visits is more efficient in the sense that it saves time and increases productivity. Kemp et al. [47], Koch & Guhres [13], as well as Saiyed et al. [48] studies have shown that digital visits took less time than in-person visits, according to healthcare professionals. A total of eight of twenty-eight selected studies indicated that digital care visits increase productivity or efficiency. Healthcare professionals felt that by using telemedicine, they could be more productive [41,54], more structured and efficient due to greater focus during the sessions [49]. The use of technology made the visits more efficient [35], meaning that the patients were prepared, and physicians could easily end the video calls after the consultation and consult another patient right away [47,59]. Some physicians reported that it was easier to consult patients via digital care visits as their cases were simpler than those in the physical visits, making it possible to provide consultation to more patients [37,39].

3.2.3. Satisfaction

Overall, healthcare professionals, regardless of specialization or location, had mostly positive experiences with digital care visits, ranging from at least slightly [45] to highly satisfied [35,40,44], as stated in nearly half of the selected publications. Professionals felt that digital care visits have a positive impact on their work environment [13], were generally happy with their experience [41,43,55,58,59], and enjoyed it [48].

The usability of digital visits varied due to the use of different platforms. Some studies stated that systems used for digital care visits were easy [41,44], and twelve out of twenty-eight studies found that they were quite straightforward and easy to use [13,39,41,43,46,48,50,55,59–61]. Several publications revealed that healthcare professionals were satisfied with the system/platform itself [13,47,58] and/or appreciated its features [46].

Regarding the interaction between healthcare professionals and patients, healthcare professionals in six studies expressed having a positive experience regarding patient-professional interaction. Clinicians could discuss patients' issues, assess their condition, and offer treatment advice effectively. In many cases, an in-person visit was unnecessary [39,43,48,55,58].

3.2.4. Convenient, Accessible Care and Saved Resources for Patients

Increased flexibility and greater accessibility are some of the benefits of digital care visits. Digital care visits allow patients to schedule visits at their convenience [13,39,46,58,59]. Patients with responsibilities for, e.g., caring for their children, do not need to organize childcare for visiting healthcare professionals [46,51,52,57]. Digital care visits offer high-quality medical care for patients from rural or remote areas where such care is inaccessible [13,39,41,52,58].

In addition, patients who are homebound due to their medical conditions or those who simply do not have the means or wish to travel to a health care facility benefit from having their visit digitally [46,51,57,59]. Saving travel costs was mentioned multiple times in eight reviewed articles [13,46,51,52,57–59]. Moreover, patients feel more at ease when they are surrounded by the environment and people that they are used to, such as their family or their pets [39,45,58,59]. Digital care visits were emphasized in mental health-related studies as well. Studies showed that patients tended to be more open, feel more secure, and often shared more with their therapists during the online sessions from their homes [45,46,49]. Having remote visits from patients' homes allows for better family member inclusion [51], knowing more about their condition, and caring for them [51]. Hinman et al. study on remote physiotherapy found that patients felt empowered when doing exercises at home. Digital care visits increased their adherence to the program and allowed them to learn correct and safe exercise techniques [59]. Patients gained more confidence in performing rehabilitation exercises at home [59] and took more initiative to care for themselves and be more self-reliant [60]. They even could form stronger therapeutic alliances or cooperation with therapists [45,46].

Furthermore, by having remote health care visits, patients avoided transmission of and exposure to communicable diseases, which enhanced their safety and contributed to controlling the spread of infectious diseases [39,51]. Overall, clinicians from several studies indicated they felt their patients were satisfied with digital care visits, their complaints were addressed, and they got the necessary care [37,38,48,56].

3.3. *Facilitators*

3.3.1. New Perspectives in Remote Care

Digital care visits employ video-conferencing technology, opening new perspectives in remote care. The ability to get instant non-verbal feedback through video, i.e., seeing the patients' facial reactions and body language, enables healthcare professionals to get more unspoken information from the visit [59]. Seeing a patient's symptoms during the video consultation allows health care professionals to intervene in real-time [57]. Some clinicians thought that caring for patients remotely made them focus more on what was the most important in the treatment [49,59].

Clinicians noted that digital care visits felt more personal in Hinman's study because physical therapists had to listen to their patients to provide good service [59]. Levy et al. stated that in a therapeutic setting with a close-up video of the patient's face, the session could be as intimate as in-person [52]. Being able to see the patient was one of the reasons for healthcare professionals' preference for digital care visits over phone consultations [13,60]. Interestingly, another new perspective brought by digital care visits, which was not present in traditional visits, is a possibility to get insight into a patient's home environment. This allows clinicians to get a better overview of the patient's life and gives valuable insight into how they communicate, e.g., with their relatives or pets if they are in the picture, which creates a unique possibility to "get closer" to the patient and many healthcare professionals appreciated that [35,39,45,46,57]. According to clinicians, digital care visits allow for shorter and more frequent visits [46,49,51] and ensure continuity of care [45,46], supporting access to care for patients.

3.3.2. Technical Qualities

Eight out of twenty-eight studies indicated that technical features such as audio and video were of good quality or that there were no issues during the visits [13,37,38,43,48,55,59,61]. Clinicians thought they could hear and see patients well enough to provide healthcare services.

3.3.3. Possibilities of Digital Care Visits

Using video-conferencing technology for digital care visits, it is possible to consult, examine, and diagnose patients, as stated in eight reviewed studies [35,39,43,45,48,57,58,61]. Digital care visits seemed like a suitable form of care for some clinicians [58]. A total of

57% of therapists in Becevic's et al. study reported that they could treat patients via digital care visits [61]. Furthermore, 83% of couple therapists in Hardy et al. study replied that they could at least somewhat solve the conflicts as effectively as in in-person visits [45]. In the Kirby et al. survey, surgeons were fairly confident in their diagnoses and assessments [43]. Other studies showed that clinicians were comfortable treating their patients or that their patients were appropriate subjects for getting treatment via telemedicine [35,39,48]. Several studies indicated that clinicians felt they could establish a connection with patients, an imperative part of patient-clinician interaction [45,46]. Some even stated that the relationship with patients was as authentic as face-to-face visits [48,56].

3.3.4. Suitability

A total of four of the selected studies explained that digital care visits are best suitable for follow-up visits, as it is easier to consult a patient who is known and whose condition is not completely new for the healthcare professional [38,40,44,55]. Some other studies showed that, in clinicians' opinion, digital care visits are suitable for treating mental health problems as no physical examination is required [38,39]. In addition, digital visits are appropriate for some less complicated skin conditions if the video quality is good enough [39,46]. This type of visit is also suitable for palliative and pediatric care [35], for chronic disease management [40], and for administrative purposes such as extending a sick leave for working patients [38].

3.4. *Negative Experiences/Challenges of Using Digital Care Visits*

Despite numerous advantages of digital care visits, multiple drawbacks and challenges are reported in the literature. Clinicians have encountered decision-making issues, workload and workflow problems, patient-professional relationship-related considerations, patient-related challenges, or low satisfaction. These negative experiences and challenges will be presented further on.

3.4.1. Complicated Decision Making

Seven studies out of twenty-eight declared that clinicians experience difficulties making decisions regarding a patient's diagnosis, treatment, or referrals [13,37–39,46,50,55]. In Koch & Guhres's paper, physicians reported that "information for decision making is limited" in digital care visits [13]. The Johnsen et al. study revealed that 15% of GPs were worried that they had possibly missed signs of serious disease. In addition, more than half of the physicians considered the inability to perform a physical examination was a serious disadvantage [38]. In another publication, it was explained that physicians think digital care visits will never be able to replace hands-on examination [55]. Sugarman et al. articulated that according to therapists' experience, it was complicated to easily treat distracted patients, to visualize their psychomotor symptoms, measure vital signs, and prescribe medication based on their observations and discussions during the digital care visit [46]. Other authors got similar findings regarding these difficulties [37,39,51,52,59]. Physicians and therapists also saw disadvantages in having to rely on the patients' observations and descriptions to diagnose, assess, or prescribe proper treatment [35,39,51,52,59]. GPs were sometimes hesitant about trusting a patient's complaints without an examination when extending their sick leaves or prescribing medication [39].

Several studies revealed that healthcare professionals had difficulties guiding the right patients to digital care visits. It was complicated for physicians to sort the patients whose conditions were appropriate to be treated via digital care visits, who needed to have an in-person visit, and who could have their problems solved by other health professionals, e.g., by nurses to utilize the limited healthcare resources efficiently [13,37,39].

3.4.2. Professional Development and Work Environment

Clinicians' concerns, such as lack of medical skills practice and loss of competence, were raised in the Fernemark et al. paper [37]. GPs worried that digital care visits often deal with simpler cases where physical examinations and more complicated medical manipulations are unnecessary. They were concerned that by working exclusively with digital care visits, they would lose some of their skills and competence [37].

3.4.3. Workload and Workflow

Clinicians in several studies reported that digital care visits require a higher level of concentration compared to traditional visits [45,54,57]. Over 55% of therapists said digital care visits require a higher concentration level in the Cioffi et al. study [54]. Therapists from another study pointed out that 30% of them experienced less engagement, that they had to work harder because they needed to monitor technical aspects of the session, that it tended to non-verbal communication (difficult for 80% of the therapists), and that disruptions during the visits occurred for 92% of the respondents [57]. Therapists in the Hardy et al. study experienced "clinician fatigue—lethargy, tiredness, and discomfort" and claimed digital care visits were more tiring [45]. Similarly, other studies reported that treating patients online is more tiring as the clinician needs to compensate for the absence of physical presence, focus harder, and use senses other than touch to assess patients, as well as often helping patients with technology issues and dealing with a sometimes higher workload [45,49,54,56,57]. Healthcare professionals from the Cioffi et al. and Fernemark et al. studies felt it was more difficult to structure their time when working from home, and they were unsure as to when and if they should take breaks [37,54].

Around one-fifth of the physicians who participated in the Gold et al. survey replied that they experience increased stress while working with digital care visits [50]. Some identified that the type of digital care visits conflicted with their views on how care should be delivered [50]. Johnsson et al. and Paulik et al. discovered that clinicians feel administration and preparation for digital care visits takes a lot of time, because they must adapt certain treatment techniques to a new setting [36,49]. Other authors found that clinicians experience a lack of administrative support, and they need to schedule visits and do other tasks, that a secretary or a nurse could take over, instead of focusing on treating patients [35,36,41].

3.4.4. Dissatisfaction

One study from the USA reported that 58% of the participating physicians were generally neutral or dissatisfied with digital care visits. Nearly half were concerned that the healthcare professional-patient relationship was compromised because of digital care visits [53].

Two other studies discovered that clinicians felt their patients' needs were not adequately addressed as some patients could not get the necessary care online, or wished for a physical presence and social interaction with the clinician [53,60]. Overall, five studies reported that healthcare professionals consider digital care visits inferior to face-to-face visits and prefer traditional visits over digital ones [36,45,50,52,53].

3.4.5. Patient-Professional Relationship

Several studies addressed the issue of the healthcare professional's difficulty fostering rapport with their patients [44,56]. Bekes et al. and Tenforde et al. reported that healthcare professionals felt they had difficulties connecting emotionally to the patient [44,56]. Mental health professionals expressed that it was difficult to deal with emotional situations in digital care visits [45,49,52,56] mainly due to the inability to properly see patients' body language and facial expressions and the inability to use certain conflict management techniques from a distance [45,49,52,56].

Another concern regarding the patient-professional relationship was that digital care visits are less personal than face-to-face visits. This concern was reported in two studies that explained it happened due to a lack of physical presence and being there for the patient [52,60]. Similarly, therapists from the Wade et al. study felt that therapy sessions via digital care visits were less intimate [57]. Mental health workers experienced difficulty maintaining patients' attention and engagement due to their condition or distractions at home [45,46,56,57].

3.4.6. Unmet Patients' Expectations

A few studies reported that healthcare professionals felt their patients' desire for physical consultation was unmet, especially oncological and geriatric patients [51,60]. At times, according to clinicians, patients desired to be examined physically to feel more secure regarding their diagnosis [51] or because social interaction was important for homebound patients [60]. Koch & Guhres found that physicians reported patients having unrealistic expectations or poor understanding of what could be done during digital care visits, resulting in dissatisfaction [13].

3.4.7. Technical Challenges

The fact that patients have different socioeconomic statuses was related to their access to technology such as smartphones, tablets, and computers [51]. This meant that not all patients got access to digital care visits. In addition, patients' lack of technological skills [45,51] or patients' lack of comfort in using technology [45,49,51,57] often prevented successful interaction via digital care visits.

3.4.8. Complications from the Patients' Side

Other challenges clinicians had to deal with were more visit cancellations or rescheduling by patients due to increased flexibility, as reported by Hinman et al. and Wade et al. [57,59]. Disruptions when patients get distracted by their family members or daily chores also had a negative effect on the overall experience [45,57]. Moreover, Heyer et al. stated in their study that sometimes patients do not feel they should pay for digital care visits the same as they do for traditional ones [51].

3.4.9. Safety and Privacy

Several studies have covered clinicians' concerns regarding patient safety, privacy, confidentiality, and informed consent. Patients' immediate safety due to acute conditions and the need for emergency hospitalization [37,38] or safety regarding conflicts at home and risks posed by their mental state [45] concerned clinicians. Privacy was an issue discussed in three studies. In therapy sessions, privacy is important, and it is severely compromised when a patient is unable to find a place in their homes where they feel secure and cannot be overheard by family members [45,52,56].

3.5. *Barriers*

3.5.1. Physical Barriers

Using digital care visits for treatment sometimes poses barriers, such as the inability to apply certain treatment techniques that could otherwise be used in a face-to-face visit. Cioffi et al. showed that 50.69% of responding therapists felt digital care visits restrict or prevent applying certain techniques [54], and surgeons from Kirby et al. had similar experiences [43]. Allied health specialists reported that the medical interventions were limited to those who did not require a trained occupational therapist's presence. Therefore, patients were less successful in reaching motor goals [36].

A similar problem occurred to healthcare professionals trying to examine the patients. In 13 studies, clinicians reported that lack of physical examination was problematic [36,38,39,43,44,46,48,50,51,55,58,59,61]. Surgeons in Kirby et al. pointed out that they had difficulty measuring sensation and tenderness [43]. In addition, occupational therapists sometimes struggle to evaluate motor skills [36]. Mammen et al. found occasional technical problems and the inability to touch sometimes hindered the physician's ability to conduct the examination [58]. Physical therapists experienced discomfort without hands-on assessment [59]. Other studies showed that clinicians thought not having a physical examination was a loss, and digital care visits cannot replace hands-on examination [38,39,44,46,48,50,51,55,61]. Mental health workers [45,49,52,57] and physicians [48,50] considered the inability to see non-verbal cues as a disadvantage.

3.5.2. Suitability

Clinicians expressed that digital care visits were not always a suitable form of care for some patients. Studies have found that digital care visits were less applicable for new patients [38,50]. This also applies to patients that have musculoskeletal, skin [38,50], pediatric problems, acute and severe health issues [38,51], or conditions that certainly require physical examination [35]. They were also unsuitable for patients with severe mental problems such as paranoia, psychosis, etc. [46]. Medical oncology professionals noted that digital care visits were inappropriate for sensitive conversations with the patient, such as for delivering bad news [51].

3.5.3. Technical Issues

Technical issues may become a serious barrier to providing quality care. In ten of twenty-eight studies, authors reported that healthcare professionals had encountered connectivity issues. Lost connection [36,45,51,52,60], difficulty logging on [46], patients not being able to connect [42], poor or unstable internet connection [50,60], over half the clinicians in Mammen et al. and Yu et al. indicated they experienced audio, video, and connectivity issues [53,58]. Healthcare professionals from other studies also expressed they had problems with poor quality sound or video during the visits, which affected the overall quality of the consultation as it was more difficult to communicate and assess the patient [42,46,52,60,61].

3.5.4. Reimbursement Issues

Healthcare professionals mentioned problems related to reimbursement for digital care visits. Due to rapid telehealth adoption, the insurance companies have not adapted their policies for coverage regarding digital care visits, which is problematic for healthcare professionals [51]. Allied health therapists experienced they had to spend a lot of time training to use digital care visits and administrate them. At the same time, they were only compensated for the factual duration of the visits, not the preparation, which posed a risk of job dissatisfaction [36]. Other authors suggested the reimbursement models should be adapted to offer fair pay for healthcare professionals providing care via this technology [13,41,44]. Negative experiences and challenges are presented in Table 4.

3.6. Suggestions for Improvement

Lastly, a theme about possible suggestions for improving digital visits emerged from the reviewed literature. The findings suggest that main improvements could be done in training and education, improving tools, and adapting the system, as well as offering greater support for clinicians.

3.6.1. Training and Education

Providing proper training for healthcare professionals on how to use the technology and train them in providing health care services remotely would be beneficial and improve clinicians' experience, as well as increase their confidence in using digital care visits, as reported in six studies [13,36,42,44,47,56]. Preparing tutorial materials such as video clips or booklets concerning how to use the digital care visit platform to support both professionals and patients was indicated as a potential benefit [55,59]. Promotion and education about digital care visits could raise awareness and encourage and support healthcare professionals in using the technology [41,44].

3.6.2. System and Tools

Standardized equipment for providers would ensure that digital care visit platforms are supported by all used devices, and it would be easier to use, as stated in Kemp et al. [47].

Clinicians would also benefit from the video-conferencing tool being integrated into the EHR system they use routinely to grant easy access to patient records [37]. Additionally, healthcare professionals find it difficult to guide appropriate patients to digital care; therefore, implementing a triage system would be helpful [13,37,50]. In response to safety and confidentiality concerns, improvements could be made to enhance security by setting session passwords, end-to-end encryption, and ensuring GDPR compliance [42]. Paulik et al. suggested using two cameras for the patients—one showing a close-up image of the face and another capturing the whole body to improve visibility and understanding of non-verbal cues shared by the patient during therapy sessions [49].

3.6.3. Clinician Support

Providing clinicians with properly functioning devices and ensuring they have a suitable work environment that is private, quiet, and well-lit would contribute to the professional's comfort and the quality of the consultation [45]. Not putting a burden of excessive administration and coordination tasks that could be done by other staff on the clinicians [42,44,47] could help them better cope with the workload. Lastly, another important aspect is the promotion of self-care. It has been reported that digital care visits may be more tiring than regular sessions, and professionals caring for themselves to cope with fatigue caused by digital care visits is crucial [46]. Suggestions for improvement are presented in Table 5.

4. Discussion

This study aimed to explore the literature and determine the user experience of digital care visits from different healthcare professionals' points of view [62]. This study showed that healthcare professionals mostly had positive experiences with digital care visits. Many authors stated that healthcare professionals believe that digital care visits are advantageous for the professionals considering benefits such as remote work, efficiency, satisfaction with this type of consultation, and new perspectives in remote care. Clinicians particularly appreciated the ability to be flexible in terms of work hours, choosing the work environment, increased productivity and efficiency, as well as the ease of use of the technology. Similarly, when exploring patients' points of view, a systematic review has shown that the patients had overall high satisfaction with information sharing and consumer focus [6].

Many healthcare professionals agreed that digital care visits significantly increased the accessibility of health care services to patients who live in remote locations, are not able to travel to a health facility due to various reasons (limited or restricted mobility, social phobias, lack of resources, etc.,), or even those with responsibilities at home such as caring for young children or sick relatives. These findings align with other studies that explored the caregivers' and patients' points of view toward remote care [63,64]. A significant portion of studies declared clinicians found their patients became more confident in themselves, felt more at ease, and cooperated in their treatment better when they had the chance to stay in their home environment. Specifically, this was mentioned not only by mental health professionals, who rarely need to apply hands-on techniques in their treatment, but also by physical therapists who were teaching their patients exercise techniques and managed to achieve good outcomes [39,45,58,59]. Facilitators found in the literature were related to new perspectives and features of remote care, such as real-time video aspects that added visual information compared to consultations over the phone.

Health care professionals had mixed experiences with technical quality. Eight of the selected papers reported the quality as being good with no issues at all. In ten of the other studies, it was reported that clinicians experienced technical issues related to video/audio quality or connectivity issues from their or their patients' side [13,37,38,43,48,55,59,61].

The possibilities of digital care visits were rated positively among healthcare professionals. Physicians and therapists thought it was possible to consult, examine, diagnose, and treat patients via digital care visits. However, in almost half of the reviewed papers, healthcare professionals expressed that the lack of physical examination was at least somewhat problematic. Particularly, general practitioners and healthcare professionals who work with musculoskeletal disorders and oncologic patients found the inability to physically examine patients to be an obstacle in some cases [36,38,39,43,44,46,48,50,51,55,58,59,61]. Overall, there were mixed opinions on whether digital care visits could replace face-to-face visits. Health care professionals reported that digital care visits are suitable to assess some conditions, such as simpler skin conditions, mental health issues, and other conditions that did not require touch to assess as well as follow-up visits for chronically ill patients [38–40,46]. On the other hand, when the conditions were more complicated or the patient was new, clinicians reported that a physical visit would be more suitable [35,38,46,50,51]. Compared with another study, health personnel found both benefits and disadvantages of treating patients remotely. Some found it advantageous because the patients did not need to wait long to receive care; others expressed it was easier for them to write a referral rather than have a digital care visit [64].

The findings suggest that digital care visits are suitable for visits that involve treatment of rather simple conditions. Those that do not require a physical examination or do not involve sensitive conversations would be better managed in face-to-face visits. Naturally, selecting the right kind of patients for remote care would decrease the complexity of decision-making when a professional must rely on other senses and information collected without being physically present with a patient. This could be achieved by employing a triage system as suggested in two Swedish studies [13,37] and one American study [50]. By implementing a triage system that would filter the patients and direct them to the right type of care, limited medical resources could be utilized more efficiently. Other suggestions for improvement include training and educational materials, which could potentially improve healthcare professionals' experience in using digital care visits, as well as raise awareness among those who have not started to use it yet and encourage them to employ the technology [13,36,42,44,47,56]. Conversely, general practitioners from another study noted that even though reading manuals on how to use the technology were often helpful, they rarely found the time "to read and understand the instructions" due to tight scheduling [64].

Some healthcare professionals expressed digital care visits were not well integrated into their workflow. They felt unsure when prescribing medication to patients without knowing their health history [39]. Using separate video-conferencing tools, scheduling consultations, and coordinating remote care added to the workload [50], thus implementing the necessary tools into the EHR system could make the workflow smoother and allow healthcare professionals to access patients' health records, ensuring greater confidence for the clinicians and safety for the patient [37]. Another important aspect of remote consultations or remote work, in general, is fatigue that comes from communicating online and the feeling of isolation and loneliness from being unable to meet with peers. One study suggested that self-care should be promoted among healthcare professionals working remotely [46]. Online social activities for the healthcare teams such as communication channels, virtual social groups, peer support, and games or team challenges could be offered to healthcare professionals to provide them with an opportunity for casual and less formal communication with colleagues as a replacement for running into each other at the office [65]. This, in turn, could make them feel more connected to the team and less isolated.

Few of the studies in this review involved resident doctors or young professionals who do not have extensive work experience [50,51,53]. It was reported that they had more difficulties in using digital care visits. It was more complicated to assess and diagnose patients due to limited work experience [50,51,53], and therefore it is possible their experience with remote consultations was more negative. It is possible that clinicians with more in-person work experience would be a better fit for providing such services [39], and they would be more comfortable in such a setting. Alternatively, it could be beneficial if young professionals got mentorship or support from their more experienced peers whenever they needed to increase their confidence. Also, as mentioned in some of the reviewed articles, training and education on how to use the technology and provide remote health care would be beneficial [66].

It is worth mentioning that because of the COVID-19 pandemic, the adoption of digital care visits was extremely rapid, and many health organizations were unprepared for it. Health workers were pushed out of their comfort zones and forced to move to remote care abruptly without having the time to prepare or train for it properly, and many of the organizational changes had to be made suddenly to make the shift happen [45,46,54]. Therefore, the studies published during the pandemic were strongly influenced by these aspects, and healthcare professionals' experiences of using digital care visits were affected by the sudden shift as well as general stress and pressure caused by this unprecedented contagion. Many of the professionals have not used digital care visits prior to the pandemic, and the sudden change may have influenced their experiences more negatively. However, on the contrary, many stated an overall positive experience and would continue to use the technology to a smaller or larger extent in the post-pandemic future [35,38,41,44,45].

It is clear from the results that digital care visits will never fully replace in-person visits [67,68]. However, the studies showed that it is possible to provide health care services via digital care visits in cases that do not require a physical examination for the assessment, such as chronic disease management [40]. Patients with conditions such as diabetes who are consulted and monitored by healthcare professionals online may be able to manage their condition at home without the need for hospitalization, thus saving time and resources for both parties [68–70]. Almathami et al. performed a systematic literature review on "Barriers and facilitators that influence telemedicine-based, real-time, online consultation at patients' homes" in 2020 and found that the majority of their included studies (98 percent) proved the effectiveness of digital care visits in "improving patients' overall health conditions and in assessing patients' health conditions successfully" [71]. The same review found that more than a quarter of analyzed studies proved online consultations were as good as face-to-face visits [71]. However, digital care visits should not take over all in-person visits but act as a complement to the physical visits. It is important to note that the social interaction and physical presence facilitate better conditions for showing empathy and simply "being there" for the patient, which are essential parts of care and bear significant value to patients and professionals alike.

4.1. Implications of This Study

Admittedly, only a fraction of healthcare professionals' specializations was included in the reviewed studies. The knowledge of the experience of surgeons, midwives, dental care professionals, and specialized physicians other than those included in this study is limited and should be studied in the future to get a clearer picture of their perspectives. The COVID-19 pandemic is surely transforming remote care, and there have already been studies that described the shift towards digital care visits and the organizational changes [11,18,72,73]. However, a more detailed review of how the perspectives have changed and how the rapid adoption has affected the use and experience of clinicians could be beneficial in the future. Moreover, more research could be done on the usability of digital care visits integrated into the EHR systems because, so far, the clinicians mostly use separate platforms. In addition, more explorations of how digital care visits could be combined with in-person care and the perspectives of professionals, patients, and caregivers on this approach could be studied further. Finally, more studies regarding the use of digital care visits in self-management and follow-up of chronically ill patients are needed. Knowing how digital care visits affect patient safety is also of interest to be studied in the future. Moreover, studying the economic impact of digital care visits on health care is also of great importance.

4.2. Strengths and Limitations

One of the strengths of this study is a comprehensive search strategy in three large databases containing large amounts of healthcare and technology-related publications. The search was carefully documented. The search queries were tested and adjusted to retrieve more relevant results. MeSH terms were used to broaden the search. Many studies were retrieved from the databases, and an additional search for the grey literature was performed. Various types of publications and different study types were included in the review to ensure broad coverage of the topic. All the citations were managed using Mendeley's reference system to ensure orderly documentation.

This scoping review is not without limitations. Firstly, filtering the search results by the language (including only English papers) may have prevented getting more results and potentially missed data that could have been included in this study. Secondly, a limited time frame may have affected the quality and quantity of the findings. Additionally, lack of critical appraisal is one of the disadvantages of the scoping review type of studies and therefore poses a risk of bias [74]. The search terms and search queries were discussed in detail. The search queries were adjusted, and new search words were added several times. However, the screening and the selection of studies were performed by the first author, increasing the risk that some studies were missed [75].

5. Conclusions

To summarize, this scoping review explored the existing literature on the user experience of using digital care visits from different healthcare professionals' points of view. The themes of positive experiences/benefits, facilitators, negative experiences/challenges, barriers, and suggestions for improvement were identified. The findings suggested that overall, healthcare professionals had a positive experience with the use of digital care visits and found numerous benefits of this type of remote care for themselves as healthcare workers as well as for their patients. Despite the overall positive experience, clinicians reported challenges and issues they faced when using the technology, including decision-making difficulties, physical barriers, technical issues, suitability concerns, and others. Finally, it is suggested that digital care visits cannot replace in-person visits in full. However, they could be effectively used in combination to treat and manage suitable conditions. Further research could be done to explore the experiences of other healthcare professionals not represented in this study, as well as the effects of the COVID-19 pandemic on digital care visit use.

Author Contributions: I.L. and N.D. designed the study. I.L. and N.D. designed and discussed the search queries. I.L. undertook the data collection and screened all potentially relevant studies. I.L. performed the initial data analysis independently and discussed the categories and sub-categories with N.D. and I.L. prepared a first draft of the manuscript, and N.D. contributed to the writing. All authors have read and agreed to the published version of the manuscript.

Funding: The project was funded by the Health Informatics Center at Karolinska Institute. The design of the study, data collection, analysis, and writing of the manuscript was not affected by the funding.

Institutional Review Board Statement: Not applicable.

Informed Consent Statement: Not applicable. We only used publicly vailable datasets repared by other organizations and these datasets are standard to use for automatic diagnosis of retinal diseases.

Data Availability Statement: All relevant data are included in the article.

Conflicts of Interest: The authors declare no conflict of interest.The funders had no role in the design of the study; in the collection, analyses, or interpretation of data; in the writing of the manuscript, or in the decision to publish the results.

References

1. World Health Organization. *Telemedicine: Opportunities and Developments in Member States: Report on the Second Global Survey on eHealth*; World Health Organization: Geneva, Switzerland, 2010; pp. 8–11.
2. NCBI. Videoconferencing—MeSH. Available online: https://www.ncbi.nlm.nih.gov/mesh/?term=videoconferencing (accessed on 11 March 2021).

3. Granström, E.; Wannheden, C.; Brommels, M.; Hvitfeldt, H.; Nyström, M.E. Digital tools as promoters for person-centered care practices in chronic care? Healthcare professionals' experiences from rheumatology care. *BMC Health Serv. Res.* **2020**, *20*, 1108. [CrossRef] [PubMed]
4. Demi, S.; Hilmy, S.; Keller, C. Doctor at Your Fingertips: An Exploration of Digital Visits from Stakeholders' Perspectives. *Life* **2020**, *11*, 6. [CrossRef] [PubMed]
5. MacDonald, G.G.; Townsend, A.F.; Adam, P.; Li, L.C.; Kerr, S.; McDonald, M.; Backman, C.L. eHealth technologies, multimorbidity, and the office visit: Qualitative interview study on the perspectives of physicians and nurses. *J. Med. Internet Res.* **2018**, *20*, e8983. [CrossRef] [PubMed]
6. Orlando, J.F.; Beard, M.; Kumar, S. Systematic review of patient and caregivers' satisfaction with telehealth videoconferencing as a mode of service delivery in managing patients' health. *PLoS ONE* **2019**, *14*, e0221848. [CrossRef] [PubMed]
7. Melian, C.; Kieser, D.; Frampton, C.; Wyatt, M.C. Teleconsultation in orthopaedic surgery: A systematic review and meta-analysis of patient and physician experiences. *J. Telemed. Telecare* **2020**. [CrossRef] [PubMed]
8. Mooi, J.K.; Whop, L.J.; Valery, P.C.; Sabesan, S.S. Teleoncology for Indigenous patients: The responses of patients and health workers. *Aust. J. Rural Health* **2012**, *20*, 265–269. [CrossRef]
9. De Weger, E.; Macinnes, D.; Enser, J.; Francis, S.J.; Jones, F.W. Implementing video conferencing in mental health practice. *J. Psychiatr. Ment. Health Nurs.* **2013**, *20*, 448–454. [CrossRef]
10. Hensel, J.M.; Yang, R.; Vigod, S.N.; Desveaux, L. Videoconferencing at home for psychotherapy in the postpartum period: Identifying drivers of successful engagement and important therapeutic conditions for meaningful use. *Couns. Psychother. Res.* **2020**, *21*, 535–544. [CrossRef]
11. Santoro, S.L.; Donelan, K.; Haugen, K.; Oreskovic, N.M.; Torres, A.; Skotko, B.G. Transition to virtual clinic: Experience in a multidisciplinary clinic for Down syndrome. *Am. J. Med. Genet. Part C Semin. Med. Genet.* **2021**, *187*, 70–82. [CrossRef]
12. Dahl-Popolizio, S.; Carpenter, H.; Coronado, M.; Popolizio, N.J.; Swanson, C. Telehealth for the provision of occupational therapy: Reflections on experiences during the COVID-19 pandemic. *Int. J. Telerehabilit.* **2020**, *12*, 77–92. [CrossRef]
13. Koch, S.; Guhres, M. Physicians' experiences of patient-initiated online consultations in primary care using direct-to-consumer technology. *Stud. Health Technol. Inform.* **2020**, *270*, 643–647. [PubMed]
14. World Health Organization. Coronavirus Disease (COVID-19) Situation Reports. Available online: https://www.who.int/emergencies/diseases/novel-coronavirus-2019/situation-reports (accessed on 9 April 2021).
15. Forbes, P.; Finch, A. Redeployed Staff and Better Teamwork: How COVID-19 Has Transformed Nursing. *Nurs. Manag.* **2020**, *27*, 14–17. Available online: https://rcni.com/nursing-standard/features/redeployed-staff-and-better-teamwork-how-covid-19-has-transformed-nursing-161911 (accessed on 9 April 2021).
16. Pandey, N.; Kaushal, V.; Puri, G.D.; Taneja, S.; Biswal, M.; Mahajan, P.; Guru, R.R.; Malhotra, P.; Sehgal, I.S.; Dhooria, S.; et al. Transforming a General Hospital to an Infectious Disease Hospital for COVID-19 Over 2 Weeks. *Front. Public Health* **2020**, *8*, 382. [CrossRef] [PubMed]
17. Mehta, S.; Machado, F.; Kwizera, A.; Papazian, L.; Moss, M.; Azoulay, É.; Herridge, M. COVID-19: A heavy toll on health-care workers. *Lancet Respir. Med.* **2021**, *9*, 226–228. [CrossRef]
18. Grossman, S.N.; Han, S.C.; Balcer, L.J.; Kurzweil, A.; Weinberg, H.; Galetta, S.L.; Busis, N.A. Rapid implementation of virtual neurology in response to the COVID-19 pandemic. *Neurology* **2020**, *94*, 1077–1087. [CrossRef]
19. Keesara, S.; Jonas, A.; Schulman, K. COVID-19 and Health Care's Digital Revolution. *N. Engl. J. Med.* **2020**, *382*, e82. [CrossRef]
20. Chen, J.A.; Chung, W.-J.; Young, S.K.; Tuttle, M.C.; Collins, M.B.; Darghouth, S.L.; Longley, R.; Levy, R.; Razafsha, M.; Kerner, J.C.; et al. COVID-19 and telepsychiatry: Early outpatient experiences and implications for the future. *Gen. Hosp. Psychiatry* **2020**, *66*, 89–95. [CrossRef]
21. Lecomte, T.; Abdel-Baki, A.; Francoeur, A.; Cloutier, B.; Leboeuf, A.; Abadie, P.; Villeneuve, M.; Guay, S. Group therapy via videoconferencing for individuals with early psychosis: A pilot study. *Early Interv. Psychiatry* **2020**, *15*, 1595–1601. [CrossRef]
22. Tanaka, M.J.; Oh, L.S.; Martin, S.D.; Berkson, E.M. Telemedicine in the Era of COVID-19: The Virtual Orthopaedic Examination. *J. Bone Jt. Surg. Am.* **2020**, *102*, e57. [CrossRef]
23. Ganapathy, K. Telemedicine and Neurological Practice in the COVID-19 Era. *Neurol. India* **2020**, *68*, 555–559. [CrossRef]
24. Tarolli, C.G.; Biernot, J.M.; Creigh, P.D.; Moukheiber, E.; Salas, R.M.E.; Dorsey, E.R.; Cohen, A.B. Practicing in a pandemic. *Neurol. Clin. Pract.* **2020**, *11*, e179–e188. [CrossRef]
25. Sutherland, A.E.; Stickland, J.; Wee, B. Can video consultations replace face-to-face interviews? Palliative medicine and the COVID-19 pandemic: Rapid review. *BMJ Support. Palliat. Care* **2020**, *10*, 271–275. [CrossRef] [PubMed]
26. Margusino-Framiñán, L.; Illarro-Uranga, A.; Lorenzo-Lorenzo, K.; Monte-Boquet, E.; Márquez-Saavedra, E.; Fernández-Bargiela, N.; Gómez-Gómez, D.; Lago-Rivero, N.; Poveda-Andrés, J.L.; Díaz-Acedo, R.; et al. Pharmaceutical care to hospital outpatients during the COVID-19 pandemic. *Telepharmacy Farm. Hosp.* **2020**, *44*, 61–65. [PubMed]
27. Ghai, S. Teledentistry during COVID-19 pandemic. *Diabetes Metab. Syndr. Clin. Res. Rev.* **2020**, *14*, 933–935. [CrossRef] [PubMed]
28. Chang, C.P.; Lee, T.T.; Mills, M.E. Experience of home telehealth technology in older patients with diabetes. *CIN—Comput. Inform. Nurs.* **2017**, *35*, 530–537. [CrossRef]
29. Appireddy, R.; Khan, S.; Leaver, C.; Martin, C.; Jin, A.; Durafourt, B.A.; Archer, S.L. Home virtual visits for outpatient follow-up stroke care: Cross-sectional study. *J. Med. Internet Res.* **2019**, *21*, e13734. [CrossRef]

30. Palcu, P.; Munce, S.; Jaglal, S.B.; Allin, S.; Chishtie, J.A.; Silverstein, A.; Kim, S. Understanding patient experiences and challenges to osteoporosis care delivered virtually by telemedicine: A mixed methods study. *Osteoporos. Int.* **2020**, *31*, 351–361. [CrossRef]
31. Arksey, H.; O'Malley, L. Scoping studies: Towards a methodological framework. *Int. J. Soc. Res. Methodol. Theory Pract.* **2005**, *8*, 19–32. [CrossRef]
32. TriTricco, A.C.; Lillie, E.; Zarin, W.; O'Brien, K.K.; Colquhoun, H.; Levac, D.; Moher, D.; Peters, M.D.; Horsley, T.; Weeks, L.; et al. PRISMA Extension for Scoping Reviews (PRISMA-ScR): Checklist and Explanation. *Ann. Intern. Med.* **2018**, *169*, 467. [CrossRef]
33. Adams, J.; Hillier-Brown, F.C.; Moore, H.J.; Lake, A.A.; Araujo-Soares, V.; White, M.; Summerbell, C. Searching and synthesising "grey literature" and "grey information" in public health: Critical reflections on three case studies. *Syst. Rev.* **2016**, *5*, 164. [CrossRef]
34. Karolinska Institutet Universitetsbiblioteket. Grey Literature. Available online: https://kib.ki.se/en/search-evaluate/grey-literature (accessed on 15 April 2021).
35. Zhang, H.; Cha, E.E.; Lynch, K.; Cahlon, O.; Gomez, D.R.; Shaverdian, N.; Gillespie, E.F. Radiation Oncologist Perceptions of Telemedicine from Consultation to Treatment Planning: A Mixed-Methods Study. *Int. J. Radiat. Oncol. Biol. Phys.* **2020**, *108*, 421–429. [CrossRef]
36. Johnsson, G.; Kerslake, R.; Crook, S. Delivering allied health services to regional and remote participants on the autism spectrum via video-conferencing technology: Lessons learned. *Rural Remote Health* **2019**, *19*, 5358. [CrossRef] [PubMed]
37. Fernemark, H.; Skagerström, J.; Seing, I.; Ericsson, C.; Nilsen, P. Digital consultations in Swedish primary health care: A qualitative study of physicians' job control, demand and support. *BMC Fam. Pract.* **2020**, *21*, 241. [CrossRef] [PubMed]
38. Johnsen, T.M.; Norberg, B.L.; Kristiansen, E.; Zanaboni, P.; Austad, B.; Krogh, F.H.; Getz, L. Suitability of video consultations during the COVID-19 pandemic lockdown: Cross-sectional survey among Norwegian general practitioners. *J. Med. Internet Res.* **2021**, *23*, e26433. [CrossRef] [PubMed]
39. Björndell, C.; Premberg, Å. Physicians' experiences of video consultation with patients at a public virtual primary care clinic: A qualitative interview study. *Scand. J. Prim. Health Care* **2021**, *39*, 67–76. [CrossRef] [PubMed]
40. Lechien, J.R.; Radulesco, T.; Distinguin, L.; Chekkoury-Idrissi, Y.; Circiu, M.P.; Afia, F.E.; Michel, J.; Papon, J.F.; Hans, S. Patient and otolaryngologist perceptions of telemedicine during COVID-19 pandemic. *Eur. Arch. Otorhinolaryngol.* **2021**, *278*, 4101–4105. [CrossRef]
41. Dubin, J.M.; Wyant, W.A.; Balaji, N.C.; Ong, W.L.; Kettache, R.H.; Haffaf, M.; Zouari, S.; Santillan, D.; Gómez, A.M.A.; Sadeghi-Nejad, H.; et al. Telemedicine Usage among Urologists during the COVID-19 Pandemic: Cross-Sectional Study. *J. Med. Internet Res.* **2020**, *22*, e21875. [CrossRef]
42. Vandekerckhove, P.; Vandekerckhove, Y.; Tavernier, R.; de Jaegher, K.; de Mul, M. Leveraging user experience to improve video consultations in a cardiology practice during the COVID-19 pandemic: Initial insights. *J. Med. Internet Res.* **2020**, *22*, e19771. [CrossRef]
43. Kirby, D.J.; Fried, J.W.; Buchalter, D.B.; Moses, M.J.; Hurly, E.T.; Cardone, D.A.; Yang, S.S.; Virk, M.S.; Rokito, A.S.; Jazrawi, L.M.; et al. Patient and Physician Satisfaction with Telehealth during the COVID-19 Pandemic: Sports Medicine Perspective. *Telemed. e-Health* **2021**, *27*, 1151–1159. [CrossRef]
44. Tenforde, A.S.; Iaccarino, M.A.; Borgstrom, H.; Hefner, J.E.; Silver, J.; Ahmed, M.; Babu, A.N.; Blauwet, C.A.; Elson, L.; Eng, C.; et al. Telemedicine during COVID-19 for Outpatient Sports and Musculoskeletal Medicine Physicians. *PM&R* **2020**, *12*, 926–932.
45. Hardy, N.R.; Maier, C.A.; Gregson, T.J. Couple teletherapy in the era of COVID-19: Experiences and recommendations. *J. Marital. Fam. Ther.* **2021**, *47*, 225–243. [CrossRef]
46. Sugarman, D.E.; Horvitz, L.E.; Greenfield, S.F.; Busch, A.B. Clinicians' Perceptions of Rapid Scale-up of Telehealth Services in Outpatient Mental Health Treatment. *Telemed. e-Health* **2021**, *27*, 1399–1408. [CrossRef] [PubMed]
47. Kemp, M.T.; Liesman, D.R.; Williams, A.M.; Brown, C.S.; Iancu, A.M.; Wakam, G.K.; Biesterveld, B.E.; Alam, H.B. Surgery Provider Perceptions on Telehealth Visits during the COVID-19 Pandemic: Room for Improvement. *J. Surg. Res.* **2021**, *260*, 300–306. [CrossRef] [PubMed]
48. Saiyed, S.; Nguyen, A.; Singh, R. Physician Perspective and Key Satisfaction Indicators with Rapid Telehealth Adoption during the Coronavirus Disease 2019 Pandemic. *Telemed. e-Health* **2021**, *27*, 1225–1234. [CrossRef]
49. Paulik, G.; Maloney, G.; Arntz, A.; Bachrach, N.; Koppeschaar, A.; McEvoy, P. Delivering Imagery Rescripting via Telehealth: Clinical Concerns, Benefits, and Recommendations. *Curr. Psychiatry Rep.* **2021**, *23*, 24. [CrossRef] [PubMed]
50. Gold, K.J.; Laurie, A.R.; Kinney, D.R.; Harmes, K.M.; Serlin, D.C. Video visits: Family physician experiences with uptake during the COVID-19 pandemic. *Fam. Med.* **2021**, *53*, 207–210. [CrossRef]
51. Heyer, A.; Granberg, R.E.; Rising, K.L.; Binder, A.F.; Gentsch, A.T.; Handley, N.R. Medical Oncology Professionals' Perceptions of Telehealth Video Visits. *JAMA Netw. Open* **2021**, *4*, e2033967. [CrossRef]
52. Levy, S.; Mason, S.; Russon, J.; Diamond, G. Attachment-based family therapy in the age of telehealth and COVID-19. *J. Marital. Fam. Ther.* **2021**, *47*, 440–454. [CrossRef]
53. Yu, J.; Afridi, S.M.; Cozart, A.C.; Isea, L.; Guan, J. Evaluation and Feedback for Telehealth from Patients and Physicians during the Early Stage of COVID-19 Pandemic Period. *Cureus* **2021**, *13*, e12633. [CrossRef]

54. Cioffi, V.; Cantone, D.; Guerriera, C.; Architravo, M.; Mosca, L.L.; Sperandeo, R.; Moretto, E.; Longobardi, T.; Alfano, Y.M.; Continisio, G.I.; et al. *Satisfaction Degree in the Using of Videoconferencing Psychotherapy in a Sample of Italian Psychotherapists during COVID-19 Emergency*; Institute of Electrical and Electronics Engineers Inc.: Piscataway, NJ, USA, 2020; pp. 125–132.
55. Barsom, E.Z.; Jansen, M.; Tanis, P.J.; van de Ven, A.W.H.; van Oud-Alblas, M.B.; Buskens, C.J.; Bemelman, W.A.; Schijven, M.P. Video consultation during follow up care: Effect on quality of care and patient- and provider attitude in patients with colorectal cancer. *Surg. Endosc.* **2021**, *35*, 1278–1287. [CrossRef]
56. Békés, V.; Aafjes–van Doorn, K.; Prout, T.A.; Hoffman, L. Stretching the Analytic Frame: Analytic Therapists' Experiences with Remote Therapy During COVID-19. *J. Am. Psychoanal. Assoc.* **2020**, *68*, 437–446. [CrossRef]
57. Wade, S.L.; Moscato, E.L.; Raj, S.P.; Narad, M.E. Clinician perspectives delivering telehealth interventions to children/families impacted by pediatric traumatic brain injury. *Rehabil. Psychol.* **2019**, *64*, 298–306. [CrossRef] [PubMed]
58. Mammen, J.R.; Elson, M.J.; Java, J.J.; Beck, C.A.; Beran, D.B.; Biglan, K.M.; Boyd, C.M.; Schmidt, P.N.; Simone, R.; Willis, A.W.; et al. Patient and physician perceptions of virtual visits for Parkinson's disease: A qualitative study. *Telemed. e-Health* **2018**, *24*, 255–267. [CrossRef] [PubMed]
59. Hinman, R.S.; Nelligan, R.K.; Bennell, K.L.; Delany, C. "Sounds a Bit Crazy, But It Was Almost More Personal:" A Qualitative Study of Patient and Clinician Experiences of Physical Therapist–Prescribed Exercise For Knee Osteoarthritis Via Skype. *Arthritis Care Res.* **2017**, *69*, 1834–1844. [CrossRef] [PubMed]
60. Rykkje, L.; Hjorth, G.H.B. "Safety at Home": Experiences from Testing of Video Communication between Patients and Home Health Care Personnel. *SAGE Open* **2017**, *27*, 2158244017744900. [CrossRef]
61. Becevic, M.; Boren, S.; Mutrux, R.; Shah, Z.; Banerjee, S. User satisfaction with telehealth: Study of patients, providers, and coordinators. *Health Care Manag.* **2015**, *34*, 337–349. [CrossRef]
62. Combi, C.; Pozzani, G.; Pozzi, G. Telemedicine for developing countries: A survey and some design issues. *Appl. Clin. Inform.* **2016**, *7*, 1025–1050. [CrossRef]
63. Downs, J.; Lotan, M.; Elefant, C.; Leonard, H.; Wong, K.; Buckley, N.; Stahlhut, M. Implementing telehealth support to increase physical activity in girls and women with Rett syndrome-ActivRett: Protocol for a waitlist randomised controlled trial. *BMJ Open* **2020**, *10*, e042446. [CrossRef]
64. Johansson, A.M.; Lindberg, I.; Söderberg, S. Healthcare personnel's experiences using video consultation in primary healthcare in rural areas. *Prim. Health Care Res. Dev.* **2017**, *18*, 73–83. [CrossRef]
65. DuPont Sustainable Solutions. Managing Remote Workers: Preventing Isolation and Loneliness. Available online: https://www.consultdss.com/preventing-isolation-and-loneliness/ (accessed on 13 May 2021).
66. Bradford, N.K.; Penny, R.A. Registered nurse and midwife experiences of using videoconferencing in practice: A qualitative systematic review protocol. *JBI Database Syst. Rev. Implement. Rep.* **2016**, *14*, 3–9. [CrossRef]
67. Pidgeon, F.M. Use of telehealth videoconferencing as a supplement to visiting allied health services. *Aust. J. Rural Health* **2017**, *25*, 58–59. [CrossRef]
68. Lovo, S.; Harrison, L.; O'Connell, M.E.; Trask, C.; Bath, B. Experience of patients and practitioners with a team and technology approach to chronic back disorder management. *J. Multidiscip. Healthc.* **2019**, *12*, 855–869. [CrossRef] [PubMed]
69. Electronic Health Reporter. How Telemedicine Is Revolutionizing the Healthcare Industry. Available online: https://electronichealthreporter.com/how-telemedicine-is-revolutionizing-the-healthcare-industry/ (accessed on 30 March 2021).
70. Morris, J.; Campbell-Richards, D.; Wherton, J.; Sudra, R.; Vijayaraghavan, S.; Greenhalgh, T.; Collard, A.; Byrne, E.; O'Shea, T. Webcam consultations for diabetes: Findings from four years of experience in Newham. *Pract. Diabetes* **2017**, *34*, 45–50. [CrossRef]
71. Almathami, H.K.Y.; Than Win, K.; Vlahu-Gjorgievska, E. Barriers and facilitators that influence telemedicine-based, real-time, online consultation at patients' homes: Systematic literature review. *J. Med. Internet Res.* **2020**, *22*, e16407. [CrossRef] [PubMed]
72. Uscher-Pines, L.; Sousa, J.; Raja, P.; Mehrotra, A.; Barnett, M.L.; Huskamp, H.A. Suddenly becoming a "Virtual doctor": Experiences of psychiatrists transitioning to telemedicine during the COVID-19 pandemic. *Psychiatr. Serv.* **2020**, *71*, 1143–1150. [CrossRef] [PubMed]
73. Yang, L.; Brown-Johnson, C.G.; Miller-Kuhlmann, R.; Kling, S.M.; Saliba-Gustafsson, E.A.; Shaw, J.G.; Gold, C.A.; Winget, M. Accelerated launch of video visits in ambulatory neurology during COVID-19: Key lessons from the Stanford experience. *Neurology* **2020**, *95*, 305–311. [CrossRef]
74. Munn, Z.; Peters, M.D.J.; Stern, C.; Tufanaru, C.; McArthur, A.; Aromataris, E. Systematic review or scoping review? Guidance for authors when choosing between a systematic or scoping review approach. *BMC Med. Res. Methodol.* **2018**, *18*, 143. [CrossRef]
75. Clarke, V.; Braun, V. Thematic analysis. *J. Posit. Psychol.* **2017**, *12*, 297–298. [CrossRef]

life

Article

What Went Wrong with the IMMUNI Contact-Tracing App in Italy? A Cross-Sectional Survey on the Attitudes and Experiences among Healthcare University Students

Claudia Isonne *, Maria Roberta De Blasiis, Federica Turatto, Elena Mazzalai, Carolina Marzuillo, Corrado De Vito, Paolo Villari and Valentina Baccolini

Department of Public Health and Infectious Diseases, Sapienza University of Rome, 00185 Rome, Italy; mariaroberta.deblasiis@uniroma1.it (M.R.D.B.); federica.turatto@uniroma1.it (F.T.); elena.mazzalai@uniroma1.it (E.M.); carolina.marzuillo@uniroma1.it (C.M.); corrado.devito@uniroma1.it (C.D.V.); paolo.villari@uniroma1.it (P.V.); valentina.baccolini@uniroma1.it (V.B.)

* Correspondence: claudia.isonne@uniroma1.it

Abstract: The adoption of digital contact-tracing apps to limit the spread of SARS-CoV-2 has been sup-optimal, but studies that clearly identify factors associated with the app uptake are still limited. In April 2021, we administered a questionnaire to healthcare university students to investigate their attitudes towards and experiences of the IMMUNI app. A multivariable logistic regression model was built to identify app download predictors. Adjusted odds ratios (aORs) and 95% confidence intervals (CIs) were calculated. We surveyed 247 students. Most respondents (65.6%) had not downloaded IMMUNI, reporting as the main reason the perceived app uselessness (32.7%). In the multivariable analysis, being advised to use the app (aOR: 3.21, 95%CI: 1.80–5.73), greater fear of infecting others (aOR: 1.50, 95%CI: 1.01–2.23), and greater trust in the institutional response to the emergency (aOR: 1.33, 95%CI: 1.00–1.76) were positively associated with the outcome, whereas greater belief in the "lab-leak theory" of COVID-19 was a negative predictor (aOR: 0.75, 95%CI: 0.60–0.93). Major technical issues were reported by app users. Targeted strategies aimed at improving awareness of digital health applications should be devised. Furthermore, institutions should invest in the development of these technologies, to minimize technical issues and make them accessible to the entire population.

Keywords: digital contact tracing; IMMUNI app; COVID-19; students

Citation: Isonne, C.; De Blasiis, M.R.; Turatto, F.; Mazzalai, E.; Marzuillo, C.; De Vito, C.; Villari, P.; Baccolini, V. What Went Wrong with the IMMUNI Contact-Tracing App in Italy? A Cross-Sectional Survey on the Attitudes and Experiences among Healthcare University Students. *Life* **2022**, *12*, 871. https://doi.org/10.3390/life12060871

Academic Editor: Daniele Giansanti

Received: 16 May 2022
Accepted: 9 June 2022
Published: 10 June 2022

1. Introduction

Contact tracing has long been a key public health tool for slowing or stopping the spread of infectious diseases [1]. It allows rapid and accurate identification of individuals who have been exposed to confirmed or probable cases (contacts) and, thus, the infection's chain of transmission to be broken [2]. During the COVID-19 pandemic, contact tracing has assumed a critical role in mitigating transmission of the SARS-CoV-2 virus and limiting its dramatic effects on health systems and societies [3–5]. Nevertheless, several challenges in using the traditional contact-tracing strategy have become apparent in many countries [6]. Among these, the scarcity of previously trained personnel, the short time between infection and the onset of symptoms, as well as a possible recall bias, may have hindered the effectiveness of this surveillance system [7]. For these reasons, and because a number of digital health technologies have been implemented successfully in recent years, several mobile applications have been developed to support the traditional approach by enabling digital contact tracing [8,9].

Using Bluetooth or GPS technology together with an appropriate app, it is possible to geolocate and record every device that has been in close proximity with another [10]. This allows users to be tracked and notified when they have been near the device of a person who reports testing positive for SARS-CoV-2, and, consequently, to take preventive measures, such as quarantine [11]. Recognizing the public health potential of digital contact tracing (DCT) tools, which is underpinned by modelling studies that demonstrate how DCT could help control the spread of SARS-CoV-2 [8], several countries have introduced these systems, some of them with positive experiences, especially Eastern countries [7]. In Italy, the DCT app "IMMUNI" was launched in June 2020 and its use was on voluntary basis [12]. Briefly, when installed on a smartphone, IMMUNI emits a Bluetooth signal that includes a random code. It does this on a continuous basis. When a person approaches another one, their smartphones exchange these codes and store them in their memory, thus making note of that contact. When a person is notified of testing positive to SARS-CoV-2, with the help of healthcare personnel, the user is able to report this result to IMMUNI, sharing his or her random codes and alerting the people he or she has been in close contact with [13]. However, despite initial interest in this innovation in Italy, and in similar apps in other Western countries, the intense international debate over the ethical, legal, and societal implications has hampered efforts to implement DCT strategies [8,14,15].

Recent evidence shows a generally positive attitude towards DCT apps [16], but issues of cyber security, variable risk perception, and poor awareness of benefits have been indicated as barriers to their use [17,18]. However, studies that clearly identify factors associated with app uptake are still limited, and results are mixed [19–21]. Therefore, it is critical to further investigate the factors that may have hindered app use. This is especially pertinent among young people, who on one hand are at higher risk of SARS-CoV-2 infection given their frequent opportunities to socialize [22], while on the other hand they have a greater degree of digital literacy and acceptance of downloaded apps [21]. The objective of our study was to investigate the attitude towards the IMMUNI app in a sample of healthcare university students. We also explored their experiences of using it as well as the main barriers to its download. Specifically, we aimed to identify the key factors associated with its uptake among a cohort of people who have been trained to adopt health prevention behavior, and, thereby, to better understand what may have hampered its adoption in a population that should be receptive to DCT.

2. Materials and Methods

2.1. Setting and Participants

This cross-sectional study was conducted at Sapienza University of Rome between 14 and 19 April 2021. An online survey was administered to students enrolled in the healthcare area (i.e., three nursing science courses and one physiotherapy course). Access to the questionnaire was via a link sent by e-mail to the students' institutional e-mail addresses. The study was performed in accordance with the World Medical Association Declaration of Helsinki. Participants were asked for their consent and were guaranteed anonymity in the information collected. The institutional ethics board of the Umberto I teaching hospital/Sapienza University of Rome approved this study (protocol number 571/2021).

2.2. Questionnaire

The questionnaire was self-administered and took approximately five minutes to fill out (Supplementary Materials). It consisted of a maximum of 33 closed-ended questions grouped into three sections.

The first section aimed to collect sociodemographic information: age, gender, field of study, year of study, nationality, Italian Region, finances (i.e., with the financial resources at your disposal, how well do you get to the end of the month?), main source of health information (i.e., what is your main source of health information?), health literacy (HL) (i.e., how often do you need to have someone help when you read instructions, pamphlets, or other written material from your doctor or pharmacy? [23]), chronic pathologies, and the occurrence and symptoms of a SARS-CoV-2 infection in the past.

The second section explored students' perceptions of and attitudes towards the COVID-19 pandemic. Specifically, we asked them to rate on a 5-point scale (from 1 [very low] to 5 [very high]) how great was their fear of getting the SARS-CoV-2 infection, fear of infecting others, and their concern about the COVID-19 emergency. We also asked students to express their feelings in relation to the pandemic (i.e., depression, anxiety, and anger, from 1 [not at all] to 5 [extremely]), to self-report adherence to COVID-19 precautionary measures (i.e., compliance with social distancing and use of mask, from 1 [not at all] to 5 [extremely]), their trust in institutions (i.e., on a scale from 1 [not at all] to 5 [extremely]; how much do you trust the response of the institutions to the emergency?), and their belief that the virus originated from a laboratory in Wuhan (i.e., on a scale from 1 [not at all] to 5 [extremely]; how much do you believe in the "lab-leak theory" of the origin of COVID-19?). Finally, we asked whether someone had advised them to download and use IMMUNI, when they had actually downloaded it and whether they were still using it.

The third section was different for students who had downloaded the app and those who had not. In the first group, we investigated the main reasons for such a download and their assessment of some app features (i.e., on a scale from 1 [very poor] to 5 [excellent], how would you assess the privacy features, ease of use, usefulness, and intuitiveness?). In addition, students were asked to report their personal experience with app notifications. Two possible scenarios were investigated: (i) receipt of at least one notification as a potential contact, and the nature of their post-notification behavior, or (ii) at least one notification via the app of having a SARS-CoV-2 infection, and their assessment of the notification process (from very lacking to very good) together with the difficulties encountered in submitting the notification, if applicable. For the students reported to have not download the app, the third section explored their attitudes. We asked the main reason why they did not download IMMUNI, and to rate on a scale from 1 (not at all) to 5 (definitely) how effective some hypothetical incentives would have been in promoting app uptake: (i) receiving concrete feedback on how the app could help limit the virus spread; (ii) being informed about the app's uptake among the population; (iii) making app download mandatory; (iv) having the opportunity to give feedback on the technical aspects of the app; (v) receiving more information about personal data collection and management; and (vi) receiving an economic reward.

2.3. Statistical Analysis

Descriptive statistics were obtained using median and interquartile range, or mean and standard deviation, for continuous variables and proportions for dichotomous and categorical variables. Student age was dichotomized using 21 years as a cut-off. Participants were classified into four groups according to their year of study: first-, second-, or third-year students, and students outside prescribed courses. As for nationality, respondents were classed as Italian or non-Italian. Health literacy was categorized into two groups: adequate HL (answering never/rarely) and non-adequate HL (answering sometimes/often/always) [24]. Chronic pathologies were grouped into nine categories: none, autoimmune disease, cardiovascular disease, endocrine disease, genetic disease, gynecological disease, psychiatric disease, respiratory disease, and cancer. SARS-CoV-2 infection was categorized into four groups: no infection, asymptomatic, mild symptoms, and moderate/severe symptoms.

For the univariable analysis, the Mann–Whitney U test was used to compare continuous variables between students who had download IMMUNI and the students who had not, whereas Pearson's chi-squared test or Fisher's test was used for dichotomous and categorical variables, as appropriate. A multivariable logistic regression model was built to identify predictors of app download. Variables were included in the model based on expert opinion. Multicollinearity was checked using as threshold a variance inflation factor of 5. The Hosmer and Lemeshow test was used to evaluate the goodness of fit of the model. As a result, the final model consisted of the following variables: age (<21 vs. ≥21 years), gender (male vs. female), HL (inadequate vs. adequate), fear of getting the SARS-CoV-2 infection (continuous), fear of infecting others (continuous), concern about the COVID-19 pandemic (continuous), trust in the response of the institutions to the emergency (continuous), belief in the "lab-leak theory" of the origin of COVID-19 (continuous), and receipt of some advice to download the app (yes vs. no). Adjusted odds ratios (ORs) and 95% confidence intervals (CIs) were calculated.

All analyses were performed using Stata (StataCorp LLC, 4905 Lakeway Drive, College Station, TX, USA), version 17.0. A two-sided p-value < 0.05 was considered statistically significant.

3. Results

A total of 247 students answered the questionnaire (response rate: 72.4%). Of the 85 students who had downloaded IMMUNI (34.4%), more than half had done it immediately on launch of the app (N = 48), and the remaining participants between September and November 2020, but all of them were still using it in April 2021 [Table 1]. The two groups were of a similar age. Most were females (71.8% vs. 77.2%); almost three in every four attended undergraduate nursing courses (68.2% vs. 75.3%) and more than 90% were enrolled as first- or second- year students. Only a minority of responders were non-Italian (around 2.5%) and approximately half the Italian respondents came from the Lazio Region. More than 60% of the students in both groups reported that they got to the end of the month (financially) very well or well enough. The mass media was indicated as the main source of health information in both groups (around 40%), followed by social networks and the Internet, whereas only a limited number of students reported not looking for any health information (~1%). Most students showed adequate HL, with more than two thirds answering that they never or rarely needed help understanding medical material. The vast majority of respondents in both groups did not suffer from any chronic conditions and had never contracted the SARS-CoV-2 infection.

Table 1. Students' sociodemographic characteristics vs. IMMUNI app download. Results are expressed as frequency (percentage).

	App Download		
	Yes (N = 85)	**No (N = 162)**	***p*-Value ***
Age			0.463
<21 years	43 (50.6)	74 (45.6)	
≥21 years	42 (49.4)	88 (54.3)	
Gender			0.350
Female	61 (71.8)	125 (77.2)	
Male	24 (28.2)	37 (22.8)	
Field of study			0.235
Nursing science	58 (68.2)	122 (75.3)	
Physiotherapy	27 (31.8)	40 (24.7)	
Year of study			0.911
First	37 (43.5)	69 (42.6)	
Second	39 (45.9)	79 (48.8)	
Third	8 (9.4)	13 (8.0)	
Outside prescribed course	1 (1.2)	1 (0.6)	
Nationality			0.999
Italian	83 (97.6)	158 (97.5)	
Non-Italian	2 (2.4)	4 (2.5)	
Italian Region (N = 241)			0.049
Abruzzo	0 (0.0)	2 (1.3)	
Calabria	3 (3.6)	5 (3.1)	

Table 1. *Cont.*

	App Download		*p*-Value *
	Yes (N = 85)	No (N = 162)	
Campania	3 (3.6)	11 (6.9)	
Lazio	41 (49.4)	90 (56.9)	
Puglia	18 (21.7)	13 (8.2)	
Sardegna	1 (1.2)	0 (0.0)	
Sicilia	17 (20.5)	29 (18.4)	
Umbria	0 (0.0)	1 (0.6)	
Veneto	1 (0.0)	2 (1.3)	
Missing	2 (0.0)	5 (3.1)	
Finances			0.169
I have many difficulties	6 (7.0)	11 (6.8)	
I have some difficulties	26 (30.6)	48 (29.6)	
Managing well enough	34 (40.0)	83 (51.2)	
Managing very well	19 (22.4)	20 (12.4)	
Main source of health information			0.999
Mass media	35 (41.2)	65 (40.1)	
Web	20 (23.5)	39 (24.1)	
Social network	29 (34.1)	56 (34.6)	
None	1 (1.2)	2 (1.2)	
Health literacy			0.360
Non-adequate	24 (28.2)	55 (33.9)	
Adequate	61 (71.8)	107 (66.0)	
Chronic pathologies			0.164
None	72 (84.7)	149 (91.9)	
Autoimmune disease	3 (3.5)	1 (0.6)	
Cardiovascular disease	0 (0.0)	2 (1.2)	
Cancer	1 (1.2)	0 (0.0)	
Endocrine disease	1 (1.2)	2 (1.2)	
Genetic disease	1 (1.2)	1 (0.6)	
Gynecological disease	1 (1.2)	0 (0.0)	
Psychiatric disease	0 (0.0)	1 (0.6)	
Respiratory disease	6 (7.1)	6 (3.7)	
SARS-CoV-2 infection			0.865
No infection	78 (91.7)	149 (92.0)	
Asymptomatic	1 (1.2)	1 (0.6)	
Mild symptoms	5 (5.9)	8 (4.9)	
Moderate/severe symptoms	1 (1.2)	4 (2.5)	

* Pearson's chi-squared test or Fisher test.

No significant difference in terms of fear of contracting SARS-CoV-2 was observed between those who downloaded the app compared to those who did not [Table 2]. By contrast, although it did not reach statistical significance (mean score: 4.5 vs. 4.2), the first cohort seemed to have a slightly greater fear of infecting others. Concern about the COVID-19 pandemic did not differ (mean score: 3.9 vs. 3.8), and neither did the students' feelings in relation to the pandemic, among which, depression was the most reported in both groups (depression, mean score: 3.1 vs. 3.2; anxiety, mean score: 2.9 vs. 3.0; anger, mean score: 2.5 vs. 2.8). Self-reported adherence to COVID-19 precautionary measures (i.e., respect of social distancing and use of mask) was slightly higher in the first group, albeit not significantly (mean score: 4.8 vs. 4.6 in both items). Conversely, the group that downloaded the app had a greater trust in the response of the institutions to the emergency (mean score: 3.6 vs. 3.3). By contrast, students that did not download the app had a significantly greater belief that the virus originated from a laboratory (mean score: 2.4 vs. 1.9). Finally, a greater proportion of students among those who had downloaded the app reported they had been advised to do so (64.7% vs. 38.3%).

Table 2. Students' perceptions of and attitudes towards SARS-CoV-2 pandemic vs. IMMUNI app download. Results are expressed as mean (standard deviation) or frequency (percentage).

	App Download		
	Yes (N = 85)	**No (N = 162)**	***p*-Value ***
Fear of getting the SARS-CoV-2 infection	2.8 (1.1)	2.7 (1.2)	0.545
Fear of infecting others	4.5 (0.8)	4.2 (1.1)	0.051
Concern about the COVID-19 emergency	3.9 (1.0)	3.8 (1.1)	0.954
Feelings about the COVID-19 pandemic			
Depression	3.1 (1.3)	3.2 (1.3)	0.578
Anxiety	2.9 (1.3)	3.0 (1.4)	0.547
Anger	2.5 (1.4)	2.8 (1.4)	0.122
Adherence to COVID-19 precautionary measures			
Maintaining physical distance	4.8 (0.6)	4.6 (0.7)	0.074
Use of mask	4.8 (0.5)	4.6 (0.7)	0.137
Trust in institutional response to the emergency	3.6 (1.0)	3.3 (1.1)	0.025
Belief in the lab-leak theory of COVID-19 origin	1.9 (1.3)	2.4 (1.4)	0.003
Receipt of advice to download the app			<0.001
No	30 (35.3)	100 (61.7)	
Yes	55 (64.7)	62 (38.3)	

COVID-19: coronavirus diseases 2019. * Pearson's chi-squared test for categorical variables and Mann–Whitney U test for continuous variables.

The main reasons for uptake of the app, among those who downloaded it, were sense of duty (40.0%) and respect for others (30.6%), followed by fear of getting the infection (20.0%), and curiosity (9.4%) [Table 3]. On average, students rated as very good the privacy features of the app (mean score: 4.0), and they found it easy to use (mean score: 3.8), but also quite intuitive and useful (mean score: 3.4 for both). Overall, only 8.2% of the students who downloaded the app received at least one alert that they were a potential contact and most followed the app advice (around 70%). Similarly, only seven students (8.2%) tried to notify a positive COVID-19 test through the app, but most of them were not successful (71.4%). Of these, one student was unable to get the National Unique Code (CUN) whereas three participants were unable to enter the CUN in the app. Almost three quarters of these students rated the notification process as lacking or very lacking (71.4%).

Table 3. Attitudes and experiences of surveyed students who downloaded the IMMUNI App. Results are expressed as mean (standard deviation) or frequency (percentage).

	N = 85
Main reason for the app download	
Sense of duty	34 (40.0)
Respect for others	26 (30.6)
Fear of getting the infection	17 (20.0)
Curiosity	8 (9.4)
Assessment of app features	
Privacy	4.0 (1.1)
Ease of use	3.8 (1.1)
Usefulness	3.4 (1.3)
Intuitiveness	3.4 (1.3)
Receipt of at least one contact notification	
No	78 (91.8)
Yes	7 (8.2)
Post-notification behavior (N = 7)	
I received and followed the advice provided by the app	5 (71.4)
I received the advice, but I did not do anything	2 (28.6)
Notification of positivity (N = 7)	
No, I was not able to	5 (71.4)
Yes, I was given the CUN, and I entered the requested data on the app	1 (14.3)

Table 3. *Cont.*

	N = 85
Yes, I provided the CUN to the healthcare professional who contacted me for contact-tracing purposes	1 (14.3)
Assessment of the notification process (N = 7)	
Very lacking	4 (57.1)
Lacking	1 (14.3)
Good	1 (14.3)
Very good	1 (14.3)
Challenges/technical issues (N = 7)	
I was unable to get the CUN	1 (14.3)
I was unable to enter the CUN in the app even after calling the IMMUNI call center	1 (14.3)
I was unable to enter the CUN in the app and I did not know that I could call the IMMUNI call center	2 (28.6)
I did not had any difficulty	2 (28.6)
Missing	1 (14.3)

CUN: National Unique Code.

Students who did not download IMMUNI reported that the main reason for not doing so was the belief that it was useless (32.7%) and because they did not know they had to do it (23.5%), but also for technical issues (almost 20%) and, albeit less frequently, because of a distrust in data management (around 16%) [Table 4]. In addition, a small percentage reported hearing of negative experiences (5.6%). As for the hypothetical incentives that could increase app uptake, information on how app usage could impact virus transmission dynamics was the main driver (mean score: 3.5), followed by information on the app's uptake among the population (mean score: 3.4) and making its download mandatory (mean score: 3.4). A slightly lower importance was attributed to the opportunity to give feedback on the technical aspects of the app (mean score: 3.2) and information about personal data collection and management (mean score: 3.1). Lastly, having an economic reward seemed to be the least effective incentive (mean score: 2.4).

Table 4. Attitudes of surveyed students who did not download the IMMUNI App. Results are expressed as mean (standard deviation) or frequency (percentage).

	N = 162
Reason for not downloading the app	
I do not think it is useful	53 (32.7)
I did not know I had to download the app	38 (23.5)
Technical problems (e.g., no smartphone, operating system incompatibility, battery problems, insufficient storage on the phone, etc.)	31 (19.1)
I do not trust data management (privacy issue)	26 (16.1)
I have heard of negative personal experiences	9 (5.6)
Other reasons	5 (3.1)
Effectiveness of hypothetical incentives in increasing the app uptake	
Information on how usage can impact transmission dynamics	3.5 (1.3)
Information on the app's uptake among the population	3.4 (1.3)
Making the app download mandatory	3.4 (1.4)
Opportunity to give feedback on the technical aspects of the app	3.2 (1.4)
Information about personal data collection and management	3.1 (1.4)
Economic reward	2.4 (1.5)

In the multivariable analysis [Table 5], participants who had received some advice to download the app seemed to have the highest odds of IMMUNI uptake (aOR: 3.21, 95% CI: 1.80–5.73). Similarly, reporting a higher fear of infecting other people was associated with higher likelihood of app download (aOR: 1.50, 95% CI: 1.01–2.23), as well as a greater trust in the response of the institutions to the emergency (aOR: 1.33, 95%

CI: 1.00–1.76). On the other hand, greater belief in the "lab-leak theory" of the origin of COVID-19 was negatively associated with download (aOR: 0.75, 95% CI: 0.60–0.93). By contrast, age, gender, HL, fear of getting the SARS-CoV-2 infection, and concern about the COVID-19 pandemic did not seem to be predictors of the outcome.

Table 5. Multivariable logistic regression model for IMMUNI app download among the students surveyed between 14 and 19 April 2021, Sapienza University of Rome.

	App Download	
	aOR (95% CI)	***p*-Value**
Age		
<21 years	Ref.	
≥21 years	0.77 (0.43–1.27)	0.373
Gender		
Female	Ref.	
Male	1.48 (0.75–2.89)	0.265
Health literacy		
Adequate	Ref.	
Non-adequate	0.69 (0.36–1.30)	0.256
Fear of getting the SARS-CoV-2 infection	1.04 (0.79–1.37)	0.776
Fear of infecting others	1.50 (1.01–2.23)	0.042
Concern about the COVID-19 emergency	0.85 (0.62–1.17)	0.327
Trust in institutional response to the emergency	1.33 (1.00–1.76)	0.049
Belief in lab-leak theory of COVID-19 origin	0.75 (0.60–0.93)	0.011
Receipt of advice to download the app		
No	Ref.	
Yes	3.21 (1.80–5.73)	<0.001

aOR: adjusted Odds Ratio. CI: confidence interval. COVID-19: coronavirus diseases 2019.

4. Discussion

The usefulness of DCT apps has been a subject of intense discussion during the COVID-19 pandemic [18,25]. Most governments have struggled with low participation rates, which, in turn, have limited the effectiveness of these tools, contributing to the idea that they are useless and, thus, hindering their adoption [21]. Recently, several researchers have investigated the acceptability and use of contact tracing apps. Most studies are based on surveys assessing the uptake of DCT apps among different population subgroups with a focus on hypothetical tools or the intention to use it, but only a few collect information on the use of an existing app [21]. The majority of documents report the real uptake using data of national statistics without a scientific and theoretical background, while other studies are critical viewpoints arguing on the ethical, technical, political, and scientific impact of contact tracing apps on society [21]. In our study, we found a higher uptake rate of the IMMUNI app compared to that in the general Italian population [12], probably because our sample consisted of students attending healthcare courses, who are more likely to be committed to health prevention strategies [26]. In addition, the fact that, in our analysis, almost all students had the opportunity to download the app since they owned a smartphone, in contrast to the official data where it is more difficult to estimate the number of people eligible for the app uptake, may have contributed to such discrepancy [18]. Nevertheless, we found that use of the DCT app was relatively limited, albeit at a similar rate to uptake of comparable apps in other European nations [21]. This is a concern, however, as these students are the healthcare workforce of tomorrow, and, therefore, it is imperative to implement educational programs that further encourage the adoption of preventive strategies [27]. Moreover, it should be mentioned that the current increase in virus transmission rates due to the omicron variant, and the concomitant abolition of restrictive measures at both the national and regional level, could make it difficult to promptly identify the transmission chain using traditional methods [28]. In this scenario, a high uptake rate of the app would have some advantages, including the support to trace contacts, but also would make the population autonomous in the timely application of the preventive measures and create an environment in which citizens are effectively engaged in maintaining their personal and community health [29].

As for determinants, IMMUNI uptake was not associated with any socio-demographic characteristic, including HL, probably because of the healthcare curricula of our students, but some attitudes towards the pandemic seemed slightly different between the two groups. Risk perception was confirmed to be a key driver, but it changed over time; thus, people may have become used to a high level of risk as the pandemic continued, consequently reducing their motivation to act and use DCT tools [18]. Such changes in risk perception could explain the app download trend in Italy, which consisted of an initial peak when the app was launched in June 2020 (up to 600,000 downloads in a single day), followed by another massive increase in downloads at the beginning of the second wave, reaching more than 200,000 per day. This then tailed off to around 2000 downloads per day until April 2021, when the number of cases was limited and the vaccination campaign was at full deployment [30]. Among other factors explored, a few studies have already documented how high levels of trust in governments and health authorities can motivate people to adhere to prevention strategies [18,19,31]. It is fundamental that institutions convey official messages clearly and coherently, and combating disinformation from other sources as much as possible [18]. In addition, good communication seems important for increasing the acceptability of the app in our study population: the strongest predictor of app uptake in our analyses was being advised to download it, while a reason for non-adoption was a lack of awareness of the app. Lastly, our participants belonged to an age group that may be characterized by sociability, the importance of self-identifying with a peer group and the influence of peers on the adoption of health behavior [32]; therefore, exploiting these social mechanisms by implementing targeted communication strategies is likely to be effective at reaching a large fraction of this population [33].

As aforementioned, at the time of the survey (April 2021), Italy was at the end of the second wave, which had been characterized by a high incidence of SARS-CoV-2 infections during the fall and winter of 2020–2021 [34]. Hence, in the low transmission risk scenario of April 2021, it was not unexpected that we found a low perception of the utility of the DCT tool, similarly to other studies [19,21]. Within this context, communication policies that help people understand the importance of such measures in safeguarding their own and community health should be devised [33]. Such campaigns are most effective when risk perception is high, because people are motivated to take action to protect themselves, which potentiates DCT acceptance [18]. In fact, it is well known that a low adoption rate is the main barrier to the effectiveness of these apps [35] and the poor uptake may be responsible for the limited number of app notifications that our students reported. However, while potential contacts mostly followed the health recommendations provided by the app, which is encouraging because it highlights their awareness of the need to adopt preventive measures promptly, a substantial proportion of our students claimed they were hampered by technical issues with the notification process. This underlines the importance of investing in technical improvements of these apps and making them easy to use for the entire population, most of whom are less digitally literate than young people [35,36]. It is, in fact, important to highlight that several technical skill challenges could occur, such as some people not knowing how to download and install an app, or how to interact with it, thus limiting its acceptance and usage [18].

Interestingly, our findings contrast with a few international studies that report how concerns about data privacy can negatively impact DCT app adoption [16,31,37,38]. This could be explained by the fact that, compared to other, similar apps, IMMUNI collects relatively few data [39]. Additionally, our cohort was composed of university students, who may be accustomed to sharing their data on the web and not be particularly concerned about privacy issues [40]. Lastly, as for incentives that might promote the adoption of DCT apps, despite their recognized importance [41], few studies have investigated this aspect and available evidence focuses only on financial incentives [42,43]. In our study, it was the app's actual utility (or otherwise) that seemed to influence its adoption rate. This highlights how the feeling of being engaged may motivate people to participate in a DCT system and confirms the importance of investing in communication policies that point out the potential health benefits of using such technologies [44].

This study has some limitations. Firstly, the cross-sectional design hindered the opportunity to draw causal conclusions between app uptake and associated factors. Secondly, the relatively low number of participants might have limited the statistical power. Thirdly, since we investigated students enrolled in healthcare degree courses, results are not generalizable to all university students. For these reasons, further analyses should be conducted comparing students of both medical and non-medical subjects to highlight possible differences between the two groups. However, to the best of our knowledge, this is the first study that investigates how Italian students relate to IMMUNI by analyzing factors that affect its adoption. Since these factors are specific and different across population subgroups, it is fundamental to assess and monitor them over time, so that they can be addressed in the development of similar technologies. In addition, we were able to examine the experience of students that used the app and also to explore possible incentives to encourage reluctant or disinterested users. The data provided in this study may support policymakers in developing effective strategies for the promotion of app uptake and, more broadly, to facilitate engagement of people with digital health prevention measures.

5. Conclusions

The results of our study suggest that more efforts should be made aimed at raising population awareness on the usefulness of health digital technologies, restoring their confidence in health authorities, and limiting the spread of disinformation. To maximize the active engagement of the population, stakeholders should implement strategies that provide quality, clear, targeted, and straightforward information. Furthermore, institutions should invest in the development of these technologies, minimizing technical issues and facilitating their use in the population. Since university students represent an amenable target audience, because they are undergoing (often relevant) training and are, therefore, particularly receptive to educational campaigns, interventions should focus on improving their knowledge and awareness of how adhering to these strategies can contribute to safeguarding individual and public health.

Supplementary Materials: The following supporting information can be downloaded at: https://www.mdpi.com/article/10.3390/life12060871/s1, Survey questionnaire.

Author Contributions: Conceptualization, V.B. and C.I.; methodology, V.B. and C.I.; formal analysis, C.I. and M.R.D.B.; investigation, C.I., M.R.D.B., F.T. and E.M.; resources, C.I., M.R.D.B. and F.T. and E.M.; data curation, C.I., M.R.D.B., F.T. and E.M.; writing—original draft preparation, C.I. and M.R.D.B.; writing—review and editing, V.B., C.D.V., C.M. and P.V.; visualization, C.I. and M.R.D.B.; supervision, V.B.; project administration, C.I. and V.B. All authors have read and agreed to the published version of the manuscript.

Funding: This research received no external funding.

Institutional Review Board Statement: The study was conducted according to the guidelines of the Declaration of Helsinki and approved by the Ethics Committee of Sapienza University/Umberto I teaching hospital of Rome (protocol code 571/2021).

Informed Consent Statement: Informed consent was obtained from all subjects involved in the study.

Data Availability Statement: The data presented in this study are available on request from the corresponding author.

Acknowledgments: We would like to thank Sapienza University students for their participation in the survey.

Conflicts of Interest: The authors declare no conflict of interest.

References

1. Hossain, A.D.; Jarolimova, J.; Elnaiem, A.; Huang, C.X.; Richterman, A.; Ivers, L.C. Effectiveness of contact tracing in the control of infectious diseases: A systematic review. *Lancet Public Health* **2022**, *7*, e259–e273. [CrossRef]
2. Shelby, T.; Schenck, C.; Weeks, B.; Goodwin, J.; Hennein, R.; Zhou, X.; Spiegelman, D.; Grau, L.E.; Niccolai, L.; Bond, M.; et al. Lessons Learned From COVID-19 Contact Tracing During a Public Health Emergency: A Prospective Implementation Study. *Front. Public Health* **2021**, *9*, 1958. [CrossRef] [PubMed]
3. World Health Organization (WHO) Regional Office for Europe. Contact tracing in the context of COVID-19. Interim guidance. *Pediatr. Med. Rodz.* **2020**, *16*, 33–39. [CrossRef]
4. Mazza, C.; Girardi, D.; Gentile, L.; Gaeta, M.; Signorelli, C.; Odone, A. Public health effectiveness of digital contact tracing in the COVID-19 pandemic: A systematic review of available data. *Acta Biomed.* **2021**, *92*, e2021439. [CrossRef]
5. European Centre for Disease Prevention and Control (ECDC). *Analysis of COVID-19 Contact Tracing Data from Ireland, Italy and Spain—2020 Data*; European Centre for Disease Prevention and Control: Stockholm, Sweden, 2022.
6. Anglemyer, A. Digital contact tracing technologies in epiDemics: A rapid review. *Saudi Med. J.* **2020**, *41*, 1028. [CrossRef]
7. O'Connell, J.; O'Keeffe, D.T. Contact Tracing for Covid-19—A Digital Inoculation against Future Pandemics. *N. Engl. J. Med.* **2021**, *385*, 481–484. [CrossRef]
8. Blasimme, A.; Ferretti, A.; Vayena, E. Digital Contact Tracing Against COVID-19 in Europe: Current Features and Ongoing Developments. *Front. Digit. Health* **2021**, *3*, 823. [CrossRef]
9. World Health Organization; European Centre for Disease Prevention and Control. *Indicator Framework to Evaluate the Public Health Effectiveness of Digital Proximity Tracing Solutions*; World Health Organization: Geneva, Switzerland, 2021; pp. 1–14.
10. Zeng, K.; Bernardo, S.N.; Havins, W.E. The use of digital tools to mitigate the COVID-19 pandemic: Comparative retrospective study of six countries. *JMIR Public Health Surveill.* **2020**, *6*, e24598. [CrossRef]
11. Hernández-Quevedo, C.; Scarpetti, G.; Webb, E. How Do Countries Structure Contact Tracing Operations and What Is the Role of Apps? Available online: https://analysis.covid19healthsystem.org/index.php/2020/06/18/how-do-countries-structure-contact-tracing-operations-and-what-is-the-role-of-apps/ (accessed on 10 May 2022).
12. Scrivano, N.; Gulino, R.A.; Giansanti, D. Digital Contact Tracing and COVID-19: Design, Deployment, and Current Use in Italy. *Healthcare* **2022**, *10*, 67. [CrossRef]
13. Ministero Della Salute. IMMUNI—Hai Qualche Domanda? Available online: https://www.immuni.italia.it/faq.html (accessed on 10 May 2022).
14. World Health Organization. *Ethical Considerations to Guide the Use of Digital Proximity Tracking Technologies for COVID-19 Contact Tracing*; World Health Organization: Geneva, Switzerland, 2020; p. 6.
15. Ranisch, R.; Nijsingh, N.; Ballantyne, A.; van Bergen, A.; Buyx, A.; Friedrich, O.; Hendl, T.; Marckmann, G.; Munthe, C.; Wild, V. Digital contact tracing and exposure notification: Ethical guidance for trustworthy pandemic management. *Ethics Inf. Technol.* **2021**, *23*, 285–294. [CrossRef]
16. Altmann, S.; Milsom, L.; Zillessen, H.; Blasone, R.; Gerdon, F.; Bach, R.; Kreuter, F.; Nosenzo, D.; Toussaert, S.; Abeler, J. Acceptability of app-based contact tracing for COVID-19: Cross-country survey study. *JMIR mHealth uHealth* **2020**, *8*, e19857. [CrossRef]
17. Walrave, M.; Waeterloos, C.; Ponnet, K. Adoption of a contact tracing app for containing COVID-19: A health belief model approach. *JMIR Public Health Surveill.* **2020**, *6*, e20572. [CrossRef]
18. Chen, A.T.-Y.; Thio, K.W. Exploring the drivers and barriers to uptake for digital contact tracing. *Soc. Sci. Humanit. Open* **2021**, *4*, 100212. [CrossRef]
19. Von Wyl, V.; Höglinger, M.; Sieber, C.; Kaufmann, M.; Moser, A.; Serra-Burriel, M.; Ballouz, T.; Menges, D.; Frei, A.; Puhan, M.A. Drivers of acceptance of COVID-19 proximity tracing apps in Switzerland: Panel survey analysis. *JMIR Public Health Surveill.* **2021**, *7*, e25701. [CrossRef]
20. Blom, A.G.; Wenz, A.; Cornesse, C.; Rettig, T.; Fikel, M.; Friedel, S.; Möhring, K.; Naumann, E.; Reifenscheid, M.; Krieger, U. Barriers to the large-scale adoption of a COVID-19 contact tracing app in Germany: Survey study. *J. Med. Internet Res.* **2021**, *23*, e23362. [CrossRef]
21. Montagni, I.; Roussel, N.; Thiébaut, R.; Tzourio, C. Health care students' knowledge of and attitudes, beliefs, and practices toward the french covid-19 app: Cross-sectional questionnaire study. *J. Med. Internet Res.* **2021**, *23*, e26399. [CrossRef]
22. Baccolini, V.; Renzi, E.; Isonne, C.; Migliara, G.; Massimi, A.; De Vito, C.; Marzuillo, C.; Villari, P. COVID-19 vaccine hesitancy among italian university students: A cross-sectional survey during the first months of the vaccination campaign. *Vaccines* **2021**, *9*, 1292. [CrossRef]
23. Bonaccorsi, G.; Grazzini, M.; Pieri, L.; Santomauro, F.; Ciancio, M.; Lorini, C. Assessment of Health Literacy and validation of single-item literacy screener (SILS) in a sample of Italian people. *Ann. Dell'istituto Super. Sanità* **2017**, *53*, 205–212. [CrossRef]
24. Baccolini, V.; Rosso, A.; Di Paolo, C.; Isonne, C.; Salerno, C.; Migliara, G.; Prencipe, G.; Massimi, A.; Marzuillo, C.; De Vito, C.; et al. What is the Prevalence of Low Health Literacy in European Union Member States? A Systematic Review and Meta-analysis. *J. Gen. Intern. Med.* **2021**, *36*, 753–761. [CrossRef]
25. O'Connell, J.; Abbas, M.; Beecham, S.; Buckley, J.; Chochlov, M.; Fitzgerald, B.; Glynn, L.; Johnson, K.; Laffey, J.; McNicholas, B.; et al. Best practice guidance for digital contact tracing apps: A cross-disciplinary review of the literature. *JMIR mHealth uHealth* **2021**, *9*, e27753. [CrossRef]

26. Tempski, P.; Arantes-Costa, F.M.; Kobayasi, R.; Siqueira, M.A.M.; Torsani, M.B.; Amaro, B.Q.R.C.; Nascimento, M.E.F.M.; Siqueira, S.L.; Santos, I.S.; Martins, M.A. Medical students' perceptions and motivations during the COVID-19 pandemic. *PLoS ONE* **2021**, *16*, e0248627. [CrossRef] [PubMed]
27. Baccolini, V.; Isonne, C.; Salerno, C.; Giffi, M.; Migliara, G.; Mazzalai, E.; Turatto, F.; Sinopoli, A.; Rosso, A.; De Vito, C.; et al. The association between adherence to cancer screening programs and health literacy: A systematic review and meta-analysis. *Prev. Med.* **2022**, *155*, 106927. [CrossRef] [PubMed]
28. European Centre for Disease Prevention and Control. Assessment of the further spread and potential impact of the SARS-CoV-2 Omicron variant of concern in the EU/EEA, 19th update Risk assessed. *ECDC Stock.* **2022**, *19th updat*, 1–36.
29. Megnin-Viggars, O.; Carter, P.; Melendez-Torres, G.J.; Weston, D.; Rubin, G.J. Facilitators and barriers to engagement with contact tracing during infectious disease outbreaks: A rapid review of the evidence. *PLoS ONE* **2020**, *15*, e0241473. [CrossRef] [PubMed]
30. Ministero della Salute. I Numeri di IMMUNI. Available online: https://www.immuni.italia.it/dashboard.html (accessed on 10 May 2022).
31. Jones, K.; Thompson, R. To use or not to use a COVID-19 contact tracing app: Mixed methods survey in Wales. *JMIR mHealth uHealth* **2021**, *9*, e29181. [CrossRef]
32. Overbeek, G.; Bot, S.M.; Meeus, W.H.J.; Sentse, M.; Knibbe, R.A.; Engels, R. Where it's at! the role of best friends and peer group members in young adults' alcohol use. *J. Res. Adolesc.* **2011**, *21*, 631–638. [CrossRef]
33. Shopova, T. Digital literacy of students and its improvement at the university. *J. Effic. Responsib. Educ. Sci.* **2014**, *7*, 26–32. [CrossRef]
34. The COVID-19 Task Force of the Department of Infectious Diseases and the IT Service Istituto Superiore di Sanità COVID-19 Integrated Surveillance Data in Italy. Available online: https://www.epicentro.iss.it/coronavirus/sars-cov-2-dashboard (accessed on 10 May 2022).
35. Akinbi, A.; Forshaw, M.; Blinkhorn, V. Contact tracing apps for the COVID-19 pandemic: A systematic literature review of challenges and future directions for neo-liberal societies. *Health Inf. Sci. Syst.* **2021**, *9*, 1–15. [CrossRef] [PubMed]
36. Leslie, M. COVID-19 Fight Enlists Digital Technology: Contact Tracing Apps. *Engineering* **2020**, *6*, 1064–1066. [CrossRef] [PubMed]
37. Meier, Y.; Meinert, J.; Krämer, N.C. Investigating factors that affect the adoption of COVID-19 contact-tracing apps: A privacy calculus perspective. *Technol. Mind Behav.* **2021**, *2*, 1–10. [CrossRef]
38. Park, S.; Choi, G.J.; Ko, H. Information technology-based tracing strategy in response to COVID-19 in South Korea-privacy controversies. *JAMA J. Am. Med. Assoc.* **2020**, *323*, 2129–2130. [CrossRef]
39. Elkhodr, M.; Mubin, O.; Iftikhar, Z.; Masood, M.; Alsinglawi, B.; Shahid, S.; Alnajjar, F. Technology, privacy, and user opinions of COVID-19 mobile apps for contact tracing: Systematic search and content analysis. *J. Med. Internet Res.* **2021**, *23*, e23467. [CrossRef]
40. Madden, M.; Lenhart, A.; Cortesi, S.; Gasser, U.; Duggan, M.; Smith, A.; Beaton, M. Teens, Social Media, and Privacy. *Pew Res. Cent. Internet Am. Life Proj.* **2022**, 113–147. [CrossRef]
41. Munzert, S.; Selb, P.; Gohdes, A.; Stoetzer, L.F.; Lowe, W. Tracking and promoting the usage of a COVID-19 contact tracing app. *Nat. Hum. Behav.* **2021**, *5*, 247–255. [CrossRef]
42. Frimpong, J.A.; Helleringer, S. Strategies to increase downloads of COVID–19 exposure notification apps: A discrete choice experiment. *PLoS ONE* **2021**, *16*, e0258945. [CrossRef]
43. Fast, V.; Schnurr, D. Incentivising the Adoption of COVID-19 Contact-Tracing Apps: A Randomised Controlled Online Experiment on the German Corona-Warn-App. In Proceedings of the 2021 on Computers and People Research Conference, Virtual Event, 30 June 2021; ACM: New York, NY, USA, 2021; pp. 19–21.
44. Albouy-Llaty, M.; Martin, C.; Benamouzig, D.; Bothorel, E.; Munier, G.; Simonin, C.; Guéant, J.L.; Rusch, E. Positioning digital tracing applications in the management of the COVID-19 pandemic in France. *J. Med. Internet Res.* **2021**, *23*, e27301. [CrossRef]

life

MDPI

Editorial

A Deep Dive into the Nexus between Digital Health and Life Sciences Amidst the COVID-19 Pandemic: An Editorial Expedition

Daniele Giansanti

Centre Tisp, Istituto Superiore di Sanità, 00161 Rome, Italy; daniele.giansanti@iss.it; Tel.:+39-06-49902701

I am proposing this editorial to briefly trace the evidences that emerged from the *Special Issue* (SI)—*The Digital Health in the Pandemic Era*—[1] that I had the pleasure of following in Life.

The idea of developing this Special Issue was born when the COVID-19 pandemic was still having a major impact on the health of the planet [2].

As is well known, *digital health* (DH) encompasses a diverse range of technologies, including wearable and internal devices, various types of sensors, and innovative solutions. DH can facilitate the identification of health risks and provide assistance in the diagnosis, treatment, and monitoring of various health conditions. Generally, DH presents immense potential for both the general population and healthcare professionals.

Since the beginning of the pandemic [2], this technology has been applied to the *health domain*, and it has played a crucial role in providing remote assistance and continuity of care at home, thereby protecting patients, healthcare workers, limiting the spread of the virus, and reducing the need for hospitalization.

For instance, digital measurement of oxygen saturation at home has provided key decision-making data for patients' health, such as choosing between hospitalization and respiratory support.

Additionally, remote monitoring of frail patients with underlying conditions, such as diabetes, cardiovascular disease, or oncological problems has improved the continuity of care and reduced pressure on hospitals. Digital contributions have affected the fight against the pandemic in numerous ways, such as managing digital contact tracing and vaccination processes using smart technology.

The SI:

- Explored innovations in the field of DH stimulated by the COVID-19 pandemic, and the acceptance of this revisited DH by all, including stakeholders, healthcare professionals, and citizens.
- Also, analyzed the successes and failures of DH applications during the pandemic, highlighting the critical role played by remote health monitoring systems, the contribute of the Artificial Intelligence, and the potentials they could offer for post-pandemic healthcare delivery.

At the time of writing this editorial contribution, 18 papers have been published [1], including one editorial, one opinion, one review, one systematic review, one comment and one reply, and twelve scientific articles.

A quick overview of the contents of the published works demonstrates how the topic of DH, as faced and revisited during the pandemic, has played a connecting role for many of the branches of the life sciences [3], shown in Table 1.

Citation: Giansanti, D. A Deep Dive into the Nexus between Digital Health and Life Sciences Amidst the COVID-19 Pandemic: An Editorial Expedition. *Life* **2023**, *13*, 1154. https://doi.org/10.3390/life13051154

Received: 28 April 2023
Accepted: 9 May 2023
Published: 10 May 2023

Table 1. Branches of the life sciences.

Branches of Life Sciences
Anatomy, Biochemistry, Bioengineering, Bioinformatics, Biophysics, Biotechnology, Botany, Cell biology, Developmental biology, Ecology, Entomology, Epidemiology, Ethology, Evolutionary biology, Genetics, Hematology (also known as Haematology), Microbiology, Molecular biology, Neuroscience, Physiology, Population biology, Structural biology, Toxicology, Zoology

Five scientific papers were published, including the editorial [2,4–7], equal to 27.8% of all the papers contained in the SI, which dealt with the topic of DH applied to contact tracing, the so-called digital contact tracing (DCT). These studies reported [4,5] and/or discussed [6,7] the impact of the DCT on the spread of the pandemic in different populations, together with the factors that influenced its use. The DCT, a new application of the DH, uses apps and various methods of tracking position and proximity, and it had a wide diffusion during the pandemic [2]. It has been consolidated as a technological method with great potential, applying digital solutions for automatic contact tracing, for both a punctual and a global tracking of the evolution of an epidemic. A strategic epidemiology activity was developed for tracking communicable diseases, as in the case of the COVID-19 pandemic. The DCT highlights an important role of DH in epidemiology as a major component of public health research in the branch of the life sciences (Table 1), studying factors affecting the health of populations.

All the other studies [8–20] have dealt with the use of DH in the life sciences, which, in addition to bioengineering, touched by all, have examined, from time to time, other categories (Table 1), such as Anatomy [8,11], Bioinformatics [14,16], Cell biology [8,11], Neuroscience [8,12,15,17,20], Physiology [9,10,12,13], Population biology [14,16,18,19], and others that are shown in Table 1.

The SI, in particular, highlights how the contribution of DH to the life sciences during the pandemic also took place with the support of artificial intelligence (AI), both in applications related to COVID-19 specific diagnostics, diagnostics in general, and on population surveys regarding biomedical aspects during the pandemic.

In [8], an application of AI in brain tumors integrated in diagnostic imaging is reported. Two studies, based on AI, have addressed aspects connected with the fitness and wellness of the population.

The first study applied AI, integrated with wearable devices, to monitor physical activity (which plays an important role in controlling obesity and maintaining healthy living) during the pandemic [9].

The second study applied IoT and AI [10] in investigating the risk factors and the ratio of obesity, proposing an approach for obesity diagnosis in its initial stages, significantly increasing the patient's chances of effective treatment.

An AI algorithm for the early detection of abnormalities in chest X-rays, for COVID-19 diagnostics, was proposed in [11]. It used a deep hybrid learning-based framework for the detection of COVID-19 using chest X-ray images.

The importance of DH application in the physiological issues related to physical activity was investigated in [12,13], through population surveys. In [12], the authors tested the feasibility of virtually delivering an *exergame-based* physical activity intervention to older breast cancer survivors, while the authors of the study reported, in [13], that they investigated the changes experienced by Austrian therapists when switching to psychotherapy at a distance.

Large-scale population surveys have been conducted, both to analyze the impact of biomedical parameters and health determinants on the population, as well as digital literacy, key factors in using DH [14–16].

The study in [14] addressed the importance of *Big-medical-data classification and image detection* as crucial tasks in the field of healthcare. They proposed a specific algorithm for medical data classification and image detection in the COVID-19 era that may have significant implications in the *health domain*.

The outcome from "Understanding COVID" (a public health campaign designed in 2020 and launched in 2021 in Asturias in Spain to provide reliable and comprehensive information oriented towards vulnerable populations, which is also related to digital literacy) was reported in [15].

The study reported in [16], using AI, and particular feature selection approaches, evaluated the aspects affecting the health of students throughout the COVID-19 lockdown time period.

The importance and the impact of *mental health* monitoring through self-monitoring apps was addressed in a study reported in [17].

Two specific reviews addressed and overviewed the impact of DH in remote health care and healthcare interventions [18,19]. The impact of *eHealth* interventions on the improvement of self-care in chronic patients was overviewed in [18], while the healthcare professionals' experience of reforming digital care visits was investigated in [19].

Finally, an opinion piece investigated the impact of chatbots in the health domain [20], reporting an increasing of their use during the pandemic, also thanks to the AI.

In conclusion:

The published works on the DH, reported in the collection [1,2,4–20], demonstrate the usefulness of DH in connecting various branches of life sciences, especially during the pandemic. The use of DH in contact tracing has been particularly significant, with several studies reporting on its impact and factors influencing its use [2,4–7]. In fact, the application of DH in digital contact tracing (DCT) has emerged as a major component of public health research, showcasing its potential role in epidemiology, a major component of the life sciences (Table 1).

Additionally, DH applications have been used in biomedical diagnostics [8,11], physical activity [12,13], mental health monitoring [17], and remote healthcare interventions [12–19].

Large-scale population surveys have also been conducted to analyze the impact of biomedical parameters, health determinants, and healthcare literacy on the population [14–16], while chatbots have increasingly been used in the health domain [20].

Moreover, the support of Artificial Intelligence (AI) has further amplified the impact of DH in diagnostics [8,11], physical activity monitoring [12], obesity diagnosis [12], and healthcare interventions and determinants on the population [12–19].

DH has effectively aided public health research related to studying factors affecting the health of populations, while also contributing to advances in various fields of the life sciences (Table 1), such as, Biongineering [8–20], Anatomy [8,11], Bioinformatics [14,16], Cell biology [8,11], Neuroscience [8,12,15,17,20], Physiology [9,10,12,13], and Population biology [14,16,18].

Overall, this highlights that:

- DH has played and plays the role of a connector among different branches of life sciences, particularly in times of crisis.
- The high-quality research in this area remarks the usefulness of DH as a powerful tool for scientific and medical research.

Institutional Review Board Statement: Not applicable.

Informed Consent Statement: Not applicable.

Data Availability Statement: Not applicable.

Conflicts of Interest: The authors declare no conflict of interest.

References

1. Available online: https://www.mdpi.com/journal/life/special_issues/DigitalHealth_Pademic (accessed on 27 April 2023).
2. Giansanti, D. The digital health: From the experience of the COVID-19 pandemic onwards. *Life* **2022**, *12*, 78. [CrossRef] [PubMed]
3. What Are the Branches of Life Sciences and Their Meanings? Go Life Sciences. Available online: https://golifescience.com/life-sciences-branches/ (accessed on 27 April 2023).

4. Cao, J.; Liu, D.; Zhang, G.; Shang, M. The Impact of Digital Contact Tracing Apps Overuse on Prevention of COVID-19: A Normative Activation Model Perspective. *Life* **2022**, *12*, 1371. [CrossRef] [PubMed]
5. Isonne, C.; De Blasiis, M.R.; Turatto, F.; Mazzalai, E.; Marzuillo, C.; De Vito, C.; Villari, P.; Baccolini, V. What Went Wrong with the IMMUNI Contact-Tracing App in Italy? A Cross-Sectional Survey on the Attitudes and Experiences among Healthcare University Students. *Life* **2022**, *12*, 871. [CrossRef] [PubMed]
6. Cao, J.; Liu, D.; Zhang, G.; Shang, M. Reply to Giansanti, D., Why Has Digital Contact Tracing Worked Differently in Different Countries? Comment on "Cao et al. The Impact of Digital Contact Tracing Apps Overuse on Prevention of COVID-19: A Normative Activation Model Perspective. *Life* **2022**, *12*, 1371". *Life* **2022**, *12*, 1593. [CrossRef] [PubMed]
7. Giansanti, D. Why Has Digital Contact Tracing Worked Differently in Different Countries? Comment on Cao et al. The Impact of Digital Contact Tracing Apps Overuse on Prevention of COVID-19: A Normative Activation Model Perspective. *Life* **2022**, *12*, 1371. [CrossRef] [PubMed]
8. Almalki, Y.E.; Ali, M.U.; Ahmed, W.; Kallu, K.D.; Zafar, A.; Alduraibi, S.K.; Irfan, M.; Basha, M.A.A.; Alshamrani, H.A.; Alduraibi, A.K. Robust Gaussian and Nonlinear Hybrid Invariant Clustered Features Aided Approach for Speeded Brain Tumor Diagnosis. *Life* **2022**, *12*, 1084. [CrossRef] [PubMed]
9. Alsareii, S.A.; Awais, M.; Alamri, A.M.; AlAsmari, M.Y.; Irfan, M.; Aslam, N.; Raza, M. Physical Activity Monitoring and Classification Using Machine Learning Techniques. *Life* **2022**, *12*, 1103. [CrossRef] [PubMed]
10. Alsareii, S.A.; Shaf, A.; Ali, T.; Zafar, M.; Alamri, A.M.; AlAsmari, M.Y.; Irfan, M.; Awais, M. IoT Framework for a Decision-Making System of Obesity and Overweight Extrapolation among Children, Youths, and Adults. *Life* **2022**, *12*, 1414. [CrossRef] [PubMed]
11. Alqahtani, A.; Zahoor, M.M.; Nasrullah, R.; Fareed, A.; Cheema, A.A.; Shahrose, A.; Irfan, M.; Alqhatani, A.; Alsulami, A.A.; Zaffar, M.; et al. Computer Aided COVID-19 Diagnosis in Pandemic Era Using CNN in Chest X-ray Images. *Life* **2022**, *12*, 1709. [CrossRef] [PubMed]
12. Swartz, M.C.; Robertson, M.C.; Christopherson, U.; Wells, S.J.; Lewis, Z.H.; Bai, J.; Swartz, M.D.; Silva, H.C.; Martinez, E.; Lyons, E.J. Assessing the Suitability of a Virtual 'Pink Warrior' for Older Breast Cancer Survivors during COVID-19: A Pilot Study. *Life* **2023**, *13*, 574. [CrossRef] [PubMed]
13. Stadler, M.; Jesser, A.; Humer, E.; Haid, B.; Stippl, P.; Schimböck, W.; Maaß, E.; Schwanzar, H.; Leithner, D.; Pieh, C.; et al. Remote Psychotherapy during the COVID-19 Pandemic: A Mixed-Methods Study on the Changes Experienced by Austrian Psychotherapists. *Life* **2023**, *13*, 360. [CrossRef] [PubMed]
14. Awad, F.H.; Hamad, M.M.; Alzubaidi, L. Robust Classification and Detection of Big Medical Data Using Advanced Parallel K-Means Clustering, YOLOv4, and Logistic Regression. *Life* **2023**, *13*, 691. [CrossRef] [PubMed]
15. López-Ventoso, M.; Pisano González, M.; Fernández García, C.; Diez Valcarce, I.; Rey Hidalgo, I.; Rodríguez Nachón, M.J.; Menéndez García, A.M.; Perello, M.; Avagnina, B.; Zanutto, O.; et al. Understanding COVID: Collaborative Government Campaign for Citizen Digital Health Literacy in the COVID-19 Pandemic. *Life* **2023**, *13*, 589. [CrossRef] [PubMed]
16. Saeed, A.; Zaffar, M.; Abbas, M.A.; Quraishi, K.S.; Shahrose, A.; Irfan, M.; Huneif, M.A.; Abdulwahab, A.; Alduraibi, S.K.; Alshehri, F.; et al. A Turf-Based Feature Selection Technique for Predicting Factors Affecting Human Health during Pandemic. *Life* **2022**, *12*, 1367. [CrossRef] [PubMed]
17. Aziz, M.; Erbad, A.; Almourad, M.B.; Altuwairiqi, M.; McAlaney, J.; Ali, R. Did Usage of Mental Health Apps Change during COVID-19? A Comparative Study Based on an Objective Recording of Usage Data and Demographics. *Life* **2022**, *12*, 1266. [CrossRef] [PubMed]
18. Renzi, E.; Baccolini, V.; Migliara, G.; De Vito, C.; Gasperini, G.; Cianciulli, A.; Marzuillo, C.; Villari, P.; Massimi, A. The Impact of eHealth Interventions on the Improvement of Self-Care in Chronic Patients: An Overview of Systematic Reviews. *Life* **2022**, *12*, 1253. [CrossRef] [PubMed]
19. Lampickiene, I.; Davoody, N. Healthcare Professionals' Experience of Performing Digital Care Visits—A Scoping Review. *Life* **2022**, *12*, 913. [CrossRef] [PubMed]
20. Giansanti, D. The chatbots are invading us: A map point on the evolution, applications, opportunities and emerging problems in the health domain. *Life* **2023**, *13*, 1130. [CrossRef]

MDPI
St. Alban-Anlage 66
4052 Basel
Switzerland
Tel. +41 61 683 77 34
Fax +41 61 302 89 18
www.mdpi.com

Life Editorial Office
E-mail: life@mdpi.com
www.mdpi.com/journal/life

www.ingramcontent.com/pod-product-compliance
Lightning Source LLC
Chambersburg PA
CBHW061814210926
51850CB00006B/701
9783036577333